T0280408

A Primer on Linear Models

CHAPMAN & HALL/CRC
Texts in Statistical Science Series

Series Editors
Bradley P. Carlin, *University of Minnesota, USA*
Julian J. Faraway, *University of Bath, UK*
Martin Tanner, *Northwestern University, USA*
Jim Zidek, *University of British Columbia, Canada*

Analysis of Failure and Survival Data
P. J. Smith

The Analysis and Interpretation of Multivariate Data for Social Scientists
D.J. Bartholomew, F. Steele, I. Moustaki, and J. Galbraith

The Analysis of Time Series — An Introduction, Sixth Edition
C. Chatfield

Applied Bayesian Forecasting and Time Series Analysis
A. Pole, M. West and J. Harrison

Applied Nonparametric Statistical Methods, Fourth Edition
P. Sprent and N.C. Smeeton

Applied Statistics — Handbook of GENSTAT Analysis
E.J. Snell and H. Simpson

Applied Statistics — Principles and Examples
D.R. Cox and E.J. Snell

Bayes and Empirical Bayes Methods for Data Analysis, Second Edition
B.P. Carlin and T.A. Louis

Bayesian Data Analysis, Second Edition
A. Gelman, J.B. Carlin, H.S. Stern and D.B. Rubin

Beyond ANOVA — Basics of Applied Statistics
R.G. Miller, Jr.

Computer-Aided Multivariate Analysis, Fourth Edition
A.A. Afifi and V.A. Clark

A Course in Categorical Data Analysis
T. Leonard

A Course in Large Sample Theory
T.S. Ferguson

Data Driven Statistical Methods
P. Sprent

Decision Analysis — A Bayesian Approach
J.Q. Smith

Elementary Applications of Probability Theory, Second Edition
H.C. Tuckwell

Elements of Simulation
B.J.T. Morgan

Epidemiology — Study Design and Data Analysis, Second Edition
M. Woodward

Essential Statistics, Fourth Edition
D.A.G. Rees

Extending the Linear Model with R: Generalized Linear, Mixed Effects and Nonparametric Regression Models
J.J. Faraway

A First Course in Linear Model Theory
N. Ravishanker and D.K. Dey

Generalized Additive Models: An Introduction with R
S. Wood

Interpreting Data — A First Course in Statistics
A.J.B. Anderson

An Introduction to Generalized Linear Models, Second Edition
A.J. Dobson

Introduction to Multivariate Analysis
C. Chatfield and A.J. Collins

Introduction to Optimization Methods and Their Applications in Statistics
B.S. Everitt

Introduction to Probability with R
K. Baclawski

Introduction to Randomized Controlled Clinical Trials, Second Edition
J.N.S. Matthews

Introduction to Statistical Methods for Clinical Trials
Thomas D. Cook and David L. DeMets

Large Sample Methods in Statistics
P.K. Sen and J. da Motta Singer

Linear Models with R
J.J. Faraway

Markov Chain Monte Carlo — Stochastic Simulation for Bayesian Inference, Second Edition
D. Gamerman and H.F. Lopes

Mathematical Statistics
K. Knight

Modeling and Analysis of Stochastic Systems
V. Kulkarni

Modelling Binary Data, Second Edition
D. Collett

Modelling Survival Data in Medical Research, Second Edition
D. Collett

Multivariate Analysis of Variance and Repeated Measures — A Practical Approach for Behavioural Scientists
D.J. Hand and C.C. Taylor

Multivariate Statistics — A Practical Approach
B. Flury and H. Riedwyl

Practical Data Analysis for Designed Experiments
B.S. Yandell

Practical Longitudinal Data Analysis
D.J. Hand and M. Crowder

Practical Statistics for Medical Research
D.G. Altman

A Primer on Linear Models
J.F. Monahan

Probability — Methods and Measurement
A. O'Hagan

Problem Solving — A Statistician's Guide, Second Edition
C. Chatfield

Randomization, Bootstrap and Monte Carlo Methods in Biology, Third Edition
B.F.J. Manly

Readings in Decision Analysis
S. French

Sampling Methodologies with Applications
P.S.R.S. Rao

Statistical Analysis of Reliability Data
M.J. Crowder, A.C. Kimber, T.J. Sweeting, and R.L. Smith

Statistical Methods for Spatial Data Analysis
O. Schabenberger and C.A. Gotway

Statistical Methods for SPC and TQM
D. Bissell

Statistical Methods in Agriculture and Experimental Biology, Second Edition
R. Mead, R.N. Curnow, and A.M. Hasted

Statistical Process Control — Theory and Practice, Third Edition
G.B. Wetherill and D.W. Brown

Statistical Theory, Fourth Edition
B.W. Lindgren

Statistics for Accountants
S. Letchford

Statistics for Epidemiology
N.P. Jewell

Statistics for Technology — A Course in Applied Statistics, Third Edition
C. Chatfield

Statistics in Engineering — A Practical Approach
A.V. Metcalfe

Statistics in Research and Development, Second Edition
R. Caulcutt

Survival Analysis Using S—Analysis of Time-to-Event Data
M. Tableman and J.S. Kim

The Theory of Linear Models
B. Jørgensen

Time Series Analysis
H. Madsen

Texts in Statistical Science

A Primer on Linear Models

John F. Monahan

North Carolina State Univesity

Raleigh, North Carolina, U.S.A.

CRC Press
Taylor & Francis Group
Boca Raton London New York

CRC Press is an imprint of the
Taylor & Francis Group, an **informa** business
A CHAPMAN & HALL BOOK

Chapman & Hall/CRC
Taylor & Francis Group
6000 Broken Sound Parkway NW, Suite 300
Boca Raton, FL 33487-2742

© 2008 by Taylor & Francis Group, LLC
Chapman & Hall/CRC is an imprint of Taylor & Francis Group, an Informa business

No claim to original U.S. Government works

ISBN 13: 978-1-4200-6201-4 (pbk)

This book contains information obtained from authentic and highly regarded sources. Reprinted material is quoted with permission, and sources are indicated. A wide variety of references are listed. Reasonable efforts have been made to publish reliable data and information, but the author and the publisher cannot assume responsibility for the validity of all materials or for the consequences of their use.

Except as permitted under U.S. Copyright Law, no part of this book may be reprinted, reproduced, transmitted, or utilized in any form by any electronic, mechanical, or other means, now known or hereafter invented, including photocopying, microfilming, and recording, or in any information storage or retrieval system, without written permission from the publishers.

For permission to photocopy or use material electronically from this work, please access www.copyright.com (http://www.copyright.com/) or contact the Copyright Clearance Center, Inc. (CCC) 222 Rosewood Drive, Danvers, MA 01923, 978-750-8400. CCC is a not-for-profit organization that provides licenses and registration for a variety of users. For organizations that have been granted a photocopy license by the CCC, a separate system of payment has been arranged.

Trademark Notice: Product or corporate names may be trademarks or registered trademarks, and are used only for identification and explanation without intent to infringe.

Library of Congress Cataloging-in-Publication Data

Monahan, John F.
A primer on linear models / John Monahan.
p. cm. -- (Chapman & Hall/CRC texts in statistical science series)
Includes bibliographical references and index.
ISBN-13: 978-1-4200-6201-4 (Hardcover)
ISBN-10: 1-4200-6201-8 (Hardcover)
1. Linear models (Statistics) I. Title.

QA279.M659 2008
519.5--dc22 2007037035

Visit the Taylor & Francis Web site at
http://www.taylorandfrancis.com

and the CRC Press Web site at
http://www.crcpress.com

To Carol, whose patience, love, and understanding made this possible.

Contents

Preface

The linear model forms the foundation of the practice of statistics. Understanding the theory upon which this foundation rests is a critical step in graduate training in statistics. This book arose from the need for an appropriate text for the graduate course entitled *Linear Models and Variance Components* at North Carolina State University. The philosophy of this course has two distinguishing features: the use of non-full-rank design matrices to seamlessly incorporate regression, analysis of variance (ANOVA), and various mixed models, and attention to the exact, finite sample theory supporting common statistical methods.

The primary goal in writing this book is to provide a brief, yet complete foundation for the understanding of basic linear models. The intention is not a comprehensive text on regression methods, but the focus is on the theory behind the methodology students learn in other courses. As a result, the style follows the route of theorem/proof/corollary/example/exercise with discussions of the motivations for the assumptions from theory and practice. A secondary goal for this book is to serve as a basic reference book for statisticians for exact, finite sample results. Viewed in this role, the reader will notice a steady progression in both difficulty and sophistication of the proofs, discussions, and exercises, paralleling the growth of the students as they master the subject.

The subject matter also masks another subsidiary goal. As the study of geometry has a paramount role in secondary education because of its practice in the tools of logic, the study of linear models also becomes a convenient vehicle for teaching students the tools for statistical theory.

I have tried to avoid leaps in the development and proofs, or referring to other sources for the details of proofs, in order that the students see the complete development of this material. While the sophisticated reader may find the inclusion of these small steps superfluous, students learning the material for the first time suffer from crippling myopia, and shrink from leaping to the next step.

Because students' learning styles vary, I have attempted to include proofs and discussions from both algebraic and geometric viewpoints. Since I believe that drill and practice is a valuable learning style, I have included many exercises for each chapter at varying levels of difficulty. The multipart exercises were taken from past quizzes and final exams; the shorter ones served as homework problems. Certain exercises, marked by an asterisk (*), should not be skipped by the reader; the results of many of these exercises are used in other parts of the book. I have not referred to an exercise for the completion of a proof without having tested it on a class.

In teaching this course, I usually begin with the review of linear algebra. Yes, that means I start with the appendix. Since geometry is often passed over in most introductory courses on linear algebra, I spend considerable time covering vector spaces and projections. The reward for this effort is a simpler development of the concept of estimability. The order of the topics is designed to parallel the second of a two-semester course in statistical theory that uses Casella and Berger's *Statistical Inference*. As a result, the statistical theory of linear model is postponed, and the basic algebra and geometry of the linear model are introduced as an approximation method—the linear least squares problem as it is known in applied mathematics. Progressively adding statistical assumptions leads to estimability and then, with variances, the Gauss–Markov model. Once the students have mastered the statistical tools of hypothesis tests and confidence intervals in their statistical theory course, they then apply those concepts to the linear model. The mixed model makes subtle but important changes in the goals of estimation and inference. In a typical year, I am only able to cover three-fourths of this material in a one-semester course; I prefer to sacrifice coverage to ensure depth.

I owe a great debt of gratitude to Sastry Pantula. We began this book as a collaboration, and he set the order of topics as well as the level and tone of the discussion. Because of limitations imposed by his administrative duties, he encouraged me to venture alone with this book. Nonetheless, vestiges of his words and ideas remain. His support and advice have been invaluable.

I also thank colleagues whose comments and discussion have contributed to this effort. Robert Hartwig gave me valuable ideas on the presentation of Gram–Schmidt orthogonalization. I leaned on Bibhuti Bhattacharyya for advice on multivariate issues. Discussions with Jason Osborne and Dave Dickey on applications, especially split plots and mixed models, and Dennis Boos and Len Stefanski on theory were all a great help. I also want to thank the students who contributed in many ways: some whose research problems became homework exercises, including Amy Nail and Hyunjung Cho; many students who suffered through error-riddled earlier versions of the manuscript and found and corrected typos, especially Yu-Cheng Ku; and the many students whose indirect feedback—learning successes and struggles—has aided the development of this work.

Chapter 1

Examples of the General Linear Model

1.1 Introduction

The focus of this book is the analysis of the general linear model of the form

$$\mathbf{y} = \mathbf{X}\mathbf{b} + \mathbf{e} \tag{1.1}$$

where $\mathbf{y}$ is a $N \times 1$ vector of observed responses, $\mathbf{X}$ is a $N \times p$ matrix of fixed constants, $\mathbf{b}$ is a $p \times 1$ vector of fixed but unknown parameters, and $\mathbf{e}$ is a $N \times 1$ vector of (unobserved) errors with zero mean. This model is called a *linear model* because the mean of the response vector $\mathbf{y}$ is linear in the unknown parameters $\mathbf{b}$. Our interest is to estimate the parameters of this model and test hypotheses regarding linear combinations of the parameters.

Several models commonly used in statistical methods are examples of the general linear model (1.1). As further described in this chapter, these include simple linear regression and multiple regression models, one-way analysis of variance (ANOVA), two-way crossed analysis with or without interaction, the analysis of covariance (ANACOVA) model, mixed effects models, and some time series models. We will discuss these models and give some examples of models that are not special cases of (1.1).

1.2 One-Sample Problem

The simplest version of the linear model arises with one of the basic problems in elementary statistics, where y_i are randomly sampled from a population with unknown mean μ and variance σ^2. In this case, $\mathbf{X}\mathbf{b}$ takes a very simple form

$$\mathbf{X}\mathbf{b} = \mathbf{1}(\mu)$$

so that the unknown coefficient vector $\mathbf{b}$ is just the scalar μ. This very simple model will be extended to handle cases such as heteroskedasticity in Chapter 4 or multivariate responses in Chapter 9.

1.3 Simple Linear Regression

Consider the model where the response variable y_i is linearly related to an independent variable x_i, given by

$$y_i = \beta_0 + \beta_1 x_i + e_i, \quad i = 1, \ldots, N \tag{1.2}$$

where $e_1, e_2, \ldots, e_n$ are typically assumed to be uncorrelated random variables with mean zero and constant variance σ^2. If we assume that $x_1, x_2, \ldots, x_n$ are fixed constants, observed without error, then (1.2) is a special case of (1.1) with

$$\mathbf{y} = \begin{bmatrix} y_1 \\ y_2 \\ \ldots \\ y_N \end{bmatrix}, \quad \mathbf{Xb} = \begin{bmatrix} 1 & x_1 \\ 1 & x_2 \\ \ldots & \ldots \\ 1 & x_{N-1} \\ 1 & x_N \end{bmatrix} \begin{bmatrix} \beta_0 \\ \beta_1 \end{bmatrix}, \quad \mathbf{e} = \begin{bmatrix} e_1 \\ e_2 \\ \ldots \\ e_N \end{bmatrix}$$

so that $\mathbf{X}$ is $N \times 2$, $\mathbf{b}$ is 2×1, and $\mathbf{e}$ is $N \times 1$. Note that if the x_i's were measured with error, then the model (1.2) is not a special case of (1.1), since, in this case, the matrix $\mathbf{X}$ is random, not fixed.

Example 1.1: Measurement Error
For example, consider a study of the relationship between crop yield y_i and soil nutrients, such as nitrogen n_i. Suppose we expect the relationship between yield and nitrogen to follow

$$y_i = \beta_0 + \beta_1 n_i + e_i.$$

Typically, soil nitrogen is estimated by taking soil samples in the plot and performing a chemical analysis of the soil samples. In practice, we really observe $x_i = n_i + u_i$, where u_i is the measurement error in the sampling and chemical analysis. See, for example, Fuller [10] for examples and analysis of measurement error models or a case of errors in the variables. See also Example 5.4.

1.4 Multiple Regression

Consider the model where the response variable y_i is linearly related to several independent variables $x_{i1}, x_{i2}, \ldots, x_{ik}$, given by

$$y_i = \beta_0 + \beta_1 x_{i1} + \beta_2 x_{i2} + \cdots + \beta_k x_{ik} + e_i, \quad i = 1, \ldots, N \tag{1.3}$$

where again $e_1, e_2, \ldots, e_n$ are typically assumed to be uncorrelated random variables with mean zero and constant variance σ^2. If we assume that $x_{i1}, x_{i2}, \ldots, x_{ik}$ are fixed constants, observed without error, then the multiple regression model (1.3) is a special case of the general linear model (1.1), with $p = k + 1$,

$$\mathbf{y} = \begin{bmatrix} y_1 \\ y_2 \\ \dots \\ y_N \end{bmatrix}, \quad \mathbf{Xb} = \begin{bmatrix} 1 & x_{11} & x_{12} & \dots & x_{1k} \\ 1 & x_{21} & x_{22} & \dots & x_{2k} \\ 1 & x_{31} & x_{32} & \dots & x_{3k} \\ \dots & & \dots & \dots & \\ 1 & x_{N1} & x_{N2} & \dots & x_{Nk} \end{bmatrix} \begin{bmatrix} \beta_0 \\ \beta_1 \\ \beta_2 \\ \dots \\ \beta_k \end{bmatrix}, \quad \mathbf{e} = \begin{bmatrix} e_1 \\ e_2 \\ \dots \\ e_N \end{bmatrix}$$

so that $\mathbf{X}$ is $N \times (k+1)$, $\mathbf{b}$ is $(k+1) \times 1$, and $\mathbf{e}$ is $n \times 1$.

For example,

$$y_i = \beta_0 + \beta_1 i + \beta_2 i^2 + e_i, \quad i = 1, \dots, N$$

and

$$y_i = \beta_0 + \beta_1 cos(2\pi i/7) + \beta_2 log(i) + \beta_3 \sqrt{i^2 + 1} + e_i, \quad i = 1, \dots, N$$

are each special cases of (1.3). Note again that the general linear model requires linearity in terms of the unknown parameters, and not in terms of the independent variables. In contrast, the nonlinear models

$$y_i = \beta_0 + \beta_1 e^{-\beta_2 x_i} + e_i \tag{1.4}$$

and

$$y_i = \frac{\beta_0}{\beta_1 + \beta_2 x_i} + e_i \tag{1.5}$$

are not linear in the unknown parameters.

A multiple regression model as described here always includes an intercept. A general linear model, however, may or may not have an intercept parameter β_0; more generally, it may or may not include a column of ones in the column space of $\mathbf{X}$. For example, the quadratic with zero intercept model

$$y_i = \beta_1 i + \beta_2 i^2 + e_i$$

is an example of (1.1), but not (1.3).

1.5 One-Way ANOVA

One of the most fundamental tools in science is an experiment designed to compare a treatments. For example, we may have a different methods for determining nitrogen concentration in the soil. For the ith treatment level, n_i experimental units are selected at random and assigned to the ith treatment. Consider the model

$$y_{ij} = \mu + \alpha_i + e_{ij} \tag{1.6}$$

where $i = 1, \ldots, a$ (treatments); $j = 1, \ldots, n_i$ (experimental units or subjects), and the e_{ij}'s are assumed to be uncorrelated random errors with zero mean and constant, unknown variance σ^2. If we assume that the treatment effects α_i are fixed, then the model (1.6) is a special case of the general linear model (1.1) with

$$\mathbf{y} = \begin{bmatrix} y_{11} \\ y_{12} \\ \cdots \\ y_{1n_1} \\ y_{21} \\ y_{22} \\ \cdots \\ y_{2n_2} \\ \cdots \\ y_{a1} \\ \cdots \\ y_{an_a} \end{bmatrix}, \quad \mathbf{Xb} = \begin{bmatrix} \mathbf{1}_{n_1} & \mathbf{1}_{n_1} & \mathbf{0} & \cdots & \mathbf{0} \\ \mathbf{1}_{n_2} & \mathbf{0} & \mathbf{1}_{n_2} & \cdots & \mathbf{0} \\ \mathbf{1}_{n_3} & \mathbf{0} & \mathbf{0} & \cdots & \mathbf{0} \\ \cdots & & \cdots & \cdots & \\ \mathbf{1}_{n_a} & \mathbf{0} & \mathbf{0} & \cdots & \mathbf{1}_{n_a} \end{bmatrix} \begin{bmatrix} \mu \\ \alpha_1 \\ \alpha_2 \\ \cdots \\ \alpha_a \end{bmatrix}, \quad \mathbf{e} = \begin{bmatrix} e_{11} \\ e_{12} \\ \cdots \\ e_{1n_1} \\ e_{21} \\ e_{22} \\ \cdots \\ e_{2n_2} \\ \cdots \\ e_{a1} \\ \cdots \\ e_{an_a} \end{bmatrix} \tag{1.7}$$

so that $\mathbf{X}$ is $N \times (a + 1)$, $\mathbf{b}$ is $(a + 1) \times 1$, $p = 1 + a$, and $N = \sum_{i=1}^{a} n_i$.

If we assume that the a treatments $(\alpha_1, \alpha_2, \ldots, \alpha_a)$ were randomly selected from a population of treatments, then (1.6) is an example of a random effects (or mixed effects) model, but still may be viewed as a special case of the general linear model (1.1), but with $\mathbf{X} = \mathbf{1}_n$ and correlated errors. Random effects and mixed models are examined in Chapter 8.

Also note that the first column in the matrix $\mathbf{X}$ given in (1.7) above is the sum of the last a columns. We will see that this linear dependence in the design matrix $\mathbf{X}$ leads to several interesting—and challenging—problems.

1.6 First Discussion

In discussing the linear model, often our focus is on the response vector $\mathbf{y}$ or on the mean function $\mathbf{Xb}$. And, indeed, we will begin by assuming that the difference, the error, $\mathbf{e} = \mathbf{y} - \mathbf{Xb}$, should be small and treat the problem as an approximation. The next step will be to assume that these errors are random variables with a zero mean, then to assume a pattern for the covariance matrix for $\mathbf{e}$. Chapters 3 and 4 will follow from just such a simple level of assumptions. Then in Chapter 5 we introduce the multivariate normal distribution, and from that we derive the familiar confidence intervals and tests based on normality.

While this progression of assumptions provides an outline for this book, an important subtle reminder of the power and generality of the linear model is to focus on the meaning of the phrase "error with zero mean." Consider the following example.

Example 1.2: Bernoulli Models
Let $U_1, \ldots, U_m$ be iid Bernoulli random variables with probability p, and $V_1, \ldots, V_n$ be iid Bernoulli(q). We can write this problem as a linear model in the following way by letting

$$\mathbf{y} = \begin{bmatrix} U_1 \\ U_2 \\ \cdots \\ U_m \\ V_1 \\ V_2 \\ \cdots \\ V_n \end{bmatrix}, \quad \mathbf{Xb} = \begin{bmatrix} 1 & 0 \\ 1 & 0 \\ \cdots & \cdots \\ 1 & 0 \\ 0 & 1 \\ 0 & 1 \\ \cdots & \cdots \\ 0 & 1 \end{bmatrix} \begin{bmatrix} p \\ q \end{bmatrix}.$$

Clearly, we can see that $E(\mathbf{y}) = \mathbf{Xb}$, but what is $\mathbf{e}$? In this situation, the error $\mathbf{e}$ is just the difference between the response vector $\mathbf{y}$ and its mean vector $\mathbf{Xb}$. While we may point to $\mathbf{e}$ in most applications and list what sources of variation may be captured by it, in its most general form, $\mathbf{e}$ is just the difference $\mathbf{y} - E(\mathbf{y})$.

1.7 Two-Way Nested Model

Consider an experiment where we have two factors with one of the factors B nested within the other factor A. That is, every level of B appears with exactly one level of factor A. Consider the model

$$y_{ijk} = \mu + \alpha_i + \beta_{ij} + e_{ijk} \tag{1.8}$$

where $i = 1, \ldots, a$ (factor A treatments), $j = 1, \ldots, b_i$ (factor B treatments), $k = 1, \ldots, n_{ij}$ (experimental units or subjects), and we assume the errors e_{ij} are uncorrelated random errors with zero mean and constant, unknown variance σ^2. If we assume that both treatments are fixed, then the model (1.8) is a special case of the general linear model (1.1). For example, the A factor may be states and the B factor counties within states. An identifying feature of a nested model is that the name or index of the nested factor B within one level of factor A is unrelated to the same name or index within another level of factor A: Alleghany County in North Carolina has no real relationship with Alleghany County in either New York or Maryland, or, for that matter, Allegheny County in Pennsylvania.

Example 1.3: Nested Model
For a simple case, suppose we have sampled individuals, say looking at hospital costs for Medicare patients, from two states, and are looking at state and county effects as factors A and B, respectively. The two levels of factor A are Florida (F) and

Pennsylvania (P); the levels of factor B are Dade (D) and Monroe (M) counties within Florida, and Allegheny (A), Beaver (B), and Erie (E) counties within Pennsylvania. The response vector, design matrix, and parameter vector may then take the form

$$\mathbf{y} = \begin{bmatrix} y_{F,D,1} \\ \cdots \\ y_{F,D,n_{11}} \\ y_{F,M,1} \\ \cdots \\ y_{F,M,n_{12}} \\ y_{P,A,1} \\ \cdots \\ y_{P,A,n_{21}} \\ y_{P,B,1} \\ \cdots \\ y_{P,B,n_{22}} \\ y_{P,E,1} \\ \cdots \\ y_{P,E,n_{23}} \end{bmatrix}, \quad \mathbf{Xb} = \begin{bmatrix} \mathbf{1}_{n_{11}} & \mathbf{1}_{n_{11}} & \mathbf{0} & \mathbf{1}_{n_{11}} & \mathbf{0} & \mathbf{0} & \mathbf{0} & \mathbf{0} \\ \mathbf{1}_{n_{12}} & \mathbf{1}_{n_{12}} & \mathbf{0} & \mathbf{0} & \mathbf{1}_{n_{12}} & \mathbf{0} & \mathbf{0} & \mathbf{0} \\ \mathbf{1}_{n_{21}} & \mathbf{0} & \mathbf{1}_{n_{21}} & \mathbf{0} & \mathbf{0} & \mathbf{1}_{n_{21}} & \mathbf{0} & \mathbf{0} \\ \mathbf{1}_{n_{22}} & \mathbf{0} & \mathbf{1}_{n_{22}} & \mathbf{0} & \mathbf{0} & \mathbf{0} & \mathbf{1}_{n_{22}} & \mathbf{0} \\ \mathbf{1}_{n_{23}} & \mathbf{0} & \mathbf{1}_{n_{23}} & \mathbf{0} & \mathbf{0} & \mathbf{0} & \mathbf{0} & \mathbf{1}_{n_{23}} \end{bmatrix} \begin{bmatrix} \mu \\ \alpha_F \\ \alpha_P \\ \beta_{F,D} \\ \beta_{F,M} \\ \beta_{P,A} \\ \beta_{P,B} \\ \beta_{P,E} \end{bmatrix}.$$

Note that, in general, the dimensions for these items are $\mathbf{X}$ is $N \times (1 + a + \sum b_i)$, $p = 1 + a + \sum b_i$, where $N = \sum_{i=1}^{a} \sum_{j=1}^{b_i} n_{ij}$.

If either of the factors are considered random, say randomly selected counties within states, then (1.8) would become a special case of a mixed linear model. Note that the columns of the design matrix $\mathbf{X}$ corresponding to $\beta_{i1}, \ldots, \beta_{ib_i}$ add up to the column corresponding to α_i. And, as with (1.7), the columns corresponding to $\alpha_1, \ldots, \alpha_a$ add up to the first column, corresponding to the intercept parameter μ.

1.8 Two-Way Crossed Model

Consider an experiment with two-factors A and B, with factor A having a levels and factor B with b levels. Consider the two-factor crossed model given by

$$y_{ijk} = \mu + \alpha_i + \beta_j + \gamma_{ij} + e_{ijk} \tag{1.9}$$

where $i = 1, \ldots, a$ (factor A treatments), $j = 1, \ldots, b$ (factor B treatments), and $k = 1, \ldots, n_{ij}$ (experimental units or subjects). If we assume both treatments are fixed, then the model (1.9) is a special case of the general linear model (1.1). The design matrix $\mathbf{X}$ for this model is similar to the design matrix in the nested model in

Example 1.1 with the interaction effect γ_{ij} columns corresponding to the nested factor columns, but we need to include columns for $\beta_1, \ldots, \beta_b$. If there is no interaction in the model, then we delete the last set of columns (which number ab) corresponding to γ_{ij}. If at least one of the two treatment factors is random, then (1.9) is a mixed linear effects model.

Taking the simpler balanced case without interaction and $n_{ij} = 1$ for all i and j (no replication), the two-way crossed model without interaction is given by

$$y_{ij} = \mu + \alpha_i + \beta_j + e_{ij}, \quad i = 1, \ldots, a;\ j = 1, \ldots, b \tag{1.10}$$

where the errors e_{ij} are assumed to be uncorrelated random errors with zero mean and constant, unknown variance σ^2. This model can be written as a general linear model,

$$\mathbf{y} = \mathbf{Xb} + \mathbf{e}$$

where

$$\mathbf{y} = \begin{bmatrix} y_{11} \\ y_{12} \\ \ldots \\ y_{1b} \\ y_{21} \\ y_{22} \\ \ldots \\ y_{2b} \\ \ldots \\ y_{a1} \\ \ldots \\ y_{ab} \end{bmatrix}, \text{ and } \mathbf{Xb} = \begin{bmatrix} \mathbf{1}_b & \mathbf{1}_b & \mathbf{0} & \ldots & \mathbf{0} & \mathbf{I}_b \\ \mathbf{1}_b & \mathbf{0} & \mathbf{1}_b & \ldots & \mathbf{0} & \mathbf{I}_b \\ \mathbf{1}_b & \mathbf{0} & \mathbf{0} & \ldots & \mathbf{0} & \mathbf{I}_b \\ \ldots & & \ldots & \ldots & & \\ \mathbf{1}_b & \mathbf{0} & \mathbf{0} & \ldots & \mathbf{1}_b & \mathbf{I}_b \end{bmatrix} \begin{bmatrix} \mu \\ \alpha_1 \\ \alpha_2 \\ \ldots \\ \alpha_a \\ \beta_1 \\ \ldots \\ \beta_b \end{bmatrix}$$

where $\mathbf{X}$ is $N \times (a + b + 1)$, $p = 1 + a + b$, and $N = ab$.

The model above is most commonly used for randomized block experiments where α_i represent the treatment effects and β_i represent the block effects. If the blocks effects are considered random effects, then the model (1.10) is an example of a general (mixed) linear model. In this case, the design matrix and (fixed) parameter vector are given by

$$\mathbf{Xb} = \begin{bmatrix} \mathbf{1}_b & \mathbf{1}_b & \mathbf{0} & \ldots & \mathbf{0} \\ \mathbf{1}_b & \mathbf{0} & \mathbf{1}_b & \ldots & \mathbf{0} \\ \mathbf{1}_b & \mathbf{0} & \mathbf{0} & \ldots & \mathbf{0} \\ \ldots & & \ldots & \ldots & \\ \mathbf{1}_b & \mathbf{0} & \mathbf{0} & \ldots & \mathbf{1}_b \end{bmatrix} \begin{bmatrix} \mu \\ \alpha_1 \\ \alpha_2 \\ \ldots \\ \alpha_a \end{bmatrix}.$$

where $\mathbf{X}$ is $N \times (a + 1)$, and the errors $(\beta_j + e_{ij})$ are no longer uncorrelated.

1.9 Analysis of Covariance

Consider an experiment where we wish to compare a treatments after adjusting for a covariate (x). For example, the response y_{ij} may be weight gain for individual j under treatment or diet i and the covariate x_{ij} may be initial weight or height. An analysis of covariance model that is commonly used for this situation is

$$y_{ij} = \mu + \alpha_i + \beta x_{ij} + e_{ij}, \quad i = 1, \ldots, a \text{ treatments; } j = 1, \ldots, n_i \text{ subjects} \tag{1.11}$$

where the errors e_{ij} are assumed to be uncorrelated random errors with zero mean and constant, unknown variance σ^2. Assuming that the treatments are fixed and that the covariate is measured without error, we can express this model (1.11) as a special case of the general linear model (1.1) where

$$\mathbf{y} = \begin{bmatrix} y_{11} \\ y_{12} \\ \ldots \\ y_{1n_1} \\ y_{21} \\ y_{22} \\ \ldots \\ y_{2n_2} \\ \ldots \\ y_{a1} \\ \ldots \\ y_{an_a} \end{bmatrix}, \quad \text{and} \quad \mathbf{Xb} = \begin{bmatrix} \mathbf{1}_{n_1} & \mathbf{1}_{n_1} & \mathbf{0} & \ldots & \mathbf{0} & \mathbf{x}_{1.} \\ \mathbf{1}_{n_2} & \mathbf{0} & \mathbf{1}_{n_2} & \ldots & \mathbf{0} & \mathbf{x}_{2.} \\ \mathbf{1}_{n_3} & \mathbf{0} & \mathbf{0} & \ldots & \mathbf{0} & \mathbf{x}_{3.} \\ \ldots & & \ldots & \ldots & & \\ \mathbf{1}_{n_a} & \mathbf{0} & \mathbf{0} & \ldots & \mathbf{1}_{n_a} & \mathbf{x}_{a.} \end{bmatrix} \begin{bmatrix} \mu \\ \alpha_1 \\ \alpha_2 \\ \ldots \\ \alpha_a \\ \beta \end{bmatrix}.$$

and $\mathbf{x}_{i.}$ are n_i-dimensional vectors so that $(\mathbf{x}_{i.})_j = x_{ij}$, $\mathbf{X}$ is $N \times (a+2)$, and $p = 2+a$. An analysis of covariance model where the slopes differ across treatments can be written as

$$y_{ij} = \mu + \alpha_i + \beta_i x_{ij} + e_{ij}$$

and is also a special case of the general linear model (1.1).

1.10 Autoregression

Consider a simple linear trend model for data collected over time, given by

$$y_t = \beta_0 + \beta_1 t + e_t, \quad t = 1, 2, \ldots, N \tag{1.12}$$

where e_t are random variables with zero mean. Notice the change in index from i to t to emphasize the time ordering. Data collected over time on an individual, unit, or country tend to be correlated, usually modeled here with e_t being correlated. The model (1.12) is a special case of (1.1) with

$$\mathbf{y} = \begin{bmatrix} y_1 \\ y_2 \\ \dots \\ y_n \end{bmatrix}, \quad \mathbf{Xb} = \begin{bmatrix} 1 & 1 \\ 1 & 2 \\ \dots & \dots \\ 1 & N-1 \\ 1 & N \end{bmatrix} \begin{bmatrix} \beta_0 \\ \beta_1 \end{bmatrix}$$

so that $\mathbf{X}$ is $N \times 2$ and $p = 2$. Typically, autocorrelated models, such as the class of autoregressive-moving average (ARMA) time series models (see Fuller [11], Brockwell and Davis [5], or Shumway and Stoffer [43]), are used to model the e_t process.

In its simplest form, the first-order autoregressive error model can be written as

$$e_t = \eta + \rho e_{t-1} + a_t \tag{1.13}$$

where the a_t's are uncorrelated random variables with zero mean and constant variance. With some substitutions and some algebra, we can try to write this as a linear model:

$$\begin{aligned} y_t &= \beta_0 + \beta_1 t + e_t \\ &= \beta_0 + \beta_1 t + \eta + \rho e_{t-1} + a_t \\ &= \beta_0 + \beta_1 t + \eta + \rho[y_{t-1} - \beta_0 - \beta_1(t-1)] + a_t \\ &= [\beta_0(1-\rho) + \eta + \beta_1] + \beta_1(1-\rho)t + \rho y_{t-1} + a_t. \end{aligned} \tag{1.14}$$

This model (1.14) is not a special case of the general linear model (1.1), or the multiple regression model (1.3), because one covariate y_{t-1} is a random variable. The estimators and distributional results that we derive for the general linear model are not applicable to the estimators of the parameter ρ in (1.13).

One way that the autoregressive model can be incorporated in the framework here is to model the covariance matrix of the e_t process:

$$Cov(e_t, e_s) = Var(a_t)\rho^{|t-s|}. \tag{1.15}$$

Analysis of linear models with a covariance structure will first be encountered in Chapter 4.

1.11 Discussion

From the discussion above, it should be clear that many commonly used statistical models are special cases of the general linear model (1.1). In some cases, the errors e_i are uncorrelated random variables with zero mean and constant variance σ^2, and in other cases, they may be correlated. We will develop estimation and inference procedures for both of these situations.

We will first consider (1.1) as a numerical approximation problem and find the best approximation of $\mathbf{y}$ as a linear function of the columns of $\mathbf{X}$. Next, we assume that $E[\mathbf{e}] = \mathbf{0}$ and find unbiased estimators of $\lambda^T\mathbf{b}$, a linear combination of the parameters. We will see that not all linear combinations $\lambda^T\mathbf{b}$ have an unbiased estimator for it, while other linear combinations may have several linear unbiased estimators for it. Next we include the assumption that the errors e_i are uncorrelated with constant variance, and find the best (smallest variance) unbiased estimator for $\lambda^T\mathbf{b}$ when unbiased estimators are available. We extend these results to the case where the errors may not be uncorrelated or the variances are unequal. Finally, to perform hypothesis tests or construct confidence intervals, we need to assume a distribution for the e_i's. We discuss the properties of a multivariate normal distribution. We then use these properties to derive the distributions of quadratic forms, ANOVA sums of squares, and the t- and F-statistics commonly used in statistical methods. Likelihood ratio tests and models with a restricted parameter space will also be discussed. These eight models will be cited from time to time as examples upon which to apply the theory that will be developed throughout this book.

1.12 Summary

Several examples of the linear model are given: one-sample problem, simple linear regression, multiple regression, one-way ANOVA, two-way nested model, two-way crossed model, analysis of covariance, and autoregression.

1.13 Notes

- In the Introduction, the phrase describing *linear model* was "because the mean of the response vector $\mathbf{y}$ is linear in the unknown parameters b." More technically, let $E(\mathbf{y}) = \mu(\mathbf{b})$ be the mean function of $\mathbf{y}$. Then "linear" means algebraically $\mu(a_1\mathbf{b}^{(1)} + a_2\mathbf{b}^{(2)}) = a_1\mu(\mathbf{b}^{(1)}) + a_2\mu(\mathbf{b}^{(2)})$. Using the elementary vectors $\mathbf{e}^{(j)}$, $j = 1, \ldots, p$, we can then write $\mu(\mathbf{b}) = \sum_j b_j\mu(\mathbf{e}^{(j)})$. Stacking the vectors $\mu(\mathbf{e}^{(j)})$ as columns of a design matrix $\mathbf{X}$, we have the familiar $E(\mathbf{y}) = \mathbf{Xb}$.

- The models in Equations (1.4) and (1.5) exemplify *nonlinear models*. In practice, the analysis of nonlinear models follows very much the linear model in the use of least squares estimators. However, no finite sample results are available, as the least squares problem must be solved numerically.

- The *generalized linear model* differs in two ways. First, the distribution of the response is a key part of the specification of the model, and second, the parameters of that distribution are linked by a one-to-one *link function* to a linear model in the covariates. The most common example is *logistic regression*, where the response y_i follows a Bernoulli distribution with probability p_i, where p_i is related to the covariate vector $\mathbf{x}_i$ by the link function $p_i = e^{\gamma_i}/(1 + e^{\gamma_i})$ and $\gamma_i = \mathbf{b}^T\mathbf{x}_i$. A second case is *Poisson regression*, where the response y_i follows a Poisson distribution with rate λ_i and link $log(\lambda_i) = \mathbf{b}^T\mathbf{x}_i$. In some designed experimental situations (as in Example 1.2), these models can be rewritten as linear models with different parameters and will not follow the Gauss–Markov assumptions of Chapter 4 because of heteroskedasticity.

1.14 Exercises

1.1. Let $Y_i, i = 1, \ldots, N$ be independent binomial random variables, with parameters n_i and p_i where $p_i = \beta_0 + \beta_1 x_i$. Write this as a general linear model.

1.2. Let $Y_i, i = 1, \ldots, N$ be independent Poisson random variables, with parameters λ_i where $\lambda_i = \beta_0 + \beta_1 x_i$. Write this as a general linear model.

1.3. Let $y_t = \alpha_0 + \rho cos(2\pi f t - \phi) + e_t$, for $t = 1, \ldots, N$ where $E(e_t) = 0$. Here f is known and the unknown parameters are α_0, ρ, and ϕ. Write this as a linear model with three, perhaps different, parameters. Hint: $cos(a - b) = cos(a)cos(b) - sin(a)sin(b)$.

1.4. Consider the simple linear regression model $y_i = \beta_0 + \beta_1 x_i + e_i, i = 1, \ldots, N$, and let $z_i = y_i/x_i$. Write a linear model for z_i.

1.5. For the analysis of covariance model in Section 1.9, write out $\mathbf{y}$ and $\mathbf{Xb}$ where $a = 3$ and $x_{ij} = 2i + j$.

Chapter 2

The Linear Least Squares Problem

2.1 The Normal Equations

In the previous chapter, we have seen several examples of the linear model in the form $\mathbf{y} = \mathbf{X}\mathbf{b} + \mathbf{e}$, relating N response observations in the vector $\mathbf{y}$ to the explanatory variables stored as columns in the design matrix $\mathbf{X}$. One mathematical view of the linear model is the closest or best approximation $\mathbf{X}\mathbf{b}$ to the observed vector $\mathbf{y}$. If we define closeness or distance in the familiar Euclidean manner, then finding the best approximation means minimizing the squared distance between the observed vector $\mathbf{y}$ and its approximation $\mathbf{X}\mathbf{b}$, given by the sum of squares function

$$Q(\mathbf{b}) = (\mathbf{y} - \mathbf{X}\mathbf{b})^T(\mathbf{y} - \mathbf{X}\mathbf{b}) = \|\mathbf{y} - \mathbf{X}\mathbf{b}\|^2. \tag{2.1}$$

A value of the vector $\mathbf{b}$ that minimizes $Q(\mathbf{b})$ is called a *least squares solution*. We will refrain from calling it an estimator of $\mathbf{b}$, until we discuss seriously the idea of estimation. In order to find the least squares solution, we take the partial derivatives with respect to the components of $\mathbf{b}$ to obtain the gradient vector

$$\frac{\partial Q}{\partial \mathbf{b}} = \begin{bmatrix} \frac{\partial Q}{\partial \mathbf{b}_1} \\ \cdots \\ \frac{\partial Q}{\partial \mathbf{b}_p} \end{bmatrix}, \tag{2.2}$$

a vector whose j^{th} component is the partial derivative of the function Q with respect to the jth variable b_j, also written as $(\frac{\partial Q(\mathbf{b})}{\partial \mathbf{b}})_j = \partial Q(\mathbf{b})/\partial b_j$. The minimum of the function $Q(\mathbf{b})$ will occur when the gradient is zero, so we solve $\frac{\partial Q}{\partial \mathbf{b}} = \mathbf{0}$ to find least squares solutions. We now include some results on vector derivatives that will be helpful.

Result 2.1 *Let* $\mathbf{a}$ *and* $\mathbf{b}$ *be* $p \times 1$ *vectors and* $\mathbf{A}$ *be a* $p \times p$ *matrix of constants. Then (a)* $\frac{\partial \mathbf{a}^T\mathbf{b}}{\partial \mathbf{b}} = \mathbf{a}$, *and (b)* $\frac{\partial \mathbf{b}^T\mathbf{A}\mathbf{b}}{\partial \mathbf{b}} = (\mathbf{A} + \mathbf{A}^T)\mathbf{b}$.

Proof: For (a), note that $\mathbf{a}^T\mathbf{b} = a_1b_1 + a_2b_2 + \cdots + a_pb_p$, so $(\frac{\partial \mathbf{a}^T\mathbf{b}}{\partial \mathbf{b}})_j = \frac{\partial \mathbf{a}^T\mathbf{b}}{\partial \mathbf{b}_j} = a_j$. Similarly,

$$\mathbf{b}^T\mathbf{A}\mathbf{b} = \sum_{j=1}^{p}\sum_{k=1}^{p} A_{jk}b_jb_k,$$

and to form $(\frac{\partial \mathbf{b}^T\mathbf{A}\mathbf{b}}{\partial \mathbf{b}})_j$, the terms that depend on b_j are

$$A_{jj}b_j^2 + \sum_{k\neq j}(A_{jk} + A_{kj})b_kb_j,$$

so the partial derivative with respect to b_j is

$$2A_{jj}b_j + \sum_{k\neq j} A_{jk}b_k + \sum_{k\neq j} A_{kj}b_k,$$

which is the jth element of $(\mathbf{A} + \mathbf{A}^T)\mathbf{b}$. □

Since we can write

$$Q(\mathbf{b}) = (\mathbf{y} - \mathbf{X}\mathbf{b})^T(\mathbf{y} - \mathbf{X}\mathbf{b}) = \mathbf{y}^T\mathbf{y} - 2\mathbf{y}^T\mathbf{X}\mathbf{b} + \mathbf{b}^T\mathbf{X}^T\mathbf{X}\mathbf{b},$$

using Result 2.1, we have

$$\frac{\partial Q}{\partial \mathbf{b}} = -2\mathbf{X}^T\mathbf{y} + 2\mathbf{X}^T\mathbf{X}\mathbf{b} = -2\mathbf{X}^T(\mathbf{y} - \mathbf{X}\mathbf{b}).$$

Setting this gradient to zero, we obtain the normal equations (NEs):

$$\mathbf{X}^T\mathbf{X}\mathbf{b} = \mathbf{X}^T\mathbf{y}. \tag{2.3}$$

Let us consider two examples of the normal equations.

Example 2.1: Simple Linear Regression
Consider the model

$$y_i = \beta_0 + \beta_1x_i + e_i, \quad i = 1, \ldots, N.$$

In this case, we find

$$\mathbf{X}^T\mathbf{X}\mathbf{b} = \begin{bmatrix} N & \sum_{i=1}^N x_i \\ \sum_{i=1}^N x_i & \sum_{i=1}^N x_i^2 \end{bmatrix} \begin{bmatrix} \beta_0 \\ \beta_1 \end{bmatrix} \quad \text{and} \quad \mathbf{X}^T\mathbf{y} = \begin{bmatrix} \sum_{i=1}^N y_i \\ \sum_{i=1}^N x_iy_i \end{bmatrix}$$

and the normal equations are given by

$$N\beta_0 + \sum_{i=1}^{N} x_i\beta_1 = \sum_{i=1}^{N} y_i \tag{2.4}$$

$$\sum_{i=1}^{N} x_i\beta_0 + \sum_{i=1}^{N} x_i^2\beta_1 = \sum_{i=1}^{N} x_iy_i. \tag{2.5}$$

As long as $\sum_{i=1}^{N}(x_i - \overline{x})^2 > 0$, the solution to (2.4) is

$$\hat{\beta}_1 = \sum_{i=1}^{N}(x_i - \overline{x})y_i / \sum_{i=1}^{N}(x_i - \overline{x})^2$$

$$\hat{\beta}_0 = \overline{y} - \hat{\beta}_1\overline{x}$$

where $\overline{x} = \sum_{i=1}^{N} x_i/N$ and $\overline{y} = \sum_{i=1}^{N} y_i/N$. Note that if $\sum_{i=1}^{N}(x_i - \overline{x})^2 = 0$, that is, all x_i's are equal to the same value, then $\hat{\beta}_0 = \overline{y} - c\overline{x}$, and $\hat{\beta}_1 = c$ is a solution for all values of c. That is, in this case, there are infinitely many solutions to the normal equations.

Example 2.2: Balanced One-Way ANOVA
Consider the balanced ($n_i = n$) model

$$y_{ij} = \mu + \alpha_i + e_{ij}, \quad \text{for} \quad i = 1, \ldots, a;\ j = 1, \ldots, n. \tag{2.6}$$

For this model, the design matrix is given in (1.7) with $n_i = n$ for all i, and so

$$\mathbf{X}^T\mathbf{X}\mathbf{b} = \begin{bmatrix} na & n & n & \cdots & n \\ n & n & 0 & \cdots & 0 \\ n & 0 & n & \cdots & 0 \\ \cdots & \cdots & \cdots & & \cdots \\ n & 0 & 0 & \cdots & n \end{bmatrix} \begin{bmatrix} \mu \\ \alpha_1 \\ \alpha_2 \\ \cdots \\ \alpha_a \end{bmatrix} \quad \text{and} \quad \mathbf{X}^T\mathbf{y} = \begin{bmatrix} y_{..} \\ y_{1.} \\ y_{2.} \\ \cdots \\ y_{a.} \end{bmatrix}$$

where $y_{i.} = \sum_{j=1}^{N} y_{ij}$ and $y_{..} = \sum_{i=1}^{a} y_{i.}$. Note that in this case, $\mathbf{X}^T\mathbf{X}$ is a singular matrix because the first column can be written as the sum of the other columns. However, $\hat{\mu} = c, \hat{\alpha}_i = \bar{y}_{i.} - c, i = 1, \ldots, a$, is a solution to the normal equations for any value of c, where $\bar{y}_i = \sum y_{ij}/n_i$.

From the two examples above, it should be clear that the normal equations may not have a unique solution. We will show that the normal equations are always consistent, that is, always have at least one solution. We will also show that sum of squares function $Q(\mathbf{b})$ in (2.1) is minimized at $\hat{\mathbf{b}}$ if and only if $\hat{\mathbf{b}}$ is a solution to the normal equations.

2.2 The Geometry of Least Squares

Keeping in mind the discussion in appendix A of the general problem of solving systems of linear equations, our attention turns to a particular case, namely, the normal equations:

$$\mathbf{X}^T\mathbf{X}\mathbf{b} = \mathbf{X}^T\mathbf{y}.$$

The first question, then, is whether these equations are consistent: Is the right-hand-side $\mathbf{X}^T\mathbf{y}$ in $\mathcal{C}(\mathbf{X}^T\mathbf{X})$? One (algebraic) route is to construct a solution for any response vector $\mathbf{y}$, another is to describe the geometry.

Lemma 2.1 $\mathcal{N}(\mathbf{X}^T\mathbf{X}) = \mathcal{N}(\mathbf{X})$

Proof: Let $\mathbf{w} \in \mathcal{N}(\mathbf{X})$, then $\mathbf{X}\mathbf{w} = \mathbf{0}$; premultiplying by $\mathbf{X}^T$ shows that $\mathbf{w} \in \mathcal{N}(\mathbf{X}^T\mathbf{X})$. Conversely, if $\mathbf{w} \in \mathcal{N}(\mathbf{X}^T\mathbf{X})$, then $\mathbf{X}^T\mathbf{X}\mathbf{w} = \mathbf{0}$ and also $\mathbf{w}^T\mathbf{X}^T\mathbf{X}\mathbf{w} = \|\mathbf{X}\mathbf{w}\|^2 = 0$, so that $\mathbf{X}\mathbf{w} = \mathbf{0}$, or $\mathbf{w} \in \mathcal{N}(\mathbf{X})$. ▯

Result 2.2 $\mathcal{C}(\mathbf{X}^T\mathbf{X}) = \mathcal{C}(\mathbf{X}^T)$

Proof: Combine Result A.6 and Lemma 2.1, since $\mathcal{N}(\mathbf{X})$ and $\mathcal{C}(\mathbf{X}^T)$ are orthogonal complements from Result A.5. ▯

Corollary 2.1 *The normal equations are consistent.*

Proof: First, $\mathbf{X}^T\mathbf{y}$ is obviously a vector in $\mathcal{C}(\mathbf{X}^T) = \mathcal{C}(\mathbf{X}^T\mathbf{X})$ by Result 2.2; consistency follows from Result A.9. ▯

Having established that the normal equations are consistent, then a solution, or the whole family of solutions, can be found once a generalized inverse for $\mathbf{X}^T\mathbf{X}$ is found. Finding a generalized inverse is easier if the rank is known:

Corollary 2.2 *The rank of* $\mathbf{X}^T\mathbf{X}$ *is the same as the rank of* $\mathbf{X}$.

Proof: A consequence of Result 2.2 is $rank(\mathbf{X}^T\mathbf{X}) = rank(\mathbf{X}^T)$ because of the relationship between rank and dimension of a column space, and $\mathbf{X}$ has the same rank as its transpose. ▯

The original motivation for the normal equations was to minimize an error sum of squares; we can now complete that discussion with the following result.

Result 2.3 $\hat{\mathbf{b}}$ *is a solution to the normal Equations (2.3) iff* $\hat{\mathbf{b}}$ *minimizes* $Q(\mathbf{b})$ *defined in (2.1).*

Proof: Let $\hat{\mathbf{b}}$ be a solution to the normal equations. Then

$$\begin{aligned}
Q(\mathbf{b}) &= (\mathbf{y}-\mathbf{X}\mathbf{b})^T(\mathbf{y}-\mathbf{X}\mathbf{b}) = (\mathbf{y}-\mathbf{X}\hat{\mathbf{b}}+\mathbf{X}\hat{\mathbf{b}}-\mathbf{X}\mathbf{b})^T(\mathbf{y}-\mathbf{X}\hat{\mathbf{b}}+\mathbf{X}\hat{\mathbf{b}}-\mathbf{X}\mathbf{b}) \\
&= (\mathbf{y}-\mathbf{X}\hat{\mathbf{b}})^T(\mathbf{y}-\mathbf{X}\hat{\mathbf{b}}) + 2(\hat{\mathbf{b}}-\mathbf{b})^T\mathbf{X}^T(\mathbf{y}-\mathbf{X}\hat{\mathbf{b}}) + (\mathbf{X}\hat{\mathbf{b}}-\mathbf{X}\mathbf{b})^T(\mathbf{X}\hat{\mathbf{b}}-\mathbf{X}\mathbf{b}) \\
&= Q(\hat{\mathbf{b}}) + \|\mathbf{X}(\hat{\mathbf{b}}-\mathbf{b})\|^2
\end{aligned}$$

where the cross-product term vanishes since $\hat{\mathbf{b}}$ solves the normal equations: $\mathbf{X}^T(\mathbf{y} - \mathbf{X}\mathbf{b}) = \mathbf{0}$. Now since no other value of $\mathbf{b}$ can give a smaller value of $Q(\mathbf{b})$, clearly $\hat{\mathbf{b}}$ minimizes $Q(\mathbf{b})$. Conversely, if $\tilde{\mathbf{b}}$ also minimizes $Q(\mathbf{b})$, then $\mathbf{X}(\tilde{\mathbf{b}} - \hat{\mathbf{b}})$ must be $\mathbf{0}$ or $\mathbf{X}\hat{\mathbf{b}} = \mathbf{X}\tilde{\mathbf{b}}$. Therefore, $\mathbf{X}^T\mathbf{X}\hat{\mathbf{b}} = \mathbf{X}^T\mathbf{X}\tilde{\mathbf{b}} = \mathbf{X}^T\mathbf{y}$, and hence $\tilde{\mathbf{b}}$ also solves the normal equations. □

Lost in the algebra of the proof above is a subtle result that provides the key to the geometry of the least squares problem. First, we will present this as a corollary, and then discuss its implications.

Corollary 2.3 $\mathbf{X}\hat{\mathbf{b}}$ *is invariant to the choice of a solution* $\hat{\mathbf{b}}$ *to the normal Equations* (2.3).

Proof: Let $\hat{\mathbf{b}}$ and $\tilde{\mathbf{b}}$ be two solutions to the normal equations. Following directly from Result 2.3 above, we have $Q(\hat{\mathbf{b}}) = Q(\tilde{\mathbf{b}})$, implying $\|\mathbf{X}(\hat{\mathbf{b}} - \tilde{\mathbf{b}})\|^2 = 0$ and $\mathbf{X}\hat{\mathbf{b}} = \mathbf{X}\tilde{\mathbf{b}}$. □

The geometric side to this corollary is that the difference between two solutions of the normal equations must be a vector in $\mathcal{N}(\mathbf{X}^T\mathbf{X}) = \mathcal{N}(\mathbf{X})$. Further examination of the original least squares problem reveals that the closest point—in Euclidean distance—in $\mathcal{C}(\mathbf{X})$ to the observed $\mathbf{y}$ is vector of *fitted values* $\hat{\mathbf{y}} = \mathbf{X}\hat{\mathbf{b}}$. The difference, $\hat{\mathbf{e}} = \mathbf{y} - \mathbf{X}\hat{\mathbf{b}}$, known as the *residual vector*, is in its orthogonal complement, $\mathcal{N}(\mathbf{X}^T)$, following from the normal equations:

$$\mathbf{X}^T\hat{\mathbf{e}} = \mathbf{X}^T(\mathbf{y} - \mathbf{X}\hat{\mathbf{b}}) = \mathbf{0}$$

This unique orthogonal decomposition (from Result A.4) also yields a decomposition of sums of squares from the Pythagorean Theorem:

$$\|\mathbf{y}\|^2 = \|\hat{\mathbf{y}}\|^2 + \|\hat{\mathbf{e}}\|^2. \tag{2.7}$$

That is, the total sum of squares $\|\mathbf{y}\|^2$ is the sum of the *regression sum of squares*, or $SSR = \|\mathbf{X}\hat{\mathbf{b}}\|^2$, and *error sum of squares*, or $SSE = \|\hat{\mathbf{e}}\|^2$. Another route for constructing this orthogonal decomposition is to construct the symmetric projection matrix that projects onto $\mathcal{C}(\mathbf{X})$. The main result is preceded by some simple, yet very powerful lemmas that are important in their own right.

Result 2.4 $\mathbf{X}^T\mathbf{X}\mathbf{A} = \mathbf{X}^T\mathbf{X}\mathbf{B}$ *iff* $\mathbf{X}\mathbf{A} = \mathbf{X}\mathbf{B}$.

Proof: If $\mathbf{XA} = \mathbf{XB}$, then premultiplying by $\mathbf{X}^T$ yields one direction. For the other direction, there are two approaches; the first is geometric. If $\mathbf{X}^T\mathbf{X}(\mathbf{A} - \mathbf{B}) = \mathbf{0}$, then the columns of $[\mathbf{X}(\mathbf{A} - \mathbf{B})] \in \mathcal{N}(\mathbf{X}^T)$. But the columns of $[\mathbf{X}(\mathbf{A} - \mathbf{B})]$ are also obviously in $\mathcal{C}(\mathbf{X})$. Since $\mathcal{N}(\mathbf{X}^T)$ and $\mathcal{C}(\mathbf{X})$ are orthogonal complements (from Result A.5), then the only vector in both is $\mathbf{0}$, so clearly each column of $[\mathbf{X}(\mathbf{A} - \mathbf{B})]$ is $\mathbf{0}$. Now the second, algebraic approach is to premultiply $[\mathbf{X}^T\mathbf{X}(\mathbf{A} - \mathbf{B})]$ by $(\mathbf{A} - \mathbf{B})^T$ to give

$$(\mathbf{A}-\mathbf{B})^T\mathbf{X}^T\mathbf{X}(\mathbf{A}-\mathbf{B}) = [\mathbf{X}(\mathbf{A}-\mathbf{B})]^T[\mathbf{X}(\mathbf{A}-\mathbf{B})] = \mathbf{0}.$$

Using Lemma A.1, we see that $\mathbf{X}(\mathbf{A}-\mathbf{B}) = \mathbf{0}$. ◻

Result 2.5 $(\mathbf{X}^T\mathbf{X})^g\mathbf{X}^T$ *is a generalized inverse of* $\mathbf{X}$.

Proof: From the definition of a generalized inverse, we have $(\mathbf{X}^T\mathbf{X})(\mathbf{X}^T\mathbf{X})^g(\mathbf{X}^T\mathbf{X}) = (\mathbf{X}^T\mathbf{X})$. Now apply Result 2.4 with $\mathbf{A} = (\mathbf{X}^T\mathbf{X})^g(\mathbf{X}^T\mathbf{X})$ and

$$\mathbf{B} = \mathbf{I} \text{ to get } \mathbf{X}[(\mathbf{X}^T\mathbf{X})^g\mathbf{X}^T]\mathbf{X} = \mathbf{X},$$

or $[(\mathbf{X}^T\mathbf{X})^g\mathbf{X}^T]$ is a generalized inverse of $\mathbf{X}$. ◻

Theorem 2.1 $\mathbf{P_X} = \mathbf{X}(\mathbf{X}^T\mathbf{X})^g\mathbf{X}^T$ *is the projection matrix onto* $\mathcal{C}(\mathbf{X})$*, that is,* $\mathbf{P_X}$ *is*

(a) idempotent

(b) projects onto $\mathcal{C}(\mathbf{X})$*,*

(c) invariant to the choice of generalized inverse,

(d) symmetric, and

(e) unique.

Proof: Using Result 2.5, $\mathbf{P_X}$ can be written in the form $\mathbf{AA}^g$, so that from Result A.14 we know that $\mathbf{P_X}$ projects onto $\mathcal{C}(\mathbf{X})$, providing (a) and (b). For (c), let $\mathbf{G}_1$ and $\mathbf{G}_2$ be two generalized inverses of $(\mathbf{X}^T\mathbf{X})$, so

$$(\mathbf{X}^T\mathbf{X})\mathbf{G}_1(\mathbf{X}^T\mathbf{X}) = (\mathbf{X}^T\mathbf{X})\mathbf{G}_2(\mathbf{X}^T\mathbf{X}) = \mathbf{X}^T\mathbf{X}$$

and taking $\mathbf{A} = \mathbf{G}_1\mathbf{X}^T\mathbf{X}$, $\mathbf{B} = \mathbf{G}_2\mathbf{X}^T\mathbf{X}$ and applying Result 2.4 gives

$$\mathbf{X}\mathbf{G}_1(\mathbf{X}^T\mathbf{X}) = \mathbf{X}\mathbf{G}_2(\mathbf{X}^T\mathbf{X}).$$

Now transpose this result to give

$$(\mathbf{X}^T\mathbf{X})\mathbf{G}_1^T\mathbf{X}^T = (\mathbf{X}^T\mathbf{X})\mathbf{G}_2^T\mathbf{X}^T$$

and again apply Result 2.4 with $\mathbf{A} = \mathbf{G}_1^T\mathbf{X}^T$ and $\mathbf{B} = \mathbf{G}_2^T\mathbf{X}^T$ to obtain

$$\mathbf{X}\mathbf{G}_1\mathbf{X}^T = \mathbf{X}\mathbf{G}_2\mathbf{X}^T = \mathbf{P_X}.$$

Therefore, $\mathbf{P_X}$ is invariant to the choice of the generalized inverse of the matrix $(\mathbf{X}^T\mathbf{X})$. For symmetry, notice that if $\mathbf{G}_1$ is a generalized inverse of $\mathbf{X}^T\mathbf{X}$, so is $\mathbf{G}_1^T$ (see Exercise A.22); hence $\mathbf{P_X}$ is symmetric. Uniqueness then follows from Result A.16. ◻

The uniqueness of this projection permits the notation of the capital, bold $\mathbf{P_X}$ denoting the symmetric projection onto the column space of the matrix given as its subscript, that is, $\mathbf{P_X}$ is the projection matrix onto $\mathcal{C}(\mathbf{X})$.

Result 2.6 $(\mathbf{I}-\mathbf{P_X})$ *is the unique, symmetric projection onto* $\mathcal{N}(\mathbf{X}^T)$.

Proof: Invoking Corollary A.4 is the easy route. For a direct proof, first employ Result 2.4 and Exercise A.22 to say that $\mathbf{X}(\mathbf{X}^T\mathbf{X})^g$ is a generalized inverse of $\mathbf{X}^T$. Then invoke Result A.15 with $\mathbf{A}=\mathbf{X}^T$ and $\mathbf{A}^g=\mathbf{X}(\mathbf{X}^T\mathbf{X})^g$ to say that $\mathbf{I}-\mathbf{A}^g\mathbf{A}=\mathbf{I}-\mathbf{P_X}$ projects onto $\mathcal{N}(\mathbf{X}^T)$. Symmetry follows from Theorem 2.1; uniqueness from Result A.16. □

Example 2.3: Examples of $\mathbf{P_X}$ and $\mathbf{I}-\mathbf{P_X}$

Let us construct $\mathbf{P_X}$ and $\mathbf{I}-\mathbf{P_X}$ for a simple case. Here let $\mathbf{X}=\begin{bmatrix}1&0&1\\1&0&1\\1&1&2\\1&1&2\end{bmatrix}$, so that $rank(\mathbf{X})=2$, and $\mathbf{X}^T\mathbf{X}=\begin{bmatrix}4&2&6\\2&2&4\\6&4&10\end{bmatrix}$. One choice of generalized inverse of $\mathbf{X}^T\mathbf{X}$ is $\mathbf{G}_1=\frac{1}{2}\begin{bmatrix}1&-1&0\\-1&2&0\\0&0&0\end{bmatrix}$, leading (via Result 2.4) to $\mathbf{G}_1\mathbf{X}^T=\frac{1}{2}\begin{bmatrix}1&1&0&0\\-1&-1&1&1\\0&0&0&0\end{bmatrix}$ as a generalized inverse for $\mathbf{X}$. Then $\mathbf{P_X}=\mathbf{X}\mathbf{G}_1\mathbf{X}^T=\frac{1}{2}\begin{bmatrix}1&1&0&0\\1&1&0&0\\0&0&1&1\\0&0&1&1\end{bmatrix}$. By subtraction, $\mathbf{I}-\mathbf{P_X}=\frac{1}{2}\begin{bmatrix}1&-1&0&0\\-1&1&0&0\\0&0&1&-1\\0&0&-1&1\end{bmatrix}$. Another generalized inverse, $\mathbf{G}_2=\frac{1}{2}\begin{bmatrix}5&0&-3\\0&0&0\\-3&0&2\end{bmatrix}$, leads to the same $\mathbf{P_X}=\mathbf{X}\mathbf{G}_2\mathbf{X}^T$. Note: $trace(\mathbf{P_X})=trace(\mathbf{I}-\mathbf{P_X})=2$. Also see Exercises 2.8 and 2.9.

Example 2.4: $\mathbf{P_1}$ and the Mean

One of the simplest examples of these projection matrices follows from the case where $\mathbf{X}=\mathbf{1}_N$, a column of N ones. Here $\mathbf{X}^T\mathbf{X}=\mathbf{1}^T\mathbf{1}=N$, and $\mathbf{P_1}=(1/N)\mathbf{1}\mathbf{1}^T$, an $N\times N$ matrix with each entry equal to $1/N$. If we project any vector $\mathbf{y}$, we get a vector with each component equal to the mean:

$$\mathbf{P_1}\mathbf{y}=(1/N)\mathbf{1}\mathbf{1}^T\mathbf{y}=\overline{y}\mathbf{1}.$$

Similarly, $\mathbf{I}_N-\mathbf{P_1}=\mathbf{I}-(1/N)\mathbf{1}\mathbf{1}^T$, which is a matrix whose diagonal entries are $(N-1)/N$ and other entries $-1/N$. Projecting any vector $\mathbf{y}$ produces a vector of deviations from the mean:

$$(\mathbf{I}_N-\mathbf{P_1})\mathbf{y}=\mathbf{y}-(1/N)\mathbf{1}\mathbf{1}^T\mathbf{y}=\mathbf{y}-\overline{y}\mathbf{1},$$

a vector whose i^{th} component is $y_i-\overline{y}$.

The orthogonal decomposition of the previous section can now be viewed in terms of these projections. The first piece in $\mathcal{C}(\mathbf{X})$,

$$\hat{\mathbf{y}} = \mathbf{X}\hat{\mathbf{b}} = \mathbf{X}(\mathbf{X}^T\mathbf{X})^g\mathbf{X}^T\mathbf{y} = \mathbf{P}_\mathbf{X}\mathbf{y},$$

can be viewed in terms of a solution $(\mathbf{X}^T\mathbf{X})^g\mathbf{X}^T\mathbf{y}$, but constant for all solutions, or also as $\mathbf{P}_\mathbf{X}\mathbf{y}$ projecting the vector $\mathbf{y}$ onto $\mathcal{C}(\mathbf{X})$. Then the residual vector $\hat{\mathbf{e}} = \mathbf{y} - \mathbf{X}\hat{\mathbf{b}}$ can now be rewritten in terms of the projector onto the orthogonal complement:

$$\hat{\mathbf{e}} = \mathbf{y} - \mathbf{X}\hat{\mathbf{b}} = (\mathbf{I} - \mathbf{P}_\mathbf{X})\mathbf{y}.$$

Moreover, the least squares problem can be viewed in another fashion. Notice that we parameterized $\mathcal{C}(\mathbf{X})$ naturally with $\mathbf{Xb}$, so that an orthogonal decomposition yields

$$Q(\mathbf{b}) = \|\mathbf{y} - \mathbf{Xb}\|^2 = \|(\mathbf{y} - \hat{\mathbf{y}}) - (\mathbf{Xb} - \hat{\mathbf{y}})\|^2 = \|\hat{\mathbf{y}} - \mathbf{Xb}\|^2 + \|\hat{\mathbf{e}}\|^2$$

where the second piece cannot be minimized by varying $\mathbf{b}$. The first piece can be minimized to zero, because the equations

$$\mathbf{Xb} = \mathbf{P}_\mathbf{X}\mathbf{y} = \hat{\mathbf{y}} \tag{2.8}$$

are consistent since $\mathbf{P}_\mathbf{X}\mathbf{y} \in \mathcal{C}(\mathbf{X})$.

Result 2.7 *The solutions to the normal equations are the same as the solutions to the consistent equations* $\mathbf{Xb} = \mathbf{P}_\mathbf{X}\mathbf{y}$.

Proof: If $\hat{\mathbf{b}}$ solves $\mathbf{Xb} = \mathbf{P}_\mathbf{X}\mathbf{y}$ premultiplying by $\mathbf{X}^T$, we have $\mathbf{X}^T\mathbf{Xb} = \mathbf{X}^T\mathbf{P}_\mathbf{X}\mathbf{y} = \mathbf{X}^T\mathbf{y}$, and so $\hat{\mathbf{b}}$ also solves the normal equations. If $\hat{\mathbf{b}}$ is a solution to the normal equations, then

$$\mathbf{X}^T\mathbf{X}\hat{\mathbf{b}} = \mathbf{X}^T\mathbf{y} = \mathbf{X}^T\mathbf{P}_\mathbf{X}\mathbf{y} = \mathbf{X}^T\mathbf{X}(\mathbf{X}^T\mathbf{X})^g\mathbf{X}^T\mathbf{y}.$$

Now from Result 2.4, we obtain $\mathbf{X}\hat{\mathbf{b}} = \mathbf{X}(\mathbf{X}^T\mathbf{X})^g\mathbf{X}^T\mathbf{y} = \mathbf{P}_\mathbf{X}\mathbf{y}$, and hence $\hat{\mathbf{b}}$ is a solution to $\mathbf{Xb} = \mathbf{P}_\mathbf{X}\mathbf{y}$. □

Example 2.5: Toy Example

Consider the following toy regression problem, illustrated in Figure 2.1: Let $\mathrm{y} = \begin{bmatrix} 1 \\ 2 \\ 2 \end{bmatrix}$,

$\mathrm{X} = \begin{bmatrix} 1 & 0 \\ 1 & 1 \\ 0 & 0 \end{bmatrix}$, so that $\begin{bmatrix} 2 & 1 \\ 1 & 1 \end{bmatrix} \begin{bmatrix} b_1 \\ b_2 \end{bmatrix} = \mathbf{X}^T\mathbf{Xb} = \mathbf{X}^T\mathbf{y} = \begin{bmatrix} 3 \\ 2 \end{bmatrix}$, which has the solution $\begin{bmatrix} 1 \\ 1 \end{bmatrix}$. Equivalently, the following equations yield the same solution:

$$\begin{bmatrix} 1 & 0 \\ 1 & 1 \\ 0 & 0 \end{bmatrix} \begin{bmatrix} b_1 \\ b_2 \end{bmatrix} = \mathbf{Xb} = \mathbf{P}_\mathbf{X}\mathbf{y} = \begin{bmatrix} 1 & 0 & 0 \\ 0 & 1 & 0 \\ 0 & 0 & 0 \end{bmatrix} \begin{bmatrix} 1 \\ 2 \\ 2 \end{bmatrix} = \hat{\mathbf{y}} = \begin{bmatrix} 1 \\ 2 \\ 0 \end{bmatrix}, \text{ and } \hat{\mathbf{e}} = \begin{bmatrix} 0 \\ 0 \\ 2 \end{bmatrix}.$$

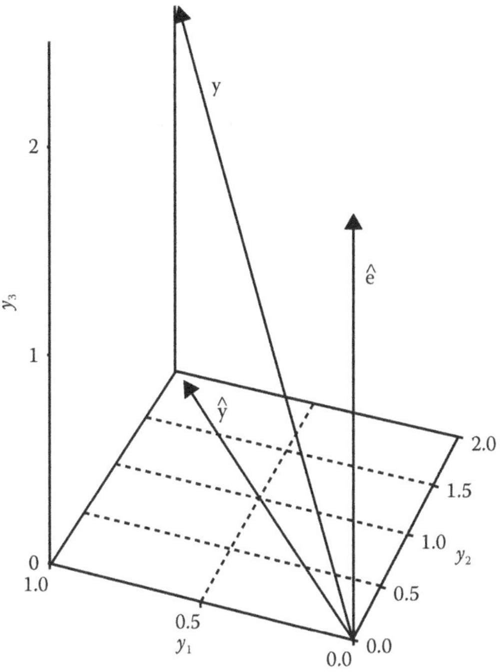

Figure 2.1: Decomposition of response vector $\mathbf{y}$ into fitted values $\hat{\mathbf{y}}$ and residuals $\hat{\mathbf{e}}$. Residual vector is orthogonal to the vector of fitted values which lies in plane of $\mathcal{C}(\mathbf{X})$ (here in plane where $y_3 = 0$).

This section concludes with a useful geometric result regarding subspaces that will arise later in discussions of subset regressions, decomposition of sums of squares, and hypothesis testing. Its importance later in this course warrants its designation as a theorem.

Theorem 2.2 *If* $\mathcal{C}(\mathbf{W}) \subseteq \mathcal{C}(\mathbf{X})$ *then* $\mathbf{P_X} - \mathbf{P_W}$ *is the projection onto* $\mathcal{C}((\mathbf{I} - \mathbf{P_W})\mathbf{X})$.

Proof: First, is $(\mathbf{P_X} - \mathbf{P_W})$ idempotent? Construct

$$(\mathbf{P_X} - \mathbf{P_W})^2 = \mathbf{P_X} - \mathbf{P_W}\mathbf{P_X} - \mathbf{P_X}\mathbf{P_W} + \mathbf{P_W}$$

since both are idempotent. Now since $\mathbf{P_W}\mathbf{z} \in \mathcal{C}(\mathbf{W}) \subseteq \mathcal{C}(\mathbf{X})$, then $\mathbf{P_X}\mathbf{P_W}\mathbf{z} = \mathbf{P_W}\mathbf{z}$ for all $\mathbf{z}$, and hence $\mathbf{P_X}\mathbf{P_W} = \mathbf{P_W}$ from Corollary A.1 (similarly for its transpose), and so $(\mathbf{P_X} - \mathbf{P_W})$ is idempotent. Second, any $\mathbf{u}$ can be decomposed into vectors in $\mathcal{C}(\mathbf{X})$ and $\mathcal{N}(\mathbf{X}^T)$, hence $\mathbf{u} = \mathbf{X}\mathbf{s} + \mathbf{t}$ for some $\mathbf{s}$ and some $\mathbf{t} \in \mathcal{N}(\mathbf{X}^T)$. Then

$$(\mathbf{P_X} - \mathbf{P_W})\mathbf{u} = (\mathbf{P_X} - \mathbf{P_W})(\mathbf{X}\mathbf{s} + \mathbf{t}) = \mathbf{X}\mathbf{s} - \mathbf{P_W}\mathbf{X}\mathbf{s} = (\mathbf{I} - \mathbf{P_W})\mathbf{X}\mathbf{s}$$

so for any $\mathbf{u}$, $(\mathbf{P_X} - \mathbf{P_W})\mathbf{u} \in \mathcal{C}((\mathbf{I} - \mathbf{P_W})\mathbf{X})$. Third, if $\mathbf{y} \in \mathcal{C}((\mathbf{I} - \mathbf{P_W})\mathbf{X})$, then $\mathbf{y} = (\mathbf{I} - \mathbf{P_W})\mathbf{X}\mathbf{c}$ for some $\mathbf{c}$, and

$$(\mathbf{P_X} - \mathbf{P_W})\mathbf{y} = (\mathbf{P_X} - \mathbf{P_W})(\mathbf{I} - \mathbf{P_W})\mathbf{X}\mathbf{c} = (\mathbf{P_X} - \mathbf{P_W})\mathbf{X}\mathbf{c} = (\mathbf{I} - \mathbf{P_W})\mathbf{X}\mathbf{c} = \mathbf{y}.$$

Finally, since its constituents are symmetric, so is $(\mathbf{P_X} - \mathbf{P_W})$, and uniqueness follows from Result A.16. ▯

Example 2.6: Decompositions for Simple Linear Regression
Suppose we have the simple linear regression model $y_i = \beta_0 + \beta_1 x_i + e_i$ with $x_i = i, i = 1, \ldots, 4 = N$. The design matrix $\mathbf{X}$ takes the form

$$\mathbf{X} = \begin{bmatrix} 1 & 1 \\ 1 & 2 \\ 1 & 3 \\ 1 & 4 \end{bmatrix},$$

then

$$\mathbf{X}^T\mathbf{X} = \begin{bmatrix} 4 & 10 \\ 10 & 30 \end{bmatrix},$$

which is nonsingular. Then the generalized inverse is unique:

$$(\mathbf{X}^T\mathbf{X})^g = \begin{bmatrix} 3/2 & -1/2 \\ -1/2 & 1/5 \end{bmatrix} = \frac{1}{10}\begin{bmatrix} 15 & -5 \\ -5 & 2 \end{bmatrix}.$$

To construct $\mathbf{P_X}$, the next step is to form

$$(\mathbf{X}^T\mathbf{X})^g\mathbf{X}^T = \frac{1}{10}\begin{bmatrix} 10 & 5 & 0 & -5 \\ -3 & -1 & 1 & 3 \end{bmatrix},$$

and finally,

$$\mathbf{P_X} = \frac{1}{10}\begin{bmatrix} 7 & 4 & 1 & -2 \\ 4 & 3 & 2 & 1 \\ 1 & 2 & 3 & 4 \\ -2 & 1 & 4 & 7 \end{bmatrix}.$$

Note that

$$\mathbf{P_1} = \frac{1}{100}\begin{bmatrix} 25 & 25 & 25 & 25 \\ 25 & 25 & 25 & 25 \\ 25 & 25 & 25 & 25 \\ 25 & 25 & 25 & 25 \end{bmatrix},$$

so that

$$\mathbf{P_X} - \mathbf{P_1} = \frac{1}{100}\begin{bmatrix} 45 & 15 & -15 & 45 \\ 15 & 5 & -5 & -15 \\ -15 & -5 & 5 & 15 \\ -45 & -15 & 15 & 45 \end{bmatrix}$$

$$= \frac{1}{20}\begin{bmatrix} 3 \\ 1 \\ -1 \\ -3 \end{bmatrix}\begin{bmatrix} 3 & 1 & -1 & -3 \end{bmatrix}.$$

Notice some aspects of these three ($\mathbf{P_1}, \mathbf{P_X}$, and $(\mathbf{P_X} - \mathbf{P_1})$) projection matrices:

1. They are symmetric and idempotent.
2. The rank is equal to the trace (see Exercise 2.7).

Notice also the cross-symmetry in $\mathbf{P_X}$, as well as the rank of $(\mathbf{P_X} - \mathbf{P_1})$ and its relationship to a linear contrast. Finally, compute the residuals from a quadratic explanatory variable:

$$(\mathbf{I} - \mathbf{P_X})\mathbf{z} = \frac{1}{10}\begin{bmatrix} 7 & 4 & 1 & -2 \\ 4 & 3 & 2 & 1 \\ 1 & 2 & 3 & 4 \\ -2 & 1 & 4 & 7 \end{bmatrix}\begin{bmatrix} 1 \\ 4 \\ 9 \\ 16 \end{bmatrix} = \begin{bmatrix} 1 \\ -1 \\ -1 \\ 1 \end{bmatrix}$$

which should remind the reader of a quadratic contrast.

2.3 Reparameterization

By *reparameterization*, we mean there are two equivalent linear models, with different design matrices and different parameters, but equivalent in the sense that they will give the same least squares fit of the data. The key idea, of course, is what we mean by equivalent, and here we mean that two models are equivalent if the column spaces of the design matrices are the same. Reparameterizations are most commonly used to take advantage of the clarity of one model and the computational ease of the other.

Definition 2.1 *Two linear models, $\mathbf{y} = \mathbf{Xb} + \mathbf{e}$, where $\mathbf{X}$ is an $N \times p$ matrix and $\mathbf{y} = \mathbf{Wc} + \mathbf{e}$, where $\mathbf{W}$ is an $N \times t$ matrix, are equivalent, or reparameterizations of each other iff the two design matrices, $\mathbf{X}$ and $\mathbf{W}$, have the same column spaces, that is, $\mathcal{C}(\mathbf{X}) = \mathcal{C}(\mathbf{W})$.*

Recalling Result A.2, if $\mathcal{C}(\mathbf{X}) = \mathcal{C}(\mathbf{W})$, then there exist matrices $\mathbf{S}$ and $\mathbf{T}$ so that $\mathbf{W} = \mathbf{XT}$ and $\mathbf{X} = \mathbf{WS}$ where $\mathbf{T}$ is $p \times t$, and $\mathbf{S}$ is $t \times p$. Clearly, both column spaces must have the same dimension, hence $rank(\mathbf{X}) = rank(\mathbf{W}) = r$. We will see that $\mathbf{P_X} = \mathbf{P_W}, \mathbf{X}\hat{\mathbf{b}} = \mathbf{W}\hat{\mathbf{c}}$, and $\hat{\mathbf{e}} = (\mathbf{I} - \mathbf{P_X})\mathbf{y} = (\mathbf{I} - \mathbf{P_W})\mathbf{y}$, and the error sum of squares and R^2 are the same for both models.

Example 2.7: Reparameterization in Simple Linear Regression
Consider the simple linear regression case as the original model, $y_i = \beta_0 + \beta_1 x_i + e_i$, say, measuring yield y against temperature x. Changing temperature units from Centigrade (x) to Fahrenheit (w) means $w_i = 1.8x_i + 32$, and now $y_i = \gamma_0 + \gamma_1 w_i + e_i$. The correspondence of the parameters is $\beta_0 = \gamma_0 + 32\gamma_1$, and $\beta_1 = 1.8\gamma_1$ can be seen

from the following algebra:

$$\begin{aligned} y_i &= \gamma_0 + \gamma_1 w_i + e_i \\ &= \gamma_0 + \gamma_1(1.8x_i + 32) + e_i \\ &= (\gamma_0 + 32\gamma_1) + (1.8\gamma_1)x_i + e_i. \end{aligned}$$

Clearly $\mathcal{C}(\mathbf{X}) = \mathcal{C}(\mathbf{W})$, where $\mathbf{X}$ has rows $[1 \quad x_i]$, and $\mathbf{W}$ has rows $[1 \quad w_i]$.

Example 2.8: One-Way ANOVA
Consider the simple one-way layout with three groups as discussed previously. The more common parameterization employs

$$\mathbf{Xb} = \begin{bmatrix} \mathbf{1}_{n_1} & \mathbf{1}_{n_1} & \mathbf{0} & \mathbf{0} \\ \mathbf{1}_{n_2} & \mathbf{0} & \mathbf{1}_{n_2} & \mathbf{0} \\ \mathbf{1}_{n_3} & \mathbf{0} & \mathbf{0} & \mathbf{1}_{n_3} \end{bmatrix} \begin{bmatrix} \mu \\ \alpha_1 \\ \alpha_2 \\ \alpha_3 \end{bmatrix}.$$

Another parameterization, using dummy variables for the first two groups, leads to a full-column-rank design matrix:

$$\mathbf{Wc} = \begin{bmatrix} \mathbf{1}_{n_1} & \mathbf{1}_{n_1} & \mathbf{0} \\ \mathbf{1}_{n_2} & \mathbf{0} & \mathbf{1}_{n_2} \\ \mathbf{1}_{n_3} & \mathbf{0} & \mathbf{0} \end{bmatrix} \begin{bmatrix} c_1 \\ c_2 \\ c_3 \end{bmatrix}.$$

Note that since the first 3 columns of $\mathbf{W}$ are the same as $\mathbf{X}$, and that the fourth column of $\mathbf{X}$ is a linear combination of those three columns, we have $\mathcal{C}(\mathbf{X}) = \mathcal{C}(\mathbf{W})$. The correspondence between these two design matrices is given by

$$\mathbf{W} = \mathbf{XT} = \begin{bmatrix} \mathbf{1}_{n_1} & \mathbf{1}_{n_1} & \mathbf{0} \\ \mathbf{1}_{n_2} & \mathbf{0} & \mathbf{1}_{n_2} \\ \mathbf{1}_{n_3} & \mathbf{0} & \mathbf{0} \end{bmatrix} = \begin{bmatrix} \mathbf{1}_{n_1} & \mathbf{1}_{n_1} & \mathbf{0} & \mathbf{0} \\ \mathbf{1}_{n_2} & \mathbf{0} & \mathbf{1}_{n_2} & \mathbf{0} \\ \mathbf{1}_{n_3} & \mathbf{0} & \mathbf{0} & \mathbf{1}_{n_3} \end{bmatrix} \begin{bmatrix} 1 & 0 & 0 \\ 0 & 1 & 0 \\ 0 & 0 & 1 \\ 0 & 0 & 0 \end{bmatrix}$$

and

$$\mathbf{X} = \mathbf{WS} = \begin{bmatrix} \mathbf{1}_{n_1} & \mathbf{1}_{n_1} & \mathbf{0} & \mathbf{0} \\ \mathbf{1}_{n_2} & \mathbf{0} & \mathbf{1}_{n_2} & \mathbf{0} \\ \mathbf{1}_{n_3} & \mathbf{0} & \mathbf{0} & \mathbf{1}_{n_3} \end{bmatrix} = \begin{bmatrix} \mathbf{1}_{n_1} & \mathbf{1}_{n_1} & \mathbf{0} \\ \mathbf{1}_{n_2} & \mathbf{0} & \mathbf{1}_{n_2} \\ \mathbf{1}_{n_3} & \mathbf{0} & \mathbf{0} \end{bmatrix} \begin{bmatrix} 1 & 0 & 0 & 1 \\ 0 & 1 & 0 & -1 \\ 0 & 0 & 1 & -1 \end{bmatrix}.$$

Example 2.9: Analysis of Covariance
Consider a simple linear regression model for two groups, $i = 1, 2$, (the assumption of common variance to be added later), that is,

$$y_{ij} = \beta_0^{(i)} + \beta_1^{(i)} x_{ij} + e_{ij},$$

leading to the linear model of the form

$$\mathbf{y} = \begin{bmatrix} y_{11} \\ y_{12} \\ \cdots \\ y_{1N_1} \\ y_{21} \\ y_{22} \\ \cdots \\ y_{2N_2} \end{bmatrix} = \mathbf{Xb} + \mathbf{e} = \begin{bmatrix} 1 & x_{11} & 0 & 0 \\ 1 & x_{22} & 0 & 0 \\ \cdots & & \cdots & \\ 1 & x_{1N_1} & 0 & 0 \\ 0 & 0 & 1 & x_{21} \\ 0 & 0 & 1 & x_{22} \\ \cdots & & \cdots & \\ 0 & 0 & 1 & x_{2N_2} \end{bmatrix} \begin{bmatrix} \beta_0^{(1)} \\ \beta_1^{(1)} \\ \beta_0^{(2)} \\ \beta_1^{(2)} \end{bmatrix} + \begin{bmatrix} e_{11} \\ e_{12} \\ \cdots \\ e_{1N_1} \\ e_{21} \\ e_{22} \\ \cdots \\ e_{2N_2} \end{bmatrix}.$$

The more commonly used form using a dummy variable for group d_{ij}, such that $d_{1j} = 0, d_{2j} = 1$, leads to the model

$$y_{ij} = \gamma_0 + \gamma_1 x_{ij} + \gamma_2 d_{ij} + \gamma_3 (d_{ij} x_{ij}) + e_{ij},$$

leading to the model of the form

$$\mathbf{y} = \begin{bmatrix} y_{11} \\ y_{12} \\ \cdots \\ y_{1N_1} \\ y_{21} \\ y_{22} \\ \cdots \\ y_{2N_2} \end{bmatrix} = \mathbf{Wc} + \mathbf{e} = \begin{bmatrix} 1 & x_{11} & 0 & 0 \\ 1 & x_{22} & 0 & 0 \\ \cdots & & \cdots & \\ 1 & x_{1N_1} & 0 & 0 \\ 1 & x_{21} & 1 & x_{21} \\ 1 & x_{22} & 1 & x_{22} \\ \cdots & & \cdots & \\ 1 & x_{2N_2} & 1 & x_{2N_2} \end{bmatrix} \begin{bmatrix} \gamma_0 \\ \gamma_1 \\ \gamma_2 \\ \gamma_3 \end{bmatrix} + \begin{bmatrix} e_{11} \\ e_{12} \\ \cdots \\ e_{1N_1} \\ e_{21} \\ e_{22} \\ \cdots \\ e_{2N_2} \end{bmatrix}.$$

See Exercise 2.16.

We now show some important results regarding the least squares solutions in the reparameterized models.

Result 2.8 *If* $\mathcal{C}(\mathbf{X}) = \mathcal{C}(\mathbf{W})$, *then* $\mathbf{P_X} = \mathbf{P_W}$.

Proof: Consider the least squares decompositions

$$\begin{aligned} \mathbf{y} &= \mathbf{P_X y} + (\mathbf{I} - \mathbf{P_X})\mathbf{y} \\ &= \mathbf{P_W y} + (\mathbf{I} - \mathbf{P_W})\mathbf{y} \end{aligned}$$

From the uniqueness of the decompositions, we have $\mathbf{P_X y} = \mathbf{P_W y}$. Since this is true for all $\mathbf{y}$, applying Corollary A.1 yields $\mathbf{P_X} = \mathbf{P_W}$. □

Corollary 2.4 *If $C(\mathbf{X}) = C(\mathbf{W})$, then $\hat{\mathbf{y}} = \mathbf{P_X y} = \mathbf{P_W y}$ and $\hat{\mathbf{e}} = (\mathbf{I} - \mathbf{P_X})\mathbf{y} = (\mathbf{I} - \mathbf{P_W})\mathbf{y}$ are the same in both parameterizations.*

Result 2.9 *If $\hat{\mathbf{c}}$ solves the normal equations in $\mathbf{W}$, that is, $\mathbf{W}^T\mathbf{W}\mathbf{c} = \mathbf{W}^T\mathbf{y}$ (for the reparameterized model $\mathbf{y} = \mathbf{W}\mathbf{c} + \mathbf{e}$), then $\hat{\mathbf{b}} = \mathbf{T}\hat{\mathbf{c}}$ solves the normal equations $\mathbf{X}^T\mathbf{X}\mathbf{b} = \mathbf{X}^T\mathbf{y}$ in $\mathbf{X}$.*

Proof: $\mathbf{X}^T\mathbf{X}\mathbf{T}\hat{\mathbf{c}} = \mathbf{X}^T\mathbf{W}\hat{\mathbf{c}} = \mathbf{X}^T\mathbf{P_W y} = \mathbf{X}^T\mathbf{P_X y} = \mathbf{X}^T\mathbf{y}$. □

Example 2.8: continued

Using the two parameterizations for the one-way ANOVA above, the following $\hat{\mathbf{b}}$ solves the normal equations using $\mathbf{X}$:

$$\hat{\mathbf{b}} = \begin{bmatrix} \overline{y}_{..} \\ \overline{y}_{1.} - \overline{y}_{..} \\ \overline{y}_{2.} - \overline{y}_{..} \\ \overline{y}_{3.} - \overline{y}_{..} \end{bmatrix} \text{ with the general solution } \begin{bmatrix} 0 \\ \overline{y}_{1.} \\ \overline{y}_{2.} \\ \overline{y}_{3.} \end{bmatrix} + z \begin{bmatrix} 1 \\ -1 \\ -1 \\ -1 \end{bmatrix}.$$

Alternatively, the following $\hat{\mathbf{c}}$ is the unique solution to the full-rank normal equations using $\mathbf{W}$:

$$\hat{\mathbf{c}} = \begin{bmatrix} \overline{y}_{3.} \\ \overline{y}_{1.} - \overline{y}_{3.} \\ \overline{y}_{2.} - \overline{y}_{3.} \end{bmatrix},$$

yielding the familiar result of the intercept corresponding the group mean of the missing dummy group, and the other coefficients as differences from that missing group. Using Result 2.8, then $\hat{\mathbf{b}} = \mathbf{T}\hat{\mathbf{c}}$ solves the normal equations for the model in $\mathbf{X}$:

$$\hat{\mathbf{b}} = \mathbf{T}\hat{\mathbf{c}} = \begin{bmatrix} 1 & 0 & 0 \\ 0 & 1 & 0 \\ 0 & 0 & 1 \\ 0 & 0 & 0 \end{bmatrix} \begin{bmatrix} \overline{y}_{3.} \\ \overline{y}_{1.} - \overline{y}_{3.} \\ \overline{y}_{2.} - \overline{y}_{3.} \end{bmatrix} = \begin{bmatrix} \overline{y}_{3.} \\ \overline{y}_{1.} - \overline{y}_{3.} \\ \overline{y}_{2.} - \overline{y}_{3.} \\ 0 \end{bmatrix},$$

which corresponds to the general solution with $z = \overline{y}_{3.}$.

Now if we reverse roles and employ the same result, we should have that $\hat{\mathbf{b}}$ solving the $\mathbf{X}$ normal equations leads to $\mathbf{S}\hat{\mathbf{b}}$ solving the $\mathbf{W}$ normal equations. Here $\mathbf{S}\hat{\mathbf{b}}$ is given by

$$\mathbf{S}\hat{\mathbf{b}} = \begin{bmatrix} 1 & 0 & 0 & 1 \\ 0 & 1 & 0 & -1 \\ 0 & 0 & 1 & -1 \end{bmatrix} \begin{bmatrix} \overline{y}_{..} \\ \overline{y}_{1.} - \overline{y}_{..} \\ \overline{y}_{2.} - \overline{y}_{..} \\ \overline{y}_{3.} - \overline{y}_{..} \end{bmatrix} = \hat{\mathbf{c}} = \begin{bmatrix} \overline{y}_{3.} \\ \overline{y}_{1.} - \overline{y}_{3.} \\ \overline{y}_{2.} - \overline{y}_{3.} \end{bmatrix}.$$

Note that if we generated all solutions $\hat{\mathbf{b}}$ to the $\mathbf{X}$ normal equations above, $\hat{\mathbf{c}}$ would not change—it is unique since the $\mathbf{W}$ design matrix has full-column rank.

2.4 Gram–Schmidt Orthonormalization

The orthogonality of the residuals $\mathbf{e}$ to the columns of the design matrix $\mathbf{X}$ is one interpretation of the term *normal* in the normal equations. This result can be applied to other purposes, such as orthogonal reparameterization. The orthogonalization part of the Gram–Schmidt algorithm produces a set of mutually orthogonal vectors from a set of linearly independent vectors by taking linear combinations sequentially, so that the span of the new set of vectors is the same as the old. For our purposes, the linearly independent vectors will be columns of the design matrix $\mathbf{X}$, denoted by $\mathbf{X}_{.1}, \mathbf{X}_{.2}, \ldots, \mathbf{X}_{.p}$. Taking just the simple case where $\mathbf{X}$ has full-column rank, the result is a set of orthogonal columns $\mathbf{U}_{.1}, \mathbf{U}_{.2}, \ldots, \mathbf{U}_{.p}$, which are linear combinations of the columns of $\mathbf{X}$, and $\mathcal{C}(\mathbf{X}) = \mathcal{C}(\mathbf{U})$. Moreover, at each step k we have the same relationship:

$$span\{\mathbf{X}_{.1}, \ldots, \mathbf{X}_{.k}\} = span\{\mathbf{U}_{.1}, \ldots, \mathbf{U}_{.k}\}.$$

The orthogonalization step of the Gram–Schmidt method can be described with an induction argument using residuals from regression to create a vector orthogonal to the design matrix. The induction begins at step $i = 1$ with $\mathbf{U}_{.1} = \mathbf{X}_{.1}$. Then at step $i+1, \mathbf{X}_{.i+1}$, column $i+1$ of $\mathbf{X}$, is regressed on the previously constructed orthogonal columns, $\mathbf{U}_{.j}$, $j = 1, \ldots, i$. So step $i+1$ consists of a regression with response vector $\mathbf{X}_{.i+1}$ and explanatory variables $\mathbf{U}_{.j}$, $j = 1, \ldots, i$. Since these explanatory variables are mutually orthogonal, the matrix in the normal equations $\mathbf{U}^T\mathbf{U}$ is an $i \times i$ diagonal matrix, so that the regression coefficients, $\hat{\mathbf{b}}^{(i+1)}$, are easily computed

$$(\hat{\mathbf{b}}^{(i+1)})_j = \mathbf{U}_{.j}^T\mathbf{X}_{.i+1}/\mathbf{U}_{.j}^T\mathbf{U}_{.j} \quad \text{for} \quad j = 1, \ldots, i. \tag{2.9}$$

For convenience that will later be obvious, store these regression coefficients as a column in the matrix $\mathbf{S}$, so that $(\hat{\mathbf{b}}^{(i+1)})_j = \mathbf{S}_{j,i+1}$. The residual vector becomes the new vector $\mathbf{U}_{.i+1}$,

$$\mathbf{U}_{.i+1} = \mathbf{X}_{.i+1} - \sum_{j=1}^{i}(\hat{\mathbf{b}}^{(i+1)})_j\mathbf{U}_{.j} = \mathbf{X}_{.i+1} - \sum_{j=1}^{i}\mathbf{S}_{j,i+1}\mathbf{U}_{.j},$$

which will be orthogonal to the previous explanatory variables $\mathbf{U}_{.j}$, $j = 1, \ldots, i$. Computing the error sum of squares $\|\mathbf{U}_{.i+1}\|^2$ completes step $i+1$.

Complete the definition of $\mathbf{S}$ with $S_{ii} = 1$ and $S_{ji} = 0$ for $j > i$, so that now $\mathbf{S}$ is unit upper triangular. Rewriting the residual equation for step i, we have

$$\text{response} = \text{residual} + \text{fitted value}$$

$$\mathbf{X}_{.i+1} = \mathbf{U}_{.i+1} + \sum_{j=1}^{i} S_{j,i+1}\mathbf{U}_{.j},$$

and stacking them side by side, we have $\mathbf{X} = \mathbf{US}$ and the columns of $\mathbf{U}$ are orthogonal: $\mathbf{U}^T\mathbf{U} = \mathbf{D}^2$ defining the diagonal matrix $\mathbf{D} = diag(\|\mathbf{U}_{.i}\|)$ and clearly $\mathcal{C}(\mathbf{X}) = \mathcal{C}(\mathbf{U})$. The normalization step of the Gram–Schmidt algorithm merely rescales each column, in matrices, postmultiplying by a diagonal matrix to form $\mathbf{Q} = \mathbf{UD}^{-1}$. In terms of the previous factorization, we have $\mathbf{X} = \mathbf{US} = (\mathbf{UD}^{-1})(\mathbf{DS}) = \mathbf{QR}$ with $\mathbf{R} = \mathbf{DS}$, so that $R_{ji} = D_j S_{ji}$. In terms of residuals, the new orthonormal vectors $\mathbf{Q}_{.i}$ are normalized versions of vectors in $\mathbf{U}$:

$$\mathbf{Q}_{.i+1} = \frac{1}{\|\mathbf{U}_{.i+1}\|}\mathbf{U}_{.i+1} = \left(\mathbf{X}_{.i+1} - \sum_{j=1}^{i} R_{j,i+1}\mathbf{Q}_{.j}\right) / R_{i+1,i+1} \tag{2.10}$$

since $R_{i+1,i+1} = \|\mathbf{U}_{.i+1}\|$. Since $\mathbf{Q}_{.i+1}$ is proportional to the residual vector $\mathbf{U}_{.i+1}$, it will be orthogonal to the columns of the design matrix, which are proportional to the previous $\mathbf{Q}_{.j}$, $j = 1, \ldots, i$, or $\mathbf{Q}^T\mathbf{Q} = \mathbf{D}^{-1}\mathbf{U}^T\mathbf{UD}^{-1} = \mathbf{D}^{-1}\mathbf{D}^2\mathbf{D}^{-1} = \mathbf{I}_p$.

Let us now rewrite the Gram–Schmidt algorithm by combining the orthogonalization and normalization steps. The induction begins again with $\mathbf{U}_{.1} = \mathbf{X}_{.1}$ so that

$$\mathbf{Q}_{.1} = \frac{1}{\|\mathbf{U}_{.1}\|}\mathbf{U}_{.1}.$$

Then the regression at step $i+1$ has the same response $\mathbf{X}_{.i+1}$ but orthonormal explanatory variables $\mathbf{Q}_{.j}$, $j = 1, \ldots, i$. The computation of the regression coefficients is merely

$$R_{j,i+1} = \mathbf{Q}_{.j}^T\mathbf{X}_{.i+1}, \quad j = 1, \ldots, i \tag{2.11}$$

since $\mathbf{Q}_{.j}^T\mathbf{Q}_{.j} = 1$. The residual vector $\mathbf{U}_{.i+1} = \mathbf{X}_{.i+1} - \sum_{j=1}^{i} R_{j,i+1}\mathbf{Q}_{.j}$ is the same since

$$S_{ji}\mathbf{U}_{.j} = (D_j S_{ji})\frac{1}{D_j}\mathbf{U}_{.j} = R_{ji}\mathbf{Q}_{.j}$$

and the step is completed by computing the sum of squares $\|\mathbf{U}_{.i+1}\|^2$ and $\mathbf{Q}_{.i+1}$ from (2.12). Solving for $\mathbf{X}_{.i+1}$ in (2.12) yields again

$$\mathbf{X}_{.i+1} = \sum_{j=1}^{i} R_{j,i+1}\mathbf{Q}_{.j} + R_{i+1,i+1}\mathbf{Q}_{.i+1}$$

or $\mathbf{X} = \mathbf{QR}$, hence the name *QR factorization*. This result also shows that $\mathcal{C}(\mathbf{X}) = \mathcal{C}(\mathbf{Q})$. Gram–Schmidt is also related to the Cholesky factorization of $\mathbf{X}^T\mathbf{X}$, that is,

$$\mathbf{X}^T\mathbf{X} = \mathbf{LL}^T = (\mathbf{QR})^T(\mathbf{QR}) = \mathbf{R}^T\mathbf{R}, \tag{2.12}$$

so $\mathbf{R}$ is the transpose of $\mathbf{L}$. For the case where $\mathbf{X}$ is full-column rank, $\mathbf{L}$ and $\mathbf{R}$ will be triangular, with $\mathbf{L}$ lower and $\mathbf{R}$ upper.

Example 2.10: Gram–Schmidt on a Simple Quadratic Design Matrix
Let

$$\mathbf{X} = \begin{bmatrix} 1 & 1 & 1 \\ 1 & 2 & 4 \\ 1 & 3 & 9 \\ 1 & 4 & 16 \end{bmatrix}, \quad \mathbf{U}_{.1} = \begin{bmatrix} 1 \\ 1 \\ 1 \\ 1 \end{bmatrix}, \quad D_1 = 2, \quad \text{and} \quad \mathbf{Q}_{.1} = \begin{bmatrix} 1/2 \\ 1/2 \\ 1/2 \\ 1/2 \end{bmatrix}.$$

Next,

$$S_{12} = 10/4 = 5/2, \quad \text{and} \quad \mathbf{U}_{.2} = \mathbf{X}_{.2} - S_{12}\mathbf{U}_{.1} = \begin{bmatrix} 1 - 5/2 \\ 2 - 5/2 \\ 3 - 5/2 \\ 4 - 5/2 \end{bmatrix} = \begin{bmatrix} -3/2 \\ -1/2 \\ 1/2 \\ 3/2 \end{bmatrix},$$

$$D_2 = \sqrt{20/4} = \sqrt{5}, \quad \text{and} \quad \mathbf{Q}_{.2} = \begin{bmatrix} -3/\sqrt{20} \\ -1/\sqrt{20} \\ 1/\sqrt{20} \\ 3/\sqrt{20} \end{bmatrix}.$$

Next step:

$S_{13} = 30/4 = 15/2$, and $S_{23} = 50/10 = 5$, so $\mathbf{U}_{.3} = \mathbf{X}_{.3} - S_{13}\mathbf{U}_{.1} - S_{23}\mathbf{U}_{.2}$

$$= \begin{bmatrix} 1 - 15/2 - 5(-3/2) \\ 4 - 15/2 - 5(-1/2) \\ 9 - 15/2 - 5(1/2) \\ 16 - 15/2 - 5(3/2) \end{bmatrix} = \begin{bmatrix} 1 - 15/2 + 15/2 \\ 4 - 15/2 + 5/2 \\ 9 - 15/2 - 5/2 \\ 16 - 15/2 - 15/2 \end{bmatrix} = \begin{bmatrix} 1 \\ -1 \\ -1 \\ 1 \end{bmatrix},$$

so $R_{33} = 2$, $\mathbf{Q}_{.3} = \begin{bmatrix} 1/2 \\ -1/2 \\ -1/2 \\ 1/2 \end{bmatrix}$,

and $\mathbf{X} = \mathbf{US} = \mathbf{QR}$ becomes

$$\begin{bmatrix} 1 & 1 & 1 \\ 1 & 2 & 4 \\ 1 & 3 & 9 \\ 1 & 4 & 16 \end{bmatrix} = \begin{bmatrix} 1 & -3/2 & 1 \\ 1 & -1/2 & -1 \\ 1 & 1/2 & -1 \\ 1 & 3/2 & 1 \end{bmatrix} \begin{bmatrix} 1 & 5/2 & 15/2 \\ 0 & 1 & 5 \\ 0 & 0 & 1 \end{bmatrix}$$

$$= \begin{bmatrix} 1/2 & -3/\sqrt{20} & 1/2 \\ 1/2 & -1/\sqrt{20} & -1/2 \\ 1/2 & 1/\sqrt{20} & -1/2 \\ 1/2 & 3/\sqrt{20} & 1/2 \end{bmatrix} \begin{bmatrix} 2 & 5 & 15 \\ 0 & \sqrt{5} & 5\sqrt{5} \\ 0 & 0 & 2 \end{bmatrix}.$$

For another approach to orthogonalization, see Exercises 2.22–2.24.

2.5 Summary of Important Results

1. In this chapter, we considered the least squares problem of finding $\mathbf{b}$ that minimizes

$$Q(\mathbf{b}) = (\mathbf{y} - \mathbf{X}\mathbf{b})^T(\mathbf{y} - \mathbf{X}\mathbf{b})$$

   This minimization problem has led to the normal equations $\mathbf{X}^T\mathbf{X}\mathbf{b} = \mathbf{X}^T\mathbf{y}$.

2. We have shown that the normal equations are always consistent. If $\mathbf{X}$ is less than full-column rank, the normal equations have infinitely many solutions.

3. $\hat{\mathbf{b}}$ minimizes $Q(\mathbf{b})$ iff $\hat{\mathbf{b}}$ is a solution to the normal equations.

4. We have seen that $\hat{\mathbf{y}} = \mathbf{X}\hat{\mathbf{b}} = \mathbf{P_X}\mathbf{y}$ is the unique point in $\mathcal{C}(\mathbf{X})$ that is closest to $\mathbf{y}$.

5. A matrix that plays an important role in the least squares problem is the projection matrix

$$\mathbf{P_X} = \mathbf{X}(\mathbf{X}^T\mathbf{X})^g\mathbf{X}^T.$$

   It is symmetric, idempotent, unique (invariant to the choice of generalized inverse), and projects onto $\mathcal{C}(\mathbf{X})$. Similarly, $\mathbf{I} - \mathbf{P_X}$ is the projection matrix onto $\mathcal{N}(\mathbf{X}^T)$.

6. The vector $\mathbf{y}$ can be decomposed uniquely as $\mathbf{y} = \hat{\mathbf{y}} + \hat{\mathbf{e}} = \mathbf{P_X}\mathbf{y} + (\mathbf{I} - \mathbf{P_X})\mathbf{y}$, where $\hat{\mathbf{y}} \in \mathcal{C}(\mathbf{X})$ and $\hat{\mathbf{e}} \in \mathcal{N}(\mathbf{X}^T)$. Applying the Pythagorean Theorem leads to the familiar decomposition of the sum of squares:

$$\|\mathbf{y}\|^2 = \|\hat{\mathbf{y}}\|^2 + \|\hat{\mathbf{e}}\|^2.$$

7. We defined $\mathbf{y} = \mathbf{W}\mathbf{c} + \mathbf{e}$ as a reparameterization of the linear model $\mathbf{y} = \mathbf{X}\mathbf{b} + \mathbf{e}$ if $\mathcal{C}(\mathbf{X}) = \mathcal{C}(\mathbf{W})$. We have shown that $\mathbf{P_X} = \mathbf{P_W}$ and $\mathbf{X}\hat{\mathbf{b}} = \mathbf{W}\hat{\mathbf{c}}$.

8. Finally, we discussed the Gram–Schmidt orthogonalization procedure.

2.6 Notes

- According to Stigler ([45], pp. 415–20), the original meaning of the name *normal equations* is unclear. Gauss used the term *Normalgleichungen* in 1822, but the connotation he intended for "normal" is not apparent. The usual interpretations are either that the residuals orthogonal to $\mathbf{X}$ or the perpendicular distance from $\mathbf{y}$ to $\mathcal{C}(\mathbf{X})$ is minimized (least squares).

- Criteria other than least squares have been proposed over the years, including other vector norms $\|\mathbf{y} - \mathbf{Xb}\|_q$. The case $q = 2$ gives us least squares; $q = 1$ and $q = \infty$ correspond to maximum likelihood estimators (see Exercise 6.6).

- In the case of simple linear regression where the covariate is measured with error, *orthogonal regression*, that is, minimizing the squared distance from the points (y_i, x_i), $i = 1, \ldots, N$, to the line $y = \beta_0 + \beta_1 x$ has been proposed as still another alternative to least squares. See Exercise 2.25.

2.7 Exercises

2.1. Let $G(\mathbf{b}) = (\mathbf{y} - \mathbf{Xb})^T \mathbf{W}(\mathbf{y} - \mathbf{Xb})$; find $\partial G/\partial \mathbf{b}$.

2.2. * Consider the problem of finding stationary points of the function $f(\mathbf{x}) = \mathbf{x}^T\mathbf{Ax}/\mathbf{x}^T\mathbf{x}$ where $\mathbf{A}$ is a $p \times 1$ symmetric matrix. Take the derivative of $f(\mathbf{x})$ using the chain rule, and show that the solutions to $\partial f(\mathbf{x})/\partial \mathbf{x} = \mathbf{0}$ are the eigenvectors of the matrix $\mathbf{A}$, leading to the useful result

$$\lambda_p = min_{\mathbf{x}} \frac{\mathbf{x}^T\mathbf{Ax}}{\mathbf{x}^T\mathbf{x}} \leq \frac{\mathbf{x}^T\mathbf{Ax}}{\mathbf{x}^T\mathbf{x}} \leq max_{\mathbf{x}} \frac{\mathbf{x}^T\mathbf{Ax}}{\mathbf{x}^T\mathbf{x}} = \lambda_1$$

where $\lambda_1 \geq \cdots \geq \lambda_p$ are the ordered eigenvalues of $\mathbf{A}$. Can the assumption of a symmetric matrix $\mathbf{A}$ be dropped to obtain the same results? See also Example B.3.

2.3. * For the simple linear regression in Example 2.1:
   a. Show that $\hat{\beta}_1 = \sum_{i=1}^N (x_i - \overline{x})y_i / \sum_{i=1}^N (x_i - \overline{x})^2$ and $\hat{\beta}_0 = \overline{y} - \hat{\beta}_1 \overline{x}$ solve the normal equations.
   b. Write both $\hat{\beta}_0$ and $\hat{\beta}_1$ as linear combinations of the y's, that is, find $\mathbf{a}$ such that $\hat{\beta}_1 = \mathbf{a}^T\mathbf{y}$, and similarly for $\hat{\beta}_0$.

2.4. From Result 2.2, $\mathcal{C}(\mathbf{X}^T\mathbf{X}) = \mathcal{C}(\mathbf{X}^T)$, so from Result A.2, there exists a matrix $\mathbf{B}$ such that $\mathbf{X}^T\mathbf{XB} = \mathbf{X}^T$. For the design matrix $\mathbf{X}$ in Example 2.3, find the matrix $\mathbf{B}$.

2.5. Repeat Exercise 2.4 with the design matrix from Example 2.9.

2.6. From Result A.13, $\hat{\mathbf{b}}(\mathbf{z}) = (\mathbf{X}^T\mathbf{X})^g\mathbf{X}^T\mathbf{y} + (\mathbf{I} - (\mathbf{X}^T\mathbf{X})^g\mathbf{X}^T\mathbf{X})\mathbf{z}$ sweeps out all solutions of the normal equations. As an alternative to Corollary 2.3, show algebraically that $\mathbf{X}\hat{\mathbf{b}}$ is the same for all solutions $\hat{\mathbf{b}}$ to the normal equations.

2.7. In Example 2.3, show that $\mathbf{G}_1\mathbf{X}^T$ is a generalized inverse for $\mathbf{X}$. Similarly, show this for $\mathbf{G}_2\mathbf{X}^T$. Find a basis for $\mathcal{N}(\mathbf{X}^T)$.

2.8. Using $\mathbf{X}$ and $\mathbf{P_X}$ from Example 2.3, and $\mathbf{y}^T = (1, 3, 5, 7)$:

a. Construct all solutions to the normal equations.
b. Construct all solutions to $\mathbf{Xb} = \mathbf{P_X y}$.

2.9. Using $\mathbf{P_X}$ from Example 2.3, show that $\mathbf{1} \in \mathcal{C}(\mathbf{X})$. Also compute $\mathbf{P_X} - \mathbf{P_1}$.

2.10. * From Example 2.4, prove the well-known identity

$$\sum_i y_i^2 = N\bar{y}^2 + \sum_i (y_i - \bar{y})^2.$$

2.11. A design matrix for a polynomial regression problem with only three design points takes the form

$$\mathbf{X} = \begin{bmatrix} 1 & -1 & 1 & -1 \\ 1 & 0 & 0 & 0 \\ 1 & 0 & 0 & 0 \\ 1 & 1 & 1 & 1 \\ 1 & 1 & 1 & 1 \end{bmatrix}.$$

a. What is the rank of $\mathbf{X}$?
b. Give the dimension of $\mathcal{C}(\mathbf{X})$ and a nonzero vector in it.
c. Give the dimension of $\mathcal{N}(\mathbf{X})$ and a nonzero vector in it.
d. Give the dimension of $\mathcal{C}(\mathbf{X}^T)$ and a nonzero vector in it.
e. Compute $\mathbf{X}^T\mathbf{X}$ and find a generalized inverse for it. (No need to check $\mathbf{AGA} = \mathbf{A}$.)
f. Which of the following vectors could be $\mathbf{P_X y}$ for an appropriate response vector $\mathbf{y}$?

$$\begin{bmatrix} 3 \\ 1 \\ 1 \\ 2 \\ 2 \end{bmatrix} \begin{bmatrix} 1 \\ -1 \\ 1 \\ -1 \\ 1 \end{bmatrix} \begin{bmatrix} 1 \\ 0 \\ 0 \\ 2 \\ 2 \end{bmatrix}$$

g. Using a right-hand side you gave above in (f), find all solutions to the equations $\mathbf{Xb} = \mathbf{P_X y}$.
h. Which of the following have the same column space as $\mathbf{X}$?

$$\mathbf{U} = \begin{bmatrix} 1 & 0 & 0 \\ 0 & 1 & 0 \\ 0 & 1 & 0 \\ 0 & 0 & 1 \\ 0 & 0 & 1 \end{bmatrix} \quad \mathbf{W} = \begin{bmatrix} 1 & 1 & 0 \\ 1 & 0 & 1 \\ 1 & 1 & 1 \\ 1 & 1 & 0 \\ 1 & 0 & 1 \end{bmatrix}$$

2.12. In Example 2.7 (Celsius to Fahrenheit), find γ_0 and γ_1 in terms of β_0 and β_1.

2.13. Consider the simple linear regression model $y_i = \beta_0 + \beta_1 x_i + e_i$. Show that if the x_i are equally spaced, that is, $x_i = s + ti$ for some values of s and t, then $y_i = \gamma_0 + \gamma_1 i + e_i$ is an equivalent parameterization. Can you extend this to a quadratic or higher degree polynomial?

2.14. For $\mathbf{X}$ and $\mathbf{W}$ below, show $\mathcal{C}(\mathbf{X}) = \mathcal{C}(\mathbf{W})$, by finding the matrices $\mathbf{S}$ and $\mathbf{T}$ so that $\mathbf{W} = \mathbf{XT}$ and $\mathbf{X} = \mathbf{WS}$:

$$\mathbf{X} = \begin{bmatrix} \mathbf{1}_{n_1} & \mathbf{1}_{n_1} & 0 & 0 \\ \mathbf{1}_{n_2} & 0 & \mathbf{1}_{n_2} & 0 \\ \mathbf{1}_{n_3} & 0 & 0 & \mathbf{1}_{n_3} \end{bmatrix}, \quad \mathbf{W} = \begin{bmatrix} \mathbf{1}_{n_1} & \mathbf{1}_{n_1} & \mathbf{1}_{n_1} \\ \mathbf{1}_{n_2} & 2\mathbf{1}_{n_2} & 4\mathbf{1}_{n_2} \\ \mathbf{1}_{n_3} & 3\mathbf{1}_{n_3} & 9\mathbf{1}_{n_3} \end{bmatrix}.$$

(Note: This shows the correspondence between ANOVA and a saturated polynomial regression.)

2.15. *Let $\mathcal{C}(\mathbf{X}) = \mathcal{C}(\mathbf{W})$ and suppose $\mathbf{W}$ has full-column rank.

a. Show that the matrix $\mathbf{S}$ such that $\mathbf{X} = \mathbf{WS}$ is unique.
b. Suppose $\mathbf{Xb} = \mathbf{Wc}$; show that $\mathbf{c}$ can be written uniquely in terms of $\mathbf{b}$.

2.16. For Example 2.9, find the components of $\mathbf{b}$ ($\beta_j^{(i)}$'s) in terms of $\mathbf{c}$ (γ's), and vice versa. The Gram–Schmidt procedure given in Section 2.4 can be extended to the non-full-rank case in the following way. Denoting the rank of the first i columns of $\mathbf{X}$ as $r(i)$, then at step $i+1$, the response vector is $\mathbf{X}_{.i+1}$ and there are $r(i)$ columns of the design matrix: $\mathbf{Q}_{.j}$, $j = 1, \ldots, r(i)$. Again store the coefficients of step i in the matrix $\mathbf{R}$:

$$\mathbf{R}_{j,i+1} = \mathbf{Q}_{.j}^T \mathbf{X}_{.i+1} \text{ for } j = 1, \ldots, r(i).$$

If $\mathbf{X}_{.i+1}$ is linearly dependent on $\mathbf{X}_{.j}$, $j = 1, \ldots, i$ (and hence the previous $\mathbf{Q}_{.j}$), then the regression will have a perfect fit, $\mathbf{U}_{.i+1} = \mathbf{X}_{.i+1} - \sum_{j=1}^{r(i)} R_{j,i+1}\,\mathbf{Q}_{.j} = \mathbf{0}$ and step $i+1$ is complete and no new $\mathbf{Q}_{.j}$ is constructed. The norm of $\mathbf{U}_{.i}$ is stored only when positive, and in $R_{r(i),i}$. The factorization still takes the form $\mathbf{X} = \mathbf{QR}$, but now $\mathbf{Q}$ is $N \times r$ where $r = rank(\mathbf{X})$. The matrix $\mathbf{R}$ will be $r \times p$, in row-echelon form; that is, the first nonzero element in each row is the positive $\|\mathbf{U}^{(i)}\|$.

2.17. Compute the Gram–Schmidt orthonormalization for the design matrix $\mathbf{X}$ in Example 2.3.

2.18. Prove Result 2.4 directly using only Lemma 2.1.

2.19. Compute the Gram–Schmidt factorization for the non-full-rank matrix below:

$$\begin{bmatrix} 1 & 1 & 3 \\ 1 & 2 & 5 \\ 1 & 3 & 7 \\ 1 & 4 & 9 \end{bmatrix}.$$

2.20. Show that $\mathbf{P_X} = \mathbf{QQ}^T$ for $\mathbf{P_X}$ from Example 2.6, and $\mathbf{Q}$ from Exercise 2.19.

2.21. * In general, let $\mathbf{X} = \mathbf{QR}$ where $\mathbf{Q}$ has orthonormal columns and $rank(\mathbf{X}) = rank(\mathbf{Q})$. Show that $\mathbf{P_X} = \mathbf{QQ}^T$.

2.22. Householder matrices (a.k.a. elementary reflectors) take the form

$$\mathbf{U} = \mathbf{I}_n - \frac{2}{\|\mathbf{u}\|^2}\mathbf{uu}^T \tag{2.13}$$

where $\mathbf{u}$ is an $n \times 1$ nonzero vector.

a. Show that $\mathbf{U}$ is a symmetric, orthogonal matrix.
b. Show that if $s = \|\mathbf{x}\|$, then choosing $\mathbf{u} = \mathbf{x} + s\mathbf{e}^{(1)}$ leads to $\mathbf{Ux} = s\mathbf{e}^{(1)}$. This can be viewed as using $\mathbf{U}$ to rotate space so that $\mathbf{x}$ corresponds to the first dimension.

2.23. Now construct a matrix $\mathbf{U}$ by partitioning as follows:

$$\mathbf{U} = \begin{bmatrix} \mathbf{I}_n & \mathbf{0} \\ \mathbf{0} & \mathbf{U}_* \end{bmatrix} \begin{matrix} p \\ n-p \end{matrix} \quad \text{and } \mathbf{x} = \begin{bmatrix} \mathbf{x}_1 \\ \mathbf{x}_2 \end{bmatrix} \tag{2.14}$$

where $\mathbf{U}_*$ is a Householder matrix constructed using (2.13) with $\mathbf{u}_* = \mathbf{x}_2 + \|\mathbf{x}_2\|\mathbf{e}^{(1)}$.

a. Show that $\mathbf{U}$ is a symmetric, orthogonal matrix.
b. Show that

$$\mathbf{Ux} = \begin{bmatrix} \mathbf{x}_1 \\ \|\mathbf{x}_2\|\mathbf{e}^{(1)} \end{bmatrix} \text{ and } \mathbf{U}\begin{bmatrix} \mathbf{v} \\ \mathbf{0} \end{bmatrix} = \begin{bmatrix} \mathbf{v} \\ \mathbf{0} \end{bmatrix}$$

for any $p \times 1$ vector $\mathbf{v}$.

2.24. Using the tools outlined in the two exercises above, for a design matrix $\mathbf{X}$ we can construct a sequence of Householder matrices $\mathbf{U}_1, \ldots, \mathbf{U}_p$ following (2.13) and (2.14) such that

$$\mathbf{U}_p \ldots \mathbf{U}_2\mathbf{U}_1\mathbf{X} = \begin{bmatrix} \mathbf{R} \\ \mathbf{0} \end{bmatrix}$$

where $\mathbf{R}$ is square and upper triangular.

a. Show that $\mathbf{U}_p \ldots \mathbf{U}_2\mathbf{U}_1$ is an orthogonal matrix.
b. Show that $\mathbf{R}^T\mathbf{R} = \mathbf{X}^T\mathbf{X}$. (Note that $\mathbf{R}$ constructed this way may differ from Gram–Schmidt by signs.)

2.25. For *orthogonal regression* (a.k.a. *total least squares*), the goal is to minimize the squared distance from the points $(y_i, x_i), i = 1, \ldots, N$, to the line $y = \beta_0 + \beta_1 x$.

a. For a point (y_i, x_i), find the closest point $(\hat{y}_i, \hat{x}_i)$ on the line $y = \beta_0 + \beta_1 x$.
b. For a given value of the slope parameter β_1, find the value of the intercept parameter β_0 that minimizes the sum of the squared distances

$$\sum_i \left[(x_i - \hat{x}_i)^2 + (y_i - \hat{y}_i)^2\right].$$

c. Using your solution to part (b), find an expression for the best-fitting slope parameter β_1.
d. Since the units of x and y may be different, or the error variances of the two variables may be different, repeat these steps with differential weight w:

$$\sum_i \left[w(x_i - \hat{x}_i)^2 + (y_i - \hat{y}_i)^2\right].$$

Chapter 3

Estimability and Least Squares Estimators

3.1 Assumptions for the Linear Mean Model

So far we have considered the linear model only as a method of mathematical approximation. In this chapter, we pose a statistical model for the data

$$\mathbf{y} = \mathbf{X}\mathbf{b} + \mathbf{e}$$

where $\mathbf{y}$ is the $(N \times 1)$ vector of responses and $\mathbf{X}$ is the $(N \times p)$ known design matrix. As before, the coefficient vector $\mathbf{b}$ is unknown and to be determined or estimated. And taking just this small step, we include the assumption that the errors $\mathbf{e}$ have a zero mean vector:

$$E(\mathbf{e}) = \mathbf{0}.$$

Following from the fact that $\mathbf{Xb}$ is fixed and $\mathbf{e}$ is random, another viewpoint is that this model merely specifies the mean vector for the response vector $\mathbf{y}$:

$$E(\mathbf{y}) = \mathbf{Xb}.$$

As discussed in Chapter 1, overparameterized models are easy to construct and interpret. But these advantages also bring obstacles. In Chapter 2, we encounter one of these obstacles where the rank of the $(N \times p)$ design matrix $\mathbf{X}$, $r = rank(\mathbf{X})$, may be less than p, leading to multiple solutions to the normal equations. The second obstacle is that we may not be able to estimate all of the parameters of the model, and that is the focus of this chapter. This problem is related to the terms *confounding* and *identifiability*.

3.2 Confounding, Identifiability, and Estimability

The formulation of a statistical model for an experiment affects the estimability of its parameters. Consider the model $y_{ij} = \mu + \alpha_i + e_{ij}$ for an experiment where we are measuring, say, blood pressure of patient j given drug protocol i. The discussion of these parameters begins by expressing the meaning of those symbols (μ, α_i, e_{ij}). Typically, we would say that μ represents the population mean blood pressure, α_i is the effect of drug i, and e_{ij} captures or subsumes the variations of an individual patient in response to drug protocol i. We include all circumstances peculiar to drug protocol i in the parameter α_i, including the color and taste of the pill, the color of the printing on the package, and the person measuring the blood pressure—all of these factors are confounded. We cannot determine whether the active ingredient in the pill or its taste may be affecting the blood pressure, since a patient given a sweet pill may think the drug is not the active drug, but the placebo, and for psychological reasons not respond to treatment. Careful design of the experimental protocol would make the features of the pill, such as taste and size, the same for all protocols. Double-blinding would keep the person measuring the blood pressure ignorant of the patient's group, and preclude that knowledge from affecting the measurement.

The related concept of identifiability, discussed more technically later in Section 6.7, assesses the ability to distinguish two parameter vectors, say $\mathbf{b}_1$ and $\mathbf{b}_2$. Given the experiment with design matrix $\mathbf{X}$, we cannot distinguish $\mathbf{b}_1$ and $\mathbf{b}_2$ if $\mathbf{X}\mathbf{b}_1 = \mathbf{X}\mathbf{b}_2$, since the two parameter vectors give the same mean for the response vector $\mathbf{y}$. In the one-way ANOVA model, notice that

$$E(y_{ij}) = \mu + \alpha_i = (\mu + c) + (\alpha_i - c),$$

for all c, so that different parameter points will give the same distribution for $\mathbf{y}$. In this case, neither μ nor α_i are identifiable, although the sum $\mu + \alpha_i$ is identifiable.

Finally, estimability of a function of the parameters corresponds to the existence of a linear unbiased estimator for it. In this book, we will focus mainly on the concept of estimability.

3.3 Estimability and Least Squares Estimators

Our primary goal is to determine whether certain functions of the parameters are estimable, and then to construct unbiased estimators for them. In Chapter 4, we will work to find the best estimators for these parameters.

Definition 3.1 *An estimator $t(\mathbf{y})$ is an unbiased estimator for the scalar $\lambda^T\mathbf{b}$ iff $E(t(\mathbf{y})) = \lambda^T\mathbf{b}$ for all $\mathbf{b}$.*

Definition 3.2 *An estimator $t(\mathbf{y})$ is a linear estimator in $\mathbf{y}$ iff $t(\mathbf{y}) = c + \mathbf{a}^T\mathbf{y}$ for constants $c, a_1, \ldots, a_N$.*

Definition 3.3 *A function $\lambda^T\mathbf{b}$ is linearly estimable iff there exists a linear unbiased estimator for it. If no such estimator exists, then the function is called* nonestimable.

Result 3.1 *Under the linear mean model, $\lambda^T\mathbf{b}$ is (linearly) estimable iff there exists a vector $\mathbf{a}$ such that $E(\mathbf{a}^T\mathbf{y}) = \lambda^T\mathbf{b}$ for all $\mathbf{b}$, or $\lambda^T = \mathbf{a}^T\mathbf{X}$ or $\lambda = \mathbf{X}^T\mathbf{a}$.*

Proof: If $\mathbf{X}^T\mathbf{a} = \lambda$, then $E(\mathbf{a}^T\mathbf{y}) = \mathbf{a}^T\mathbf{X}\mathbf{b} = \lambda^T\mathbf{b}$ for all $\mathbf{b}$, so that $\mathbf{a}^T\mathbf{y}$ is a linear unbiased estimator, and so $\lambda^T\mathbf{b}$ is estimable. Conversely, if $\lambda^T\mathbf{b}$ is estimable, then there exists c and $\mathbf{a}$ (scalar and vector) such that $c + \mathbf{a}^T\mathbf{y}$ is unbiased for $\lambda^T\mathbf{b}$. That is, $E(c+\mathbf{a}^T\mathbf{y}) = c+\mathbf{a}^T\mathbf{X}\mathbf{b} = \lambda^T\mathbf{b}$ for all $\mathbf{b}$. This can be rewritten as $c+(\mathbf{a}^T\mathbf{X}-\lambda^T)\mathbf{b} = \mathbf{0}$ for all $\mathbf{b}$. Using Result A.8, we have $c = 0$ and $\lambda^T = \mathbf{a}^T\mathbf{X}$. □

Example 3.1: Not Estimable
Consider the very simple case of ANOVA with two levels and two observations at each level,

$$E(y_{ij}) = \mu + \alpha_i, \text{ for } i = 1, 2 = a;\ j = 1, 2 = n.$$

For the nonestimable function α_1, trying to find the linear combination $\mathbf{a}$ so that $E(\mathbf{a}^T\mathbf{y}) = \alpha_1$ corresponds to trying to find a solution to the linear equations

$$\mathbf{X}^T\mathbf{a} = \lambda, \text{ where } \lambda = \begin{bmatrix} 0 \\ 1 \\ 0 \end{bmatrix}, \quad \text{or} \quad \begin{bmatrix} 1 & 1 & 1 & 1 \\ 1 & 1 & 0 & 0 \\ 0 & 0 & 1 & 1 \end{bmatrix} \begin{bmatrix} a_1 \\ a_2 \\ a_3 \\ a_4 \end{bmatrix} = \begin{bmatrix} 0 \\ 1 \\ 0 \end{bmatrix}.$$

Clearly, the right-hand-side λ is not in $\mathcal{C}(\mathbf{X}^T)$, no solution exists, and α_1 is not estimable.

In words, this result says that a particular function $\lambda^T\mathbf{b}$ is estimable if and only if it is equal to the expected value of a linear combination of observations. This result limits what parameters are estimable, or what linear combination of parameters may be estimable. To begin with, consider some obvious cases. First, the expected value of any observation is estimable: taking $\mathbf{a} = \mathbf{e}^{(i)}$ we have $\text{E}(\mathbf{y}_i) = (\mathbf{X}\mathbf{b})_i$, and hence $(\mathbf{X}\mathbf{b})_1, \ldots, (\mathbf{X}\mathbf{b})_N$ are estimable. Second, if two linear functions are estimable, then any linear combination of them is estimable; in general, if $\lambda^{(j)T}\mathbf{b}$, $j = 1, \ldots, k$, are estimable, then so is $\sum_j d_j\lambda^{(j)T}\mathbf{b}$ for any constants $d_1, \ldots, d_k$ (see Exercise 3.1). But this means that the set of vectors λ for which $\lambda^T\mathbf{b}$ is estimable forms a vector space, more specifically, following from $\lambda = \mathbf{X}^T\mathbf{a}$, this space of vectors λ for which $\lambda^T\mathbf{b}$ is estimable is $\mathcal{C}(\mathbf{X}^T)$. Since the dimension of $\mathcal{C}(\mathbf{X}^T)$ is r, then we cannot find more than $r = rank(\mathbf{X})$ linearly independent vectors $\lambda^{(j)}$ that span $\mathcal{C}(\mathbf{X}^T)$. Consequently, there are no more than r linearly independent estimable functions $\lambda^{(j)T}\mathbf{b}$. Finally, if the design matrix $\mathbf{X}$ has full-column rank, that is, $r = rank(\mathbf{X}) = p$, then $\mathbf{X}^T$ has full-row rank, $\mathcal{C}(\mathbf{X}^T) = R^p$, and so $\lambda^T\mathbf{b}$ is estimable for all λ. In words: When $\mathbf{X}$ has full-column rank, all of the components of $\mathbf{b}$ are estimable, and hence all linear combinations $\lambda^T\mathbf{b}$ are estimable and the estimability problem disappears. On the other hand, if $r < p$, then there are some linear combinations $\lambda^T\mathbf{b}$ that are not estimable.

On a practical note, many situations will arise where we will need to establish that a particular function $\lambda^T\mathbf{b}$ is estimable. There are three basic ways to do this:

Method 3.1 *Linear combinations of expected values of observations are estimable. If we can express $\lambda^T \mathbf{b}$ as a linear combination of $E(y_i)$, then $\lambda^T \mathbf{b}$ is estimable.*

Method 3.2 *If $\lambda \in \mathcal{C}(\mathbf{X}^T)$ then $\lambda^T \mathbf{b}$ is estimable. So construct a set of basis vectors for $\mathcal{C}(\mathbf{X}^T)$, say $\{\mathbf{v}^{(1)}, \mathbf{v}^{(2)}, \ldots, \mathbf{v}^{(r)}\}$, and find constants d_j so that $\lambda = \sum_j d_j \mathbf{v}^{(j)}$.*

Method 3.3 *Note that $\lambda \in \mathcal{C}(\mathbf{X}^T)$ iff $\lambda \perp \mathcal{N}(\mathbf{X})$. So find a basis for $\mathcal{N}(\mathbf{X})$, say $\{\mathbf{c}^{(1)}, \mathbf{c}^{(2)}, \ldots, \mathbf{c}^{(p-r)}\}$. Then if $\lambda \perp \mathbf{c}^{(j)}$ for all $j = 1, \ldots, p-r$, then $\lambda \in \mathcal{C}(\mathbf{X}^T)$ and $\lambda^T \mathbf{b}$ is estimable.*

The most common approaches are Methods 3.1 and 3.3 since they are often the easiest to show. The convenience of Method 3.3 must be tempered by the warning that we must show that λ is orthogonal to all basis vectors for $\mathcal{N}(\mathbf{X})$. Recall that the basis vectors for $\mathcal{N}(\mathbf{X})$ are determined in finding the rank of $\mathbf{X}$. The rank of $\mathbf{X}$ and hence the dimension of its nullspace must be known with confidence, since overstating the rank means missing a basis vector for $\mathcal{N}(\mathbf{X})$ and overstating estimability. Method 3.2 is often difficult to implement in a mechanical fashion. However, from Result A.14, $(\mathbf{X}^T\mathbf{X})(\mathbf{X}^T\mathbf{X})^g$ is a projection onto $\mathcal{C}(\mathbf{X}^T\mathbf{X}) = \mathcal{C}(\mathbf{X}^T)$. So if $(\mathbf{X}^T\mathbf{X})(\mathbf{X}^T\mathbf{X})^g\lambda = \lambda$, then $\lambda \in \mathcal{C}(\mathbf{X}^T\mathbf{X})$ and $\lambda^T\mathbf{b}$ is estimable. This approach is appealing whenever $(\mathbf{X}^T\mathbf{X})^g$ has already been constructed for solving the normal equations.

Finally, we consider estimation of a function of the parameters $\lambda^T\mathbf{b}$ using the solution to the normal equations. We first construct a linear unbiased estimator for an estimable function. Later, in Chapter 4, we will show that the estimator constructed below has desirable properties.

Definition 3.4 *The* least squares estimator *of an estimable function $\lambda^T\mathbf{b}$ is $\lambda^T\hat{\mathbf{b}}$ where $\hat{\mathbf{b}}$ is a solution to the normal equations.*

Result 3.2 *If $\lambda^T\mathbf{b}$ is estimable, then the least squares estimator $\lambda^T\hat{\mathbf{b}}$ is the same for all solutions $\hat{\mathbf{b}}$ to the normal equations.*

Proof: (Algebraic) Given a generalized inverse $(\mathbf{X}^T\mathbf{X})^g$, all solutions to the normal equations are given by $\hat{\mathbf{b}}(\mathbf{z}) = (\mathbf{X}^T\mathbf{X})^g\mathbf{X}^T\mathbf{y} + (\mathbf{I} - (\mathbf{X}^T\mathbf{X})^g\mathbf{X}^T\mathbf{X})\mathbf{z}$. If $\lambda^T\mathbf{b}$ is estimable, then we can write $\lambda^T = \mathbf{a}^T\mathbf{X}$ for some vector $\mathbf{a}$, so that

$$\begin{aligned}\lambda^T\hat{\mathbf{b}}(\mathbf{z}) &= \mathbf{a}^T\mathbf{X}(\mathbf{X}^T\mathbf{X})^g\mathbf{X}^T\mathbf{y} + \mathbf{a}^T\mathbf{X}(\mathbf{I} - (\mathbf{X}^T\mathbf{X})^g\mathbf{X}^T\mathbf{X})\mathbf{z} \\ &= \mathbf{a}^T\mathbf{P}_\mathbf{X}\mathbf{y} + \mathbf{a}^T(\mathbf{X} - \mathbf{P}_\mathbf{X}\mathbf{X})\mathbf{z} = \mathbf{a}^T\mathbf{P}_\mathbf{X}\mathbf{y}\end{aligned}$$

which does not depend on $\mathbf{z}$ or, for that matter, the choice of generalized inverse $(\mathbf{X}^T\mathbf{X})^g$. □

Proof: (Geometric) Let $\hat{\mathbf{b}}_1$ and $\hat{\mathbf{b}}_2$ be two solutions to the normal equations, hence $\mathbf{X}^T\mathbf{X}(\hat{\mathbf{b}}_1 - \hat{\mathbf{b}}_2) = \mathbf{0}$, or $(\hat{\mathbf{b}}_1 - \hat{\mathbf{b}}_2) \in \mathcal{N}(\mathbf{X}^T\mathbf{X}) = \mathcal{N}(\mathbf{X})$. If $\lambda^T\mathbf{b}$ is estimable, then $\lambda \in \mathcal{C}(\mathbf{X}^T\mathbf{X}) \perp \mathcal{N}(\mathbf{X})$, hence $\lambda \perp (\hat{\mathbf{b}}_1 - \hat{\mathbf{b}}_2)$, or $\lambda^T(\hat{\mathbf{b}}_1 - \hat{\mathbf{b}}_2) = 0$, or $\lambda^T\hat{\mathbf{b}}_1 = \lambda^T\hat{\mathbf{b}}_2$. □

The converse to Result 3.2 is also true; see Exercise 3.2.

Result 3.3 *The least squares estimator $\lambda^T\hat{\mathbf{b}}$ of an estimable function $\lambda^T\mathbf{b}$ is a linear unbiased estimator of $\lambda^T\mathbf{b}$.*

Proof: First note that for any solution $\hat{\mathbf{b}}$ of the normal equations we can write

$$\lambda^T\hat{\mathbf{b}} = \lambda^T[(\mathbf{X}^T\mathbf{X})^g\mathbf{X}^T\mathbf{y} + (\mathbf{I} - (\mathbf{X}^T\mathbf{X})^g\mathbf{X}^T\mathbf{X})\mathbf{z}] = [\lambda^T(\mathbf{X}^T\mathbf{X})^g\mathbf{X}^T]\mathbf{y}$$

□

so that $\lambda^T\hat{\mathbf{b}}$ is a linear estimator. Next, recall that if $\lambda^T\mathbf{b}$ is estimable, then we can write $\lambda^T = \mathbf{a}^T\mathbf{X}$ for some vector $\mathbf{a}$. Then to show that $\lambda^T\hat{\mathbf{b}}$ is unbiased, note that

$$\begin{aligned} E(\lambda^T\hat{\mathbf{b}}) &= \lambda^T(\mathbf{X}^T\mathbf{X})^g\mathbf{X}^T E(\mathbf{y}) = \lambda^T(\mathbf{X}^T\mathbf{X})^g\mathbf{X}^T\mathbf{X}\mathbf{b} \\ &= \mathbf{a}^T\mathbf{X}(\mathbf{X}^T\mathbf{X})^g\mathbf{X}^T\mathbf{X}\mathbf{b} = \mathbf{a}^T\mathbf{P_X}\mathbf{X}\mathbf{b} = \mathbf{a}^T\mathbf{X}\mathbf{b} = \lambda^T\mathbf{b} \end{aligned}$$

for all $\mathbf{b}$.

The results above can be used to show estimability of $\lambda^T\mathbf{b}$ and to construct linear unbiased estimators of estimable functions. We now present some examples of estimable and nonestimable functions in the two most common examples of non-full-rank linear models.

3.4 First Example: One-Way ANOVA

Consider the simple one-way ANOVA model:

$$y_{ij} = \mu + \alpha_i + e_{ij}$$

for $i = 1, 2, 3 = a$, and $j = 1, \ldots, n_i = 4 - i$, say, with $E(e_{ij}) = 0$, where μ, α_1, α_2, and α_3 are fixed, unknown parameters. Writing this in the form $\mathbf{y} = \mathbf{X}\mathbf{b} + \mathbf{e}$, we have

$$\mathbf{X} = \begin{bmatrix} 1 & 1 & 0 & 0 \\ 1 & 1 & 0 & 0 \\ 1 & 1 & 0 & 0 \\ 1 & 0 & 1 & 0 \\ 1 & 0 & 1 & 0 \\ 1 & 0 & 0 & 1 \end{bmatrix} \quad \text{and} \quad \mathbf{b} = \begin{bmatrix} \mu \\ \alpha_1 \\ \alpha_2 \\ \alpha_3 \end{bmatrix}, \quad \mathbf{X}^T\mathbf{X} = \begin{bmatrix} 6 & 3 & 2 & 1 \\ 3 & 3 & 0 & 0 \\ 2 & 0 & 2 & 0 \\ 1 & 0 & 0 & 1 \end{bmatrix},$$

and a generalized inverse of $\mathbf{X}^T\mathbf{X}$ is

$$(\mathbf{X}^T\mathbf{X})^g = \begin{bmatrix} 0 & 0 & 0 & 0 \\ 0 & 1/3 & 0 & 0 \\ 0 & 0 & 1/2 & 0 \\ 0 & 0 & 0 & 1 \end{bmatrix} \text{ so that } (\mathbf{X}^T\mathbf{X})(\mathbf{X}^T\mathbf{X})^g = \begin{bmatrix} 0 & 1 & 1 & 1 \\ 0 & 1 & 0 & 0 \\ 0 & 0 & 1 & 0 \\ 0 & 0 & 0 & 1 \end{bmatrix}.$$

The last three columns of $(\mathbf{X}^T\mathbf{X})(\mathbf{X}^T\mathbf{X})^g$ above can be chosen to be $\mathbf{v}^{(1)}$, $\mathbf{v}^{(2)}$, $\mathbf{v}^{(3)}$, to form a set of basis vectors for $\mathcal{C}(\mathbf{X}^T)$. Notice that when transposed they form linearly independent rows of $\mathbf{X}$. For $\mathcal{N}(\mathbf{X})$, notice that the first column of $\mathbf{X}$ is the sum of the last three columns, so a basis vector for $\mathcal{N}(\mathbf{X})$ is given as the nonzero column in the following matrix that projects onto $\mathcal{N}(\mathbf{X})$:

$$\mathbf{I} - (\mathbf{X}^T\mathbf{X})^g(\mathbf{X}^T\mathbf{X}) = \begin{bmatrix} 1 & 0 & 0 & 0 \\ -1 & 0 & 0 & 0 \\ -1 & 0 & 0 & 0 \\ -1 & 0 & 0 & 0 \end{bmatrix} \quad \text{and} \quad \mathbf{c}^{(1)} = \begin{bmatrix} 1 \\ -1 \\ -1 \\ -1 \end{bmatrix}.$$

To determine whether certain functions are estimable, consider $\lambda^{(1)T}\mathbf{b} = \mu + \alpha_1$, $\lambda^{(2)T}\mathbf{b} = \alpha_2$, and $\lambda^{(3)T}\mathbf{b} = \alpha_1 - \alpha_3$, corresponding to

$$\lambda^{(1)} = \begin{bmatrix} 1 \\ 1 \\ 0 \\ 0 \end{bmatrix}, \quad \lambda^{(2)} = \begin{bmatrix} 0 \\ 0 \\ 1 \\ 0 \end{bmatrix}, \quad \lambda^{(3)} = \begin{bmatrix} 0 \\ 1 \\ 0 \\ -1 \end{bmatrix},$$

Method 3.1 handles the first and third cases with ease, as $E(y_{1j}) = \mu + \alpha_1$ and $E(y_{1j}) - E(y_{3j}) = \alpha_1 - \alpha_3$. Therefore, both $\mu + \alpha_1$ and $\alpha_1 - \alpha_3$ are estimable. Applying Method 3.3 we find that $\lambda^{(2)T}\mathbf{c}^{(1)} = -1$ and not zero. Consequently, $\lambda^{(2)}$ is not orthogonal to the basis vectors for $\mathcal{N}(\mathbf{X})$, thus $\lambda^{(2)T}\mathbf{b} = \alpha_2$ is not estimable. Since we have already found $(\mathbf{X}^T\mathbf{X})^g$, we can compute $(\mathbf{X}^T\mathbf{X})(\mathbf{X}^T\mathbf{X})^g\lambda^{(j)} = \lambda^{(j)}$ for $j = 1$ and 3, but for $\lambda^{(2)}$, we have

$$\begin{bmatrix} 0 & 1 & 1 & 1 \\ 0 & 1 & 0 & 0 \\ 0 & 0 & 1 & 0 \\ 0 & 0 & 0 & 1 \end{bmatrix} \begin{bmatrix} 0 \\ 0 \\ 1 \\ 0 \end{bmatrix} = \begin{bmatrix} 1 \\ 0 \\ 1 \\ 0 \end{bmatrix} \neq \begin{bmatrix} 0 \\ 0 \\ 1 \\ 0 \end{bmatrix},$$

confirming that α_2 is not estimable. The solution to the normal equations, using the generalized inverse chosen above, is

$$\hat{\mathbf{b}}_1 = \begin{bmatrix} 0 \\ \overline{y}_{1.} \\ \overline{y}_{2.} \\ \overline{y}_{3.} \end{bmatrix}.$$

If we choose another generalized inverse, these two projection matrices will change, but not in spirit.

$$(\mathbf{X}^T\mathbf{X})^g = \begin{bmatrix} 1 & -1 & -1 & 0 \\ -1 & 4/3 & 1 & 0 \\ -1 & 1 & 3/2 & 0 \\ 0 & 0 & 0 & 0 \end{bmatrix}, \quad (\mathbf{X}^T\mathbf{X})(\mathbf{X}^T\mathbf{X})^g = \begin{bmatrix} 1 & 0 & 0 & 0 \\ 0 & 1 & 0 & 0 \\ 0 & 0 & 1 & 0 \\ 1 & -1 & -1 & 0 \end{bmatrix}$$

$$\text{and} \quad \mathbf{I} - (\mathbf{X}^T\mathbf{X})^g(\mathbf{X}^T\mathbf{X}) = \begin{bmatrix} 0 & 0 & 0 & -1 \\ 0 & 0 & 0 & 1 \\ 0 & 0 & 0 & 1 \\ 0 & 0 & 0 & 1 \end{bmatrix}.$$

With this generalized inverse, we get a different solution to the normal equations

$$\hat{\mathbf{b}}_2 = \begin{bmatrix} \overline{y}_{3.} \\ \overline{y}_{1.} - \overline{y}_{3.} \\ \overline{y}_{2.} - \overline{y}_{3.} \\ 0 \end{bmatrix},$$

but values of the least squares estimators for estimable functions will not change:

$$\begin{aligned} \lambda^{(1)T}\hat{\mathbf{b}}_1 &= \overline{y}_{1.} = \lambda^{(1)T}\hat{\mathbf{b}}_2 \\ \lambda^{(3)T}\hat{\mathbf{b}}_1 &= \overline{y}_{1.} - \overline{y}_{3.} = \lambda^{(3)T}\hat{\mathbf{b}}_2. \end{aligned}$$

For a nonestimable function, the estimates may change:

$$\lambda^{(2)T}\hat{\mathbf{b}}_1 = \overline{y}_{2.} \neq \overline{y}_{2.} - \overline{y}_{3.} = \lambda^{(2)T}\hat{\mathbf{b}}_2.$$

Notice that both generalized inverses chosen here are symmetric.

Now for the general case, with $y_{ij} = \mu + \alpha_i + e_{ij}$ for $i = 1, \ldots, a$, and $j = 1, \ldots, n_i$, ordering y_{ij} first by i, then by j within i and stacking into $\mathbf{y}$ leads to $E(\mathbf{y}) = \mathbf{Xb}$, with

$$\mathbf{X} = \begin{bmatrix} \mathbf{1}_{n_1} & \mathbf{1}_{n_1} & \mathbf{0} & \ldots & \mathbf{0} \\ \mathbf{1}_{n_2} & \mathbf{0} & \mathbf{1}_{n_2} & \ldots & \mathbf{0} \\ \mathbf{1}_{n_3} & \mathbf{0} & \mathbf{0} & \ldots & \mathbf{0} \\ \ldots & \ldots & \ldots & \ldots & \ldots \\ \mathbf{1}_{n_a} & \mathbf{0} & \ldots & \mathbf{0} & \mathbf{1}_{n_a} \end{bmatrix} \text{ and } \mathbf{b} = \begin{bmatrix} \mu \\ \alpha_1 \\ \alpha_2 \\ \ldots \\ \alpha_a \end{bmatrix}.$$

Here $p = a + 1$, $rank(\mathbf{X}) = r = a$, and $N = \sum_i n_i$ so that $\mathcal{N}(\mathbf{X})$ has just one dimension. A basis vector for $\mathcal{N}(\mathbf{X})$ is given by

$$\mathbf{c}^{(1)} = \begin{bmatrix} 1 \\ -\mathbf{1}_a \end{bmatrix}.$$

Hence, a linear combination $\lambda^T\mathbf{b} = \lambda_0\mu + \sum_i \lambda_i\alpha_i$ is estimable if and only if $\lambda_0 - \sum_i \lambda_i = 0$. The reader should see that $\mu + \alpha_i$ is estimable, as are $\alpha_i - \alpha_k$. Note that $\sum_i d_i\alpha_i$ will be estimable if and only if $\sum_i d_i = 0$. This function $\sum_i d_i\alpha_i$ with $\sum_i d_i = 0$ is known more commonly as a *contrast*. If we construct the normal equations and find a solution, we have

$$\mathbf{X}^T\mathbf{X} = \begin{bmatrix} N & n_1 & n_2 & \dots & n_a \\ n_1 & n_1 & 0 & & 0 \\ n_2 & 0 & n_2 & & 0 \\ \dots & & & \dots & 0 \\ n_a & 0 & 0 & 0 & n_a \end{bmatrix} \begin{bmatrix} \mu \\ \alpha_1 \\ \alpha_2 \\ \dots \\ \alpha_a \end{bmatrix} = \begin{bmatrix} N\overline{y}_{..} \\ n_1\overline{y}_{1.} \\ n_2\overline{y}_{2.} \\ \dots \\ n_a\overline{y}_{a.} \end{bmatrix} = \mathbf{X}^T\mathbf{y}$$

where an easy generalized inverse,

$$(\mathbf{X}^T\mathbf{X})^g = \begin{bmatrix} 0 & 0 & 0 & \dots & 0 \\ 0 & 1/n_1 & 0 & & 0 \\ 0 & 0 & 1/n_2 & & 0 \\ \dots & & & \dots & \\ 0 & 0 & 0 & & 1/n_a \end{bmatrix} \quad \text{and} \quad \hat{\mathbf{b}} = \begin{bmatrix} 0 \\ \overline{y}_{1.} \\ \overline{y}_{2.} \\ \dots \\ \overline{y}_{a.} \end{bmatrix} + z \begin{bmatrix} -1 \\ 1 \\ 1 \\ \dots \\ 1 \end{bmatrix},$$

gives the general solution. The least squares estimators of the estimable functions $\mu + \alpha_i$, $\alpha_i - \alpha_k$, and $\sum_i d_i\alpha_i$ with $\sum_i d_i = 0$ are $\overline{y}_{i.}$, $\overline{y}_{i.} - \overline{y}_{k.}$ and $\sum_i d_i\overline{y}_{i.}$, respectively. The projection matrix $\mathbf{P_X}$ takes on an enlightening form:

$$\begin{bmatrix} \mathbf{1}_{n_1} & \mathbf{1}_{n_1} & \mathbf{0} & \dots & \mathbf{0} \\ \mathbf{1}_{n_2} & \mathbf{0} & \mathbf{1}_{n_2} & \dots & \mathbf{0} \\ \mathbf{1}_{n_3} & \mathbf{0} & \mathbf{0} & \dots & \mathbf{0} \\ \dots & \dots & \dots & \dots & \dots \\ \mathbf{1}_{n_a} & \mathbf{0} & \dots & \mathbf{0} & \mathbf{1}_{n_a} \end{bmatrix} \begin{bmatrix} 0 & 0 & 0 & \dots & 0 \\ 0 & 1/n_1 & 0 & & 0 \\ 0 & 0 & 1/n_2 & & 0 \\ \dots & & & \dots & \\ 0 & 0 & 0 & & 1/n_a \end{bmatrix}$$

$$\times \begin{bmatrix} \mathbf{1}_{n_1}^T & \mathbf{1}_{n_2}^T & \mathbf{1}_{n_3}^T & \dots & \mathbf{1}_{n_a}^T \\ \mathbf{1}_{n_1}^T & \mathbf{0} & \mathbf{0} & \dots & \mathbf{0} \\ \mathbf{0} & \mathbf{1}_{n_2}^T & \mathbf{0} & \dots & \mathbf{0} \\ \dots & \dots & \dots & \dots & \dots \\ \mathbf{0} & \mathbf{0} & \dots & \mathbf{0} & \mathbf{1}_{n_a}^T \end{bmatrix} = \mathbf{X}(\mathbf{X}^T\mathbf{X})^g\mathbf{X}^T$$

$$= \begin{bmatrix} 1/n_1\mathbf{1}_{n_1}\mathbf{1}_{n_1}^T & \mathbf{0} & \dots & \mathbf{0} \\ \mathbf{0} & 1/n_2\mathbf{1}_{n_2}\mathbf{1}_{n_2}^T & \dots & \mathbf{0} \\ \dots & \dots & \dots & \dots \\ \mathbf{0} & \mathbf{0} & \dots & 1/n_a\mathbf{1}_{n_a}\mathbf{1}_{n_a}^T \end{bmatrix}$$

showing that for any vector $\mathbf{u}$ following the same indexing as $\mathbf{y}$, $(\mathbf{P_X u})_{ij} = (1/n_i) \sum_j u_{ij} = \overline{u}_{i.}$.

3.5 Second Example: Two-Way Crossed without Interaction

The two-way crossed model without interaction is the least complicated two-factor model, and for simplicity, we will consider first the balanced case without replication: $y_{ij} = \mu + \alpha_i + \beta_j + e_{ij}$ for $i = 1, \ldots, a$; $j = 1, \ldots, b$; so that $N = ab$ and $p = 1 + a + b$ are dimensions of the design matrix:

$$\mathbf{Xb} = \begin{bmatrix} \mathbf{1}_b & \mathbf{1}_b & \mathbf{0} & \ldots & \mathbf{0} & \mathbf{I}_b \\ \mathbf{1}_b & \mathbf{0} & \mathbf{1}_b & \ldots & \mathbf{0} & \mathbf{I}_b \\ \ldots & \ldots & \ldots & \ldots & & \\ \mathbf{1}_b & \mathbf{0} & \mathbf{0} & \ldots & \mathbf{0} & \mathbf{I}_b \\ \mathbf{1}_b & \mathbf{0} & \mathbf{0} & \ldots & \mathbf{1}_b & \mathbf{I}_b \end{bmatrix} \begin{bmatrix} \mu \\ \alpha_1 \\ \alpha_2 \\ \ldots \\ \alpha_a \\ \beta_1 \\ \beta_2 \\ \ldots \\ \beta_b \end{bmatrix}.$$

Notice that the rank of $\mathbf{X}$ is $r = a + b - 1$, so that $dim(\mathcal{N}(\mathbf{X})) = 2$. The following two vectors form a basis for $\mathcal{N}(\mathbf{X})$:

$$\mathbf{c}^{(1)} = \begin{bmatrix} 1 \\ -\mathbf{1}_a \\ \mathbf{0} \end{bmatrix} \quad \text{and} \quad \mathbf{c}^{(2)} = \begin{bmatrix} 1 \\ \mathbf{0} \\ -\mathbf{1}_b \end{bmatrix}.$$

As a result, $\lambda^T \mathbf{b} = \lambda_0 \mu + \sum_i \lambda_i \alpha_i + \sum_j \lambda_{a+j} \beta_j$ is estimable if and only if

$$\lambda_0 - \sum_i \lambda_i = 0, \text{ and } \lambda_0 - \sum_j \lambda_{a+j} = 0.$$

The most commonly used estimable functions are the following:

- $\mu + \alpha_i + \beta_j$
- $\alpha_i - \alpha_k$
- $\beta_j - \beta_k$
- $\sum d_i \alpha_i$ if $\sum d_i = 0$
- $\sum f_j \beta_j$ if $\sum f_j = 0$

Note that the individual parameters μ, α_i, β_j and the function $\alpha_i + \beta_j$ are not estimable. Computing a generalized inverse and constructing solutions for the general case is a more burdensome task, as the following analysis suggests:

$$\text{note } \mathbf{X}^T\mathbf{X} = \begin{bmatrix} ab & b\mathbf{1}_a^T & a\mathbf{1}_b^T \\ b\mathbf{1}_a & b\mathbf{I}_a & \mathbf{1}_a\mathbf{1}_b^T \\ a\mathbf{1}_b & \mathbf{1}_b\mathbf{1}_b^T & a\mathbf{I}_b \end{bmatrix} \begin{matrix} 1 \\ a \\ b \end{matrix} \text{ and let } \mathbf{G} = \begin{bmatrix} \frac{1}{ab} & \mathbf{0} & \mathbf{0} \\ -\frac{1}{ab}\mathbf{1}_a & \frac{1}{b}\mathbf{I}_a & \mathbf{0} \\ -\frac{1}{ab}\mathbf{1}_b & \mathbf{0} & \frac{1}{a}\mathbf{I}_b \end{bmatrix} \begin{matrix} 1 \\ a \\ b \end{matrix} \text{ also}$$

$$(\mathbf{X}^T\mathbf{X})\mathbf{G} = \begin{bmatrix} -1 & \mathbf{1}_a^T & \mathbf{1}_b^T \\ -\frac{1}{a}\mathbf{1}_a & \mathbf{I}_a & \frac{1}{a}\mathbf{1}_a\mathbf{1}_b^T \\ -\frac{1}{b}\mathbf{1}_b & \frac{1}{b}\mathbf{1}_b\mathbf{1}_a^T & \mathbf{I}_b \end{bmatrix} \text{ and } \mathbf{I} - \mathbf{G}^T(\mathbf{X}^T\mathbf{X}) = \begin{bmatrix} 2 & \frac{1}{a}\mathbf{1}_a^T & \frac{1}{b}\mathbf{1}_b^T \\ -\mathbf{1}_a & \mathbf{0} & -\frac{1}{b}\mathbf{1}_a\mathbf{1}_b^T \\ -\mathbf{1}_b & -\frac{1}{a}\mathbf{1}_b\mathbf{1}_a^T & \mathbf{0} \end{bmatrix}.$$

Notice that the rank of this last matrix is 2, and with a little effort, the basis for $\mathcal{N}(\mathbf{X})$ given above can be found in terms of its columns.

Now let us take a specific case of $a = 3$ and $b = 4$, which gives $\mathbf{Xb}$ as

$$\begin{bmatrix} 1 & 1 & 0 & 0 & 1 & 0 & 0 & 0 \\ 1 & 1 & 0 & 0 & 0 & 1 & 0 & 0 \\ 1 & 1 & 0 & 0 & 0 & 0 & 1 & 0 \\ 1 & 1 & 0 & 0 & 0 & 0 & 0 & 1 \\ 1 & 0 & 1 & 0 & 1 & 0 & 0 & 0 \\ 1 & 0 & 1 & 0 & 0 & 1 & 0 & 0 \\ 1 & 0 & 1 & 0 & 0 & 0 & 1 & 0 \\ 1 & 0 & 1 & 0 & 0 & 0 & 0 & 1 \\ 1 & 0 & 0 & 1 & 1 & 0 & 0 & 0 \\ 1 & 0 & 0 & 1 & 0 & 1 & 0 & 0 \\ 1 & 0 & 0 & 1 & 0 & 0 & 1 & 0 \\ 1 & 0 & 0 & 1 & 0 & 0 & 0 & 1 \end{bmatrix} \begin{bmatrix} \mu \\ \alpha_1 \\ \alpha_2 \\ \alpha_3 \\ \beta_1 \\ \beta_2 \\ \beta_3 \\ \beta_4 \end{bmatrix}$$

with two basis vectors for $\mathcal{N}(\mathbf{X})$:

$$\begin{bmatrix} 1 \\ -1 \\ -1 \\ -1 \\ 0 \\ 0 \\ 0 \\ 0 \end{bmatrix}, \begin{bmatrix} 1 \\ 0 \\ 0 \\ 0 \\ -1 \\ -1 \\ -1 \\ -1 \end{bmatrix}.$$

Notice that the rank of $\mathbf{X}$ is $8 - 2 = 6$ as expected, two basis vectors for $\mathcal{N}(\mathbf{X})$ are given above (right), and a set of six linearly independent estimable functions is $\{\mu + \alpha_1 + \beta_1, \alpha_2 - \alpha_1, \alpha_3 - \alpha_1, \beta_2 - \beta_1, \beta_3 - \beta_1, \beta_4 - \beta_1\}$.

Now suppose we have some missing cells. For example, suppose only observations in seven of the $a \times n = 3 \times 4 = 12$ cells are observed. In particular, only cells (1, 1),

(1, 2), (1, 3), (2, 1), (2, 3), (3, 3), and (3, 4) have observations. Then **Xb** looks like:

$$\begin{bmatrix} 1 & 1 & 0 & 0 & 1 & 0 & 0 & 0 \\ 1 & 1 & 0 & 0 & 0 & 1 & 0 & 0 \\ 1 & 1 & 0 & 0 & 0 & 0 & 1 & 0 \\ 1 & 0 & 1 & 0 & 1 & 0 & 0 & 0 \\ 1 & 0 & 1 & 0 & 0 & 0 & 1 & 0 \\ 1 & 0 & 0 & 1 & 0 & 0 & 1 & 0 \\ 1 & 0 & 0 & 1 & 0 & 0 & 0 & 1 \end{bmatrix} \begin{bmatrix} \mu \\ \alpha_1 \\ \alpha_2 \\ \alpha_3 \\ \beta_1 \\ \beta_2 \\ \beta_3 \\ \beta_4 \end{bmatrix}$$

with the same basis vectors for $\mathcal{N}(\mathbf{X})$:

$$\begin{bmatrix} 1 \\ -1 \\ -1 \\ -1 \\ 0 \\ 0 \\ 0 \\ 0 \end{bmatrix}, \begin{bmatrix} 1 \\ 0 \\ 0 \\ 0 \\ -1 \\ -1 \\ -1 \\ -1 \end{bmatrix}.$$

Notice that the rank of **X** is unchanged at 6, with the same estimable functions as in the case where no cells were missing. However, let us consider making one change—drop cell (3, 3) and add (2, 2), which does not change the number of observations. In this case, the rank of **X** decreases to 5, and one more vector is added to the basis for $\mathcal{N}(\mathbf{X})$. In this case, **Xb** takes the form

$$\begin{bmatrix} 1 & 1 & 0 & 0 & 1 & 0 & 0 & 0 \\ 1 & 1 & 0 & 0 & 0 & 1 & 0 & 0 \\ 1 & 1 & 0 & 0 & 0 & 0 & 1 & 0 \\ 1 & 0 & 1 & 0 & 1 & 0 & 0 & 0 \\ 1 & 0 & 1 & 0 & 0 & 1 & 0 & 0 \\ 1 & 0 & 1 & 0 & 0 & 0 & 1 & 0 \\ 1 & 0 & 0 & 1 & 0 & 0 & 0 & 1 \end{bmatrix} \begin{bmatrix} \mu \\ \alpha_1 \\ \alpha_2 \\ \alpha_3 \\ \beta_1 \\ \beta_2 \\ \beta_3 \\ \beta_4 \end{bmatrix}$$

but three vectors for $\mathcal{N}(\mathbf{X})$:

$$\begin{bmatrix} 1 \\ -1 \\ -1 \\ -1 \\ 0 \\ 0 \\ 0 \\ 0 \end{bmatrix}, \begin{bmatrix} 1 \\ 0 \\ 0 \\ 0 \\ -1 \\ -1 \\ -1 \\ -1 \end{bmatrix}, \begin{bmatrix} 0 \\ 0 \\ 0 \\ 1 \\ 0 \\ 0 \\ 0 \\ -1 \end{bmatrix}.$$

A full set of linearly independent estimable functions is $\{\mu+\alpha_1+\beta_1, \alpha_2-\alpha_1, \beta_2-\beta_1, \beta_3-\beta_1, \mu+\alpha_3+\beta_4\}$. The change from the case of no missing cells is that $\alpha_3-\alpha_1$ is no longer estimable.

While the rank of **X** is limited by the number of observations (rows) of **X**, it is not so much the number of missing cells that affects rank and estimability, but the pattern of missing cells. In terms of estimable functions, we find that $\alpha_3-\alpha_1$ and $\beta_4-\beta_1$ are no longer estimable, while $\mu+\alpha_3+\beta_4$ is still estimable. The obvious consequence is that α_3 and β_4 are confounded. To examine the effect of the pattern and to detect a loss of rank because of missing cells, graphical techniques are used to discover *connectedness*. The basic idea is that if all of the cells are in a sense connected, then there is no loss in rank of **X**. If groups of cells are left unconnected with other cells, then each additional group of cells signals that the rank of **X** will drop by 1. In the two cases above with missing cells, the former case (depicted on the left below) has all cells connected and no loss of rank, and everything estimable as before. In the latter case (on the right), six cells form one group and the remaining cell $(3, 4)$ forms a second group, as if there were two different experiments. The patterns of cells with observations are given in the following two tables, with the first one with $rank(\mathbf{X}) = 6$ on the left, and the second one with $rank(\mathbf{X}) = 5$ on the right:

	β_1	β_2	β_3	β_4
α_1	X	X	X	.
α_2	X	.	X	.
α_3	.	.	X	X

and

	β_1	β_2	β_3	β_4
α_1	X	X	X	.
α_2	X	X	X	.
α_3	.	.	.	X

.

Searle ([42] Section 7.6) uses tables similar to those above as part of a graphical technique to discover connectedness, going across rows (α's) to see what column effects (β's) are connected, then going down columns (β's) to see what row effects (α's) are connected. The concept of connectedness is important only because the two-way crossed model without interaction is so common.

3.6 Two-Way Crossed with Interaction

Adding interaction to the two-way crossed model adds some difficulty, but the intent here is to address a common misconception about interaction. The model can be written as

$$y_{ijk} = \mu + \alpha_i + \beta_j + \gamma_{ij} + e_{ijk} \text{ for } i = 1, \ldots, a;\ j = 1, \ldots, b;\ k = 1, \ldots, n_{ij}.$$

First of all, to simplify some matrices, the reader should have noticed that replication does not affect estimability—adding replicate design points only adds replicate rows to **X** or columns to $\mathbf{X}^T$—and does not change the rank. So, for investigating estimability, without loss of generality, let $n_{ij} = 1$. Next, consider the specific case of $a = 2$ and

$b = 3$; generalizations will be postponed. Writing $\mathbf{Xb}$ as

$$\mathbf{Xb} = \begin{bmatrix} 1 & 1 & 0 & 1 & 0 & 0 & 1 & 0 & 0 & 0 & 0 & 0 \\ 1 & 1 & 0 & 0 & 1 & 0 & 0 & 1 & 0 & 0 & 0 & 0 \\ 1 & 1 & 0 & 0 & 0 & 1 & 0 & 0 & 1 & 0 & 0 & 0 \\ 1 & 0 & 1 & 1 & 0 & 0 & 0 & 0 & 0 & 1 & 0 & 0 \\ 1 & 0 & 1 & 0 & 1 & 0 & 0 & 0 & 0 & 0 & 1 & 0 \\ 1 & 0 & 1 & 0 & 0 & 1 & 0 & 0 & 0 & 0 & 0 & 1 \end{bmatrix} \begin{bmatrix} \mu \\ \alpha_1 \\ \alpha_2 \\ \beta_1 \\ \beta_2 \\ \beta_3 \\ \gamma_{11} \\ \gamma_{12} \\ \gamma_{13} \\ \gamma_{21} \\ \gamma_{22} \\ \gamma_{23} \end{bmatrix}$$

we can see that we have only six estimable functions of the twelve parameters, so that $dim(\mathcal{N}(\mathbf{X})) = 6$. The following forms a list of six basis vectors for $\mathcal{N}(\mathbf{X})$:

$$\begin{bmatrix} -1 \\ 1 \\ 1 \\ 1 \\ 1 \\ 1 \\ 0 \\ 0 \\ 0 \\ 0 \\ 0 \\ 0 \end{bmatrix} \begin{bmatrix} 0 \\ -1 \\ 0 \\ 0 \\ 0 \\ 0 \\ 1 \\ 1 \\ 1 \\ 0 \\ 0 \\ 0 \end{bmatrix} \begin{bmatrix} 0 \\ 0 \\ -1 \\ 0 \\ 0 \\ 0 \\ 0 \\ 0 \\ 0 \\ 1 \\ 1 \\ 1 \end{bmatrix} \begin{bmatrix} 0 \\ 0 \\ 0 \\ -1 \\ 0 \\ 0 \\ 1 \\ 0 \\ 0 \\ 1 \\ 0 \\ 0 \end{bmatrix} \begin{bmatrix} 0 \\ 0 \\ 0 \\ 0 \\ -1 \\ 0 \\ 0 \\ 1 \\ 0 \\ 0 \\ 1 \\ 0 \end{bmatrix} \begin{bmatrix} 0 \\ 0 \\ 0 \\ 0 \\ 0 \\ -1 \\ 0 \\ 0 \\ 1 \\ 0 \\ 0 \\ 1 \end{bmatrix}.$$

One full set of linearly estimable functions just consists of the six cell means: $\mu + \alpha_i + \beta_j + \gamma_{ij}$, but this is not very interesting. To mimic the previous analysis, consider the following formed from a single mean, and then main effect differences: $\{\mu + \overline{\alpha}_. + \overline{\beta}_. + \overline{\gamma}_{..}, \alpha_1 + \overline{\gamma}_{1.} - \alpha_2 - \overline{\gamma}_{2.}, \beta_1 + \overline{\gamma}_{.1} - \beta_2 - \overline{\gamma}_{.2}, \beta_1 + \overline{\gamma}_{.1} - \beta_3 + \overline{\gamma}_{.3}\}$, which consists of just four estimable functions—interesting ones, for sure—but only four. The remaining two are functions of only the interaction parameters: $\gamma_{11} - \gamma_{12} - \gamma_{21} + \gamma_{22}$ and $\gamma_{11} - \gamma_{13} - \gamma_{21} + \gamma_{23}$. Notice that if both of these are zero, we have $\gamma_{11} - \gamma_{12} = \gamma_{21} - \gamma_{22}$ and $\gamma_{11} - \gamma_{13} = \gamma_{21} - \gamma_{23}$, or in terms of $E(y_{ij})$, we have $E(y_{11} - y_{12}) = E(y_{21} - y_{22})$ and $E(y_{11} - y_{13}) = E(y_{21} - y_{23})$, which is precisely what we mean by no interaction.

Later on in this book, when we are interested in testing the hypothesis of no interaction, we need to pay attention to the fact that the γ_{ij}'s are not estimable. When we express the concept of no interaction as $\gamma_{ij} = 0$, we express it in terms of functions that are not estimable. However, we can express no interaction as

$$\gamma_{ij} - \gamma_{ij'} - \gamma_{i'j} + \gamma_{i'j'} = 0 \tag{3.1}$$

and use only estimable functions. Taking care in this way, we are more likely to have the correct degrees of freedom for the test statistic.

3.7 Reparameterization Revisited

In Chapter 2, we introduced reparameterization, that is, two equivalent models, with different design matrices and different parameters, but equivalent in the sense that they will give the same least squares fit of the data. More precisely, we have the models $\mathbf{y} = \mathbf{X}\mathbf{b} + \mathbf{e}$, where $\mathbf{X}$ is an $N \times p$ matrix and $\mathbf{y} = \mathbf{W}\mathbf{c} + \mathbf{e}$, where $\mathbf{W}$ is an $N \times t$ matrix. The design matrices are related by $\mathcal{C}(\mathbf{X}) = \mathcal{C}(\mathbf{W})$, so that there exist matrices $\mathbf{S}$ and $\mathbf{T}$ so that $\mathbf{W} = \mathbf{X}\mathbf{T}$ and $\mathbf{X} = \mathbf{W}\mathbf{S}$. In this section, we are interested in the effect of reparameterization on estimability.

Result 3.4 *If $\lambda^T\mathbf{b}$ is estimable in the original model in $\mathbf{X}$, that is, $\mathbf{y} = \mathbf{X}\mathbf{b} + \mathbf{e}$, and $\hat{\mathbf{c}}$ solves the normal equations in $\mathbf{W}$, that is, $\mathbf{W}^T\mathbf{W}\mathbf{c} = \mathbf{W}^T\mathbf{y}$, then $\lambda^T(\mathbf{T}\hat{\mathbf{c}})$ is the least squares estimator of $\lambda^T\mathbf{b}$.*

Proof: If $\hat{\mathbf{c}}$ solves the normal equations in $\mathbf{W}$, from Result 2.9, then $\hat{\mathbf{b}} = \mathbf{T}\hat{\mathbf{c}}$ solves the normal equations in $\mathbf{X}$. If $\lambda^T\mathbf{b}$ is estimable, then $\lambda^T\hat{\mathbf{b}} = \lambda^T\mathbf{T}\hat{\mathbf{c}}$ is the least squares estimator of $\lambda^T\mathbf{b}$. □

Example 3.2: (Continuation of Example 2.8)
Consider the simple one-way layout with three groups as discussed previously. The more common parameterization employs

$$\mathbf{X}\mathbf{b} = \begin{bmatrix} \mathbf{1}_{n_1} & \mathbf{1}_{n_1} & \mathbf{0} & \mathbf{0} \\ \mathbf{1}_{n_2} & \mathbf{0} & \mathbf{1}_{n_2} & \mathbf{0} \\ \mathbf{1}_{n_3} & \mathbf{0} & \mathbf{0} & \mathbf{1}_{n_3} \end{bmatrix} \begin{bmatrix} \mu \\ \alpha_1 \\ \alpha_2 \\ \alpha_3 \end{bmatrix}.$$

Another parameterization, using dummy variables for the first two groups, leads to a full-column-rank design matrix:

$$\mathbf{W}\mathbf{c} = \begin{bmatrix} \mathbf{1}_{n_1} & \mathbf{1}_{n_1} & \mathbf{0} \\ \mathbf{1}_{n_2} & \mathbf{0} & \mathbf{1}_{n_2} \\ \mathbf{1}_{n_3} & \mathbf{0} & \mathbf{0} \end{bmatrix} \begin{bmatrix} c_1 \\ c_2 \\ c_3 \end{bmatrix}.$$

The relationships $\mathbf{W} = \mathbf{X}\mathbf{T}$ and $\mathbf{X} = \mathbf{W}\mathbf{S}$ hold for

$$\mathbf{T} = \begin{bmatrix} 1 & 0 & 0 \\ 0 & 1 & 0 \\ 0 & 0 & 1 \\ 0 & 0 & 0 \end{bmatrix} \quad \text{and} \quad \mathbf{S} = \begin{bmatrix} 1 & 0 & 0 & 1 \\ 0 & 1 & 0 & -1 \\ 0 & 0 & 1 & -1 \end{bmatrix}.$$

The following $\hat{\mathbf{c}}$ is the unique solution to the full-rank normal equations using $\mathbf{W}$:

$$\hat{\mathbf{c}} = \begin{bmatrix} \overline{y}_{3.} \\ \overline{y}_{1.} - \overline{y}_{3.} \\ \overline{y}_{2.} - \overline{y}_{3.} \end{bmatrix}.$$

Consider estimable functions of the form $\lambda^T \mathbf{b}$, say, for $\lambda^{(1)}$ and $\lambda^{(2)}$ given below, then $\lambda^T \mathbf{T}\hat{\mathbf{c}}$ will be the least squares estimator. For

$$\lambda^{(1)} = \begin{bmatrix} 1 \\ 1 \\ 0 \\ 0 \end{bmatrix},$$

$\lambda^{(1)T} \mathbf{T}\hat{\mathbf{c}} = \overline{y}_{1.}$ is the least squares estimator of $\mu + \alpha_1$, and for

$$\lambda^{(2)} = \begin{bmatrix} 0 \\ 1 \\ -1 \\ 0 \end{bmatrix},$$

$\lambda^{(2)T} \mathbf{T}\hat{\mathbf{c}} = \overline{y}_{1.} - \overline{y}_{2.}$ is the least squares estimator of $\alpha_1 - \alpha_2$, both as expected.

Now if we reverse roles and employ the same result, we have $\hat{\mathbf{b}}$ solving the $\mathbf{X}$ normal equations, leading to $\mathbf{S}\hat{\mathbf{b}}$ solving the $\mathbf{W}$ normal equations. Here $\mathbf{S}\hat{\mathbf{b}}$ is given by

$$\mathbf{S}\hat{\mathbf{b}} = \begin{bmatrix} 1 & 0 & 0 & 1 \\ 0 & 1 & 0 & -1 \\ 0 & 0 & 1 & -1 \end{bmatrix} \begin{bmatrix} \overline{y}_{..} \\ \overline{y}_{1.} - \overline{y}_{..} \\ \overline{y}_{2.} - \overline{y}_{..} \\ \overline{y}_{3.} - \overline{y}_{..} \end{bmatrix} = \hat{\mathbf{c}} = \begin{bmatrix} \overline{y}_{3.} \\ \overline{y}_{1.} - \overline{y}_{3.} \\ \overline{y}_{2.} - \overline{y}_{3.} \end{bmatrix}.$$

Note that if we generated all solutions $\hat{\mathbf{b}}$ to the $\mathbf{X}$ normal equations above, $\hat{\mathbf{c}}$ would not change—it is unique since the $\mathbf{W}$ design matrix has full-column rank. Moreover, all of the components of $\mathbf{c}$ are estimable, due to the full-column-rank parameterization with $\mathbf{W}$; hence the least squares estimators must be unique.

Result 3.5 *If $\mathbf{q}^T \mathbf{c}$ is estimable in the reparameterized model (that is, $\mathbf{q} \in \mathcal{C}(\mathbf{W}^T)$), then $\mathbf{q}^T \mathbf{S}\mathbf{b}$ is estimable in the original model, and its least squares estimator is $\mathbf{q}^T \hat{\mathbf{c}}$ where $\hat{\mathbf{c}}$ solves the normal equations in $\mathbf{W}$, $\mathbf{W}^T\mathbf{W}\mathbf{c} = \mathbf{W}^T\mathbf{y}$.*

Proof: Since $\mathbf{q} \in \mathcal{C}(\mathbf{W}^T)$, then $\mathbf{q} = \mathbf{W}^T\mathbf{a}$ for some $\mathbf{a}$, so that $\mathbf{S}^T\mathbf{q} = \mathbf{S}^T\mathbf{W}^T\mathbf{a} = \mathbf{X}^T\mathbf{a} \in \mathcal{C}(\mathbf{X}^T)$, so that $\mathbf{q}^T\mathbf{S}\mathbf{b}$ is estimable. Now to find its least squares estimator $\mathbf{q}^T\mathbf{S}\hat{\mathbf{b}}$, the algebra follows

$$\mathbf{q}^T\mathbf{S}\hat{\mathbf{b}} = \mathbf{q}^T\mathbf{S}\mathbf{T}\hat{\mathbf{c}} = \mathbf{a}^T\mathbf{X}\mathbf{T}\hat{\mathbf{c}} = \mathbf{a}^T\mathbf{W}\hat{\mathbf{c}} = \mathbf{q}^T\hat{\mathbf{c}},$$

with the first equality by Result 2.9 and the second equality since $\mathbf{q} \in \mathcal{C}(\mathbf{W}^T)$; the remaining steps are substitutions. □

Example 3.2: Continued
Each component of **c** is estimable in the reparameterized model, so each component of **Sb** is estimable:

$$\mathbf{Sb} = \begin{bmatrix} 1 & 0 & 0 & 1 \\ 0 & 1 & 0 & -1 \\ 0 & 0 & 1 & -1 \end{bmatrix} \begin{bmatrix} \mu \\ \alpha_1 \\ \alpha_2 \\ \alpha_3 \end{bmatrix} = \begin{bmatrix} \mu + \alpha_3 \\ \alpha_1 - \alpha_3 \\ \alpha_2 - \alpha_3 \end{bmatrix} \text{ with the vector } \begin{bmatrix} \overline{y}_{3.} \\ \overline{y}_{1.} - \overline{y}_{3.} \\ \overline{y}_{2.} - \overline{y}_{3.} \end{bmatrix}$$

as its least squares estimator. Reversing roles and taking examples of λ as before, $\lambda^{(1)}$ and $\lambda^{(2)}$, then

$$\begin{aligned} \lambda^{(1)T}\mathbf{Tc} &= c_1 + c_2 \\ \lambda^{(2)T}\mathbf{Tc} &= c_2 - c_3 \end{aligned}$$

and their least squares estimators are, respectively, $\overline{y}_{1.}$ and $\overline{y}_{1.} - \overline{y}_{2.}$.

Remark 3.1 There are vectors $\mathbf{q}$, such that $\mathbf{q}^T\mathbf{Sb}$ is estimable in the original model ($\mathbf{y} = \mathbf{Xb} + \mathbf{e}$), but $\mathbf{q}^T\mathbf{c}$ is not estimable in the reparameterized model ($\mathbf{y} = \mathbf{Wc} + \mathbf{e}$), and $\mathbf{q}^T\hat{\mathbf{c}}$ may not equal $\mathbf{q}^T\mathbf{S}\hat{\mathbf{b}}$.

Example 3.2: Continued
For Remark 3.1, reverse roles and take $\lambda = \mathbf{e}^{(1)}$, then $\lambda^T\mathbf{Tc} = c_1$, which is estimable in the reparameterized model, but $\lambda^T\mathbf{b} = \mu$, which is not estimable in the original model. Moreover, $\lambda^T\hat{\mathbf{b}} = z$ if we use the general solution from Example 2.8, while $\lambda^T\mathbf{T}\hat{\mathbf{c}} = \overline{y}_{3.}$.

In summary, these results are neither difficult nor deep, merely convenient. Rewriting all the parameterizations as

$$E(\mathbf{y}) = \mathbf{Xb} = \mathbf{Wc} = \mathbf{WSb} = \mathbf{XTc},$$

we can easily see that most of these results follow from simple substitution. The main point of Result 3.5—if $\mathbf{q}^T\mathbf{c}$ is estimable, then $\mathbf{q}^T\mathbf{Sb}$ is estimable—is again substitution. Remark 3.1 precludes the converse, owing to the arbitrariness of **S** or **T**. The issue is if $\mathbf{S}^T\mathbf{q} \in \mathcal{C}(\mathbf{X}^T)$, is $\mathbf{q} \in \mathcal{C}(\mathbf{W}^T)$? For some **a**, we can write $\mathbf{S}^T\mathbf{q} = \mathbf{X}^T\mathbf{a} = \mathbf{S}^T\mathbf{W}^T\mathbf{a}$, but following Corollary A.2, $\mathbf{S}^T$ cannot be removed from both sides of the equation unless it has full-column rank.

The most common goal of reparameterization is to find a full-rank parameterization, so that $r(X) = r = t$. A full-rank parameterization always exists—merely delete columns of the design matrix **X** that are linearly dependent on the others. If the design matrix **W** ($N \times r$) is full-column rank, then $\mathbf{W}^T\mathbf{W}$ is nonsingular and the solution to the normal equations is unique. Another important advantage to a full-column-rank reparameterization is that the interpretation of the new parameters is clear. If $\mathbf{X} = \mathbf{WS}$, with $\mathbf{W}$ full-column rank, then $\mathbf{S}$ is unique and the representation of $\mathbf{c}$ in terms of $\mathbf{b}$, that is, $\mathbf{c} = \mathbf{Sb}$, is unique.

Example 3.3: Other One-Way ANOVA Parameterizations
The most common parameterization for one-way ANOVA has been

$$y_{ij} = \mu + \alpha_i + e_{ij} \text{ for } i = 1, \ldots, a;\ j = 1, \ldots, n_i$$

which, if the observations are ordered by group (i) and then individual (j) within group, leads to the design matrix and coefficient vector

$$\mathbf{Xb} = \begin{bmatrix} \mathbf{1}_{n_1} & \mathbf{1}_{n_1} & \mathbf{0} & \ldots & \mathbf{0} \\ \mathbf{1}_{n_2} & \mathbf{0} & \mathbf{1}_{n_2} & \ldots & \mathbf{0} \\ \mathbf{1}_{n_3} & \mathbf{0} & \mathbf{0} & \ldots & \mathbf{0} \\ \ldots & \ldots & \ldots & \ldots & \ldots \\ \mathbf{1}_{n_a} & \mathbf{0} & \ldots & \mathbf{0} & \mathbf{1}_{n_a} \end{bmatrix} \begin{bmatrix} \mu \\ \alpha_1 \\ \alpha_2 \\ \ldots \\ \alpha_a \end{bmatrix}.$$

A commonly used full-rank reparameterization drops the first column to yield

$$\mathbf{Z}\mu = \begin{bmatrix} \mathbf{1}_{n_1} & \mathbf{0} & \mathbf{0} & \ldots & \mathbf{0} \\ \mathbf{0} & \mathbf{1}_{n_2} & \mathbf{0} & \ldots & \mathbf{0} \\ \mathbf{0} & \mathbf{0} & \mathbf{1}_{n_3} & \ldots & \mathbf{0} \\ \ldots & \ldots & \ldots & \ldots & \ldots \\ \mathbf{0} & \mathbf{0} & \mathbf{0} & \ldots & \mathbf{1}_{n_a} \end{bmatrix} \begin{bmatrix} \mu_1 \\ \mu_2 \\ \mu_3 \\ \ldots \\ \mu_a \end{bmatrix} \text{ so that } \mu = \mathbf{S}_2\mathbf{b} = \begin{bmatrix} \mu + \alpha_1 \\ \mu + \alpha_2 \\ \mu + \alpha_3 \\ \ldots \\ \mu + \alpha_a \end{bmatrix}$$

and another is

$$\mathbf{U}\delta = \begin{bmatrix} \mathbf{1}_{n_1} & \mathbf{1}_{n_1} & \mathbf{0} & \ldots & \mathbf{0} \\ \mathbf{1}_{n_2} & \mathbf{0} & \mathbf{1}_{n_2} & \ldots & \mathbf{0} \\ \mathbf{1}_{n_3} & \mathbf{0} & \mathbf{0} & \ldots & \mathbf{0} \\ \ldots & \ldots & \ldots & \ldots & \mathbf{1}_{n_a-1} \\ \mathbf{1}_{n_a} & -\mathbf{1}_{n_a} & -\mathbf{1}_{n_a} & -\mathbf{1}_{n_a} & -\mathbf{1}_{n_a} \end{bmatrix} \begin{bmatrix} \delta_1 \\ \delta_2 \\ \delta_3 \\ \ldots \\ \delta_a \end{bmatrix} \text{ so that } \delta = \mathbf{S}_3\mathbf{b} = \begin{bmatrix} \mu + \overline{\alpha} \\ \alpha_1 - \overline{\alpha} \\ \alpha_2 - \overline{\alpha} \\ \ldots \\ \alpha_a - \overline{\alpha} \end{bmatrix}.$$

It is easy to see that $\mathcal{C}(\mathbf{X}) = \mathcal{C}(\mathbf{W}) = \mathcal{C}(\mathbf{Z}) = \mathcal{C}(\mathbf{U})$, where $\mathbf{W}$ is generalized from Example 3.2. Motivating the choice of these different parameterizations is the interpretation of the parameters. Note that the parameters $\mu_1, \mu_2, \ldots, \mu_a$ represent the group means, and hence $\mathbf{y} = \mathbf{Z}\mu + \mathbf{e}$ is referred to as the *cell means model* and generalizes to more complicated designs. Comparing across full-rank designs, first for $\mathbf{Wc}$ and $\mathbf{Z}\mu$, we see that

$$\begin{aligned} c_1 &= \mu_a \\ c_2 &= \mu_1 - \mu_a \\ c_3 &= \mu_2 - \mu_a \\ &\ldots \\ c_a &= \mu_{a-1} - \mu_a. \end{aligned}$$

In this case, the last group takes the role of the reference group (e.g., control or placebo) and hence $\mathbf{y} = \mathbf{Wc} + \mathbf{e}$ is called the *cell reference* model. In practice, the column deleted from $\mathbf{X}$ to form $\mathbf{W}$ corresponds to the reference group. Similarly, in the last parameterization note that

$$\begin{aligned}
\delta_1 &= (\mu_1 + \cdots + \mu_a)/a = \mu + \overline{\alpha} \\
\delta_2 &= \mu_1 - \overline{\mu} = \alpha_1 - \overline{\alpha} \\
\delta_3 &= \mu_2 - \overline{\mu} = \alpha_2 - \overline{\alpha} \\
&\ldots \\
\delta_a &= \mu_{a-1} - \overline{\mu}.
\end{aligned},$$

and hence $\mathbf{y} = \mathbf{U}\delta + \mathbf{e}$ is called the *deviations from the mean model*. Finally, our original model $\mathbf{y} = \mathbf{Xb} + \mathbf{e}$ is an overparameterized model with $\mu_i = \mu + \alpha_i, i = 1, \ldots, a$; its parameters μ, α_i can not be uniquely represented by the cell means parameters μ_i.

Another common goal of reparameterization is to construct $\mathbf{W}$ so that the inner product matrix $\mathbf{W}^T\mathbf{W}$ is either diagonal or block diagonal, that is, an orthogonal reparameterization, for which Gram–Schmidt (Section 2.6) is often employed. The next three examples show its most common uses.

Example 3.4: Centering in Simple Linear Regression
First consider the usual simple linear regression model with explanatory variable x_i,

$$y_i = b_1 + b_2 x_i + e_i$$

and the reparameterized model with centered explanatory variable,

$$y_i = c_1 + c_2(x_i - \overline{x}) + e_i.$$

Unless x_i are all equal, both representations are full rank, and the relationship between the two can be expressed as

$$\mathbf{T} = \begin{bmatrix} 1 & -\overline{x} \\ 0 & 1 \end{bmatrix} \quad \text{and} \quad \mathbf{S} = \begin{bmatrix} 1 & \overline{x} \\ 0 & 1 \end{bmatrix}.$$

Since both representations are full rank, both $\mathbf{T}$ and $\mathbf{S}$ are nonsingular and inverses of the other. The advantage of the second parameterization is the diagonal cross-products matrix and convenient form of the least squares estimates

$$\mathbf{W}^T\mathbf{W} = \begin{bmatrix} N & 0 \\ 0 & S_{xx} \end{bmatrix}$$

so that

$$\hat{\mathbf{c}} = \begin{bmatrix} \hat{c}_1 \\ \hat{c}_2 \end{bmatrix} = \begin{bmatrix} \overline{y} \\ \sum(x_i - \overline{x})y_i/S_{xx} \end{bmatrix}$$

where $S_{xx} = \sum(x_i - \overline{x})^2$. The least squares estimates for the usual parameterization follow easily:

$$\hat{\mathbf{b}} = \mathbf{T}\hat{\mathbf{c}} = \begin{bmatrix} 1 & -\overline{x} \\ 0 & 1 \end{bmatrix} \begin{bmatrix} \overline{y} \\ S_{xy}/S_{xx} \end{bmatrix} = \begin{bmatrix} \overline{y} - \overline{x}\hat{b}_2 \\ \hat{b}_2 \end{bmatrix}$$

where $\hat{b}_2 = \sum(x_i - \overline{x})y_i / \sum(x_i - \overline{x})^2 = S_{xy}/S_{xx}$.

Example 3.5: Two Sets of Explanatory Variables
Consider the situation where the explanatory variables are divided into two groups or blocks:

$$\mathbf{y} = \mathbf{X}_1\mathbf{b}_1 + \mathbf{X}_2\mathbf{b}_2 + \mathbf{e}.$$

Sometimes it is convenient to reparameterize as follows:

$$\mathbf{y} = \mathbf{X}_1\mathbf{c}_1 + \tilde{\mathbf{X}}_2\mathbf{c}_2 + \mathbf{e} = \mathbf{W}_1\mathbf{c}_1 + \mathbf{W}_2\mathbf{c}_2 + \mathbf{e}.$$

where $\tilde{\mathbf{X}}_2 = (\mathbf{I} - \mathbf{P}_{\mathbf{X}1})\mathbf{X}_2$, so that $\mathbf{W}_2 = \tilde{\mathbf{X}}_2$ can be viewed as residuals from the second block of explanatory variables when regressed on the first block. The block diagonal normal equations for the reparameterization can be written as

$$\mathbf{W}^T\mathbf{W}\mathbf{c} = \begin{bmatrix} \mathbf{X}_1^T\mathbf{X}_1 & \mathbf{0} \\ \mathbf{0} & \mathbf{X}_2^T(\mathbf{I} - \mathbf{P}_{\mathbf{X}1})\mathbf{X}_2 \end{bmatrix} \begin{bmatrix} \mathbf{c}_1 \\ \mathbf{c}_2 \end{bmatrix} = \mathbf{W}^T\mathbf{y} = \begin{bmatrix} \mathbf{X}_1^T\mathbf{y} \\ \mathbf{X}_2^T(\mathbf{I} - \mathbf{P}_{\mathbf{X}1})\mathbf{y} \end{bmatrix}.$$

By finding $\mathbf{T}$ (Exercise 3.17), one can then show that $\hat{\mathbf{b}}_2 = \hat{\mathbf{c}}_2$. The result here is that we can get the correct coefficients for the second block ($\hat{\mathbf{b}}_2$) from a regression of the residuals $(\mathbf{I} - \mathbf{P}_{\mathbf{X}1})\mathbf{y}$ on the residuals from the second block of explanatory variables $(\mathbf{I} - \mathbf{P}_{\mathbf{X}1})\mathbf{X}_2$. Before the days of sophisticated statistical software, this technique was employed to do multiple regression using simple linear regression algorithms.

Example 3.6: Orthogonal Polynomials
Consider again the one-way ANOVA model, but balanced ($n_i = n$) and $a = 3$ for simplicity. Notice that the column spaces for $\mathbf{X}$ and $\mathbf{W}$ given below are the same:

$$\mathbf{X}\mathbf{b} = \begin{bmatrix} \mathbf{1}_n & \mathbf{1}_n & \mathbf{0} & \mathbf{0} \\ \mathbf{1}_n & \mathbf{0} & \mathbf{1}_n & \mathbf{0} \\ \mathbf{1}_n & \mathbf{0} & \mathbf{0} & \mathbf{1}_n \end{bmatrix} \begin{bmatrix} \mu \\ \alpha_1 \\ \alpha_2 \\ \alpha_3 \end{bmatrix} = \mathbf{W}\mathbf{c} = \begin{bmatrix} \mathbf{1}_n & \mathbf{1}_n & \mathbf{1}_n \\ \mathbf{1}_n & 2\mathbf{1}_n & 4\mathbf{1}_n \\ \mathbf{1}_n & 3\mathbf{1}_n & 9\mathbf{1}_n \end{bmatrix} \begin{bmatrix} c_1 \\ c_2 \\ c_3 \end{bmatrix}.$$

Now perform Gram–Schmidt orthonormalization (Section 2.6) on $\mathbf{W}$ for an orthogonal reparameterization:

$$\mathbf{X}\mathbf{b} = \mathbf{Q}\gamma = \begin{bmatrix} \frac{1}{\sqrt{3n}}\mathbf{1}_n & \frac{1}{\sqrt{2n}}\mathbf{1}_n & \frac{1}{\sqrt{6n}}\mathbf{1}_n \\ \frac{1}{\sqrt{3n}}\mathbf{1}_n & \mathbf{0}_n & \frac{-2}{\sqrt{6n}}\mathbf{1}_n \\ \frac{1}{\sqrt{3n}}\mathbf{1}_n & \frac{-1}{\sqrt{2n}}\mathbf{1}_n & \frac{1}{\sqrt{6n}}\mathbf{1}_n \end{bmatrix} \begin{bmatrix} \gamma_1 \\ \gamma_2 \\ \gamma_3 \end{bmatrix}.$$

3.8 Imposing Conditions for a Unique Solution to the Normal Equations

Consider the usual linear model $\mathbf{y} = \mathbf{Xb} + \mathbf{e}$, with $E(\mathbf{e}) = \mathbf{0}$, and $r = rank(\mathbf{X})$. If $r = p$, then $\mathbf{X}^T\mathbf{X}$ is nonsingular and $\hat{\mathbf{b}} = (\mathbf{X}^T\mathbf{X})^{-1}\mathbf{X}^T\mathbf{y}$ is the unique solution to the normal equations. If $r < p$, then the family of solutions to the normal equations can be written as $\hat{\mathbf{b}}(\mathbf{z}) = (\mathbf{X}^T\mathbf{X})^g\mathbf{X}^T\mathbf{y} + (\mathbf{I} - (\mathbf{X}^T\mathbf{X})^g\mathbf{X}^T\mathbf{X})\mathbf{z}$.

In the one-way ANOVA model, $y_{ij} = \mu + \alpha_i + e_{ij}$, we commonly impose a condition on the solution to the normal equations ($\hat{\mu}, \hat{\alpha}_i, i = 1, \ldots, a$) to obtain a particularly convenient solution. The three common choices are:

1. $\alpha_a = 0$
2. $\sum_i \alpha_i = 0$
3. $\sum_i n_i\alpha_i = 0$

with the first quite common (see Example 3.2) and the last preferred for the simplifications it brings. In an unbalanced (n_i unequal) one-way layout, the normal equations are

$$N\mu + \sum_i n_i\alpha_i = N\overline{y}_{..} = y_{..}$$
$$n_i\mu + n_i\alpha_i = n_i\overline{y}_{i.} = y_{i.} \text{ for } i = 1, \ldots, a.$$

While the easily obtained solution is $\hat{\mu} = 0$ and $\hat{\alpha}_i = \overline{y}_{i.}$, imposing the third constraint yields another solution $\hat{\mu} = \overline{y}_{..}$ and $\hat{\alpha}_i = \overline{y}_{i.} - \overline{y}_{..}$, which is both familiar and appealing. In other situations, similar conditions can be imposed to gain unique, appealing solutions to the normal equations. The general question to be addressed then is whether we can always impose a set of conditions to obtain a unique solution. In particular, we are considering equations of the form $\mathbf{Cb} = \mathbf{0}$, where $\mathbf{C}$ is $s \times p$, $s = p - r$ and $rank(\mathbf{C}) = s$.

Adding these equations to the normal equations leads to the augmented system of equations,

$$\begin{bmatrix} \mathbf{X}^T\mathbf{X} \\ \mathbf{C} \end{bmatrix} \mathbf{b} = \begin{bmatrix} \mathbf{X}^T\mathbf{y} \\ \mathbf{0} \end{bmatrix} \tag{3.2}$$

or equivalently,

$$\begin{bmatrix} \mathbf{X} \\ \mathbf{C} \end{bmatrix} \mathbf{b} = \begin{bmatrix} \mathbf{P_X y} \\ \mathbf{0} \end{bmatrix}. \tag{3.3}$$

Now the latter system (3.3) will have a unique solution if the left-hand-side matrix of (3.3) has full-column rank, or its nullspace is just the zero vector. This condition is equivalent to its transpose having full-row rank, or

$$\mathcal{C}([\mathbf{X}^T \quad \mathbf{C}^T]) = R^p.$$

So choosing $\mathbf{C}$ appropriately to gain a unique solution means adding rows to $\mathbf{X}$ to make it full-column rank (p), or columns to $\mathbf{X}^T$ to make the matrix $[\mathbf{X}^T\ \mathbf{C}^T]$ full-row rank. Clearly, if we add columns of $\mathbf{C}^T$ that are in $\mathcal{C}(\mathbf{X}^T)$, they cannot increase the rank of $[\mathbf{X}^T\ \mathbf{C}^T]$ beyond its current rank r. So columns of $\mathbf{C}^T$ must have some contribution from the basis vectors of $\mathcal{N}(\mathbf{X})$ to increase the rank. "Some contribution" means that these columns (of $\mathbf{C}^T$) cannot be orthogonal to $\mathcal{N}(\mathbf{X})$, and hence these components must be nonestimable. Not only must each component of $\mathbf{Cb}$ be nonestimable, but each must be jointly nonestimable. That is, the rows of $\mathbf{C}$ are linearly independent, and no linear combination of components of $\mathbf{Cb}$ leads to an estimable function—no combination of columns of $\mathbf{C}^T$ is in $\mathcal{C}(\mathbf{X}^T)$. This can be expressed as the condition $\mathcal{C}(\mathbf{X}^T) \cap \mathcal{C}(\mathbf{C}^T) = \{\mathbf{0}\}$. So in order that the augmented system of equations lead to a unique solution, we must add $s = p - r$ linearly independent equations, each one corresponding to a nonestimable function.

Example 3.7: Imposing Constraints

In the one-way ANOVA model $y_{ij} = \mu + \alpha_i + e_{ij}$, $i = 1, \ldots, a = 3$ (say), and balanced ($n_i = n$, say); then basis vectors for $\mathcal{C}(\mathbf{X}^T)$ are

$$\begin{bmatrix}1\\1\\0\\0\end{bmatrix} \quad \begin{bmatrix}1\\0\\1\\0\end{bmatrix} \quad \begin{bmatrix}1\\0\\0\\1\end{bmatrix}$$

and we add the constraint vector

$$\mathbf{C}^T = \begin{bmatrix}0\\1\\1\\1\end{bmatrix}$$

as the "usual constraint" to get a unique solution. Note that a set of basis vectors for R^4 is

$$\begin{bmatrix}1\\1\\0\\0\end{bmatrix} \quad \begin{bmatrix}1\\0\\1\\0\end{bmatrix} \quad \begin{bmatrix}1\\0\\0\\1\end{bmatrix} \quad \begin{bmatrix}1\\-1\\-1\\-1\end{bmatrix}$$

with the last vector in $\mathcal{N}(\mathbf{X})$ so that

$$\mathbf{C}^T = \begin{bmatrix}0\\1\\1\\1\end{bmatrix}$$

can be written as a linear combination of those four vectors with coefficients $1/4$, $1/4$, $1/4$, $-3/4$, respectively. Adding the constraint column raises the rank to 4.

There are some further interesting consequences to these conditions, specifically: given the $N \times p$ design matrix $\mathbf{X}$ with rank r and the $(p-r) \times p$ matrix $\mathbf{C}$,

$$rank(\mathbf{C}) = p - r \tag{3.4}$$

$$\mathcal{C}(\mathbf{X}^T) \cap \mathcal{C}(\mathbf{C}^T) = \{\mathbf{0}\}, \tag{3.5}$$

which imply

$$rank\left(\begin{bmatrix}\mathbf{X}\\ \mathbf{C}\end{bmatrix}\right) = p. \tag{3.6}$$

The proof of some interesting results follows more easily after a lemma.

Lemma 3.1 *For* $\mathbf{C}$ *constructed according to the conditions (3.4,3.5) above, the system of equations*

$$\begin{bmatrix}\mathbf{X}^T\mathbf{X}\\ \mathbf{C}^T\mathbf{C}\end{bmatrix}\mathbf{b} = \begin{bmatrix}\mathbf{X}^T\mathbf{y}\\ \mathbf{0}\end{bmatrix} \tag{3.7}$$

is equivalent to 3.2 and

$$(\mathbf{X}^T\mathbf{X} + \mathbf{C}^T\mathbf{C})\mathbf{b} = \mathbf{X}^T\mathbf{y}. \tag{3.8}$$

Proof: The equivalence of (3.7) and (3.2), in particular, replacing $\mathbf{Cb} = \mathbf{0}$ with $\mathbf{C}^T\mathbf{Cb} = \mathbf{0}$, follows from Lemma 2.1. Adding the two components of former (3.7) gives the latter (3.8). For the other direction, rewrite (3.8) as

$$\mathbf{C}^T\mathbf{Cb} = \mathbf{X}^T\mathbf{y} - \mathbf{X}^T\mathbf{Xb}.$$

The right-hand side is in $\mathcal{C}(\mathbf{X}^T)$ and its only intersection with $\mathcal{C}(\mathbf{C}^T\mathbf{C}) = \mathcal{C}(\mathbf{C}^T)$ is the zero vector, so that $\mathbf{C}^T\mathbf{Cb} = \mathbf{0}$ and $\mathbf{X}^T\mathbf{Xb} = \mathbf{X}^T\mathbf{y}$. □

Result 3.6 *If* $\mathbf{C}$ *is constructed according to the conditions (3.4,3.5) above, then:*

(*i*) *The matrix* $(\mathbf{X}^T\mathbf{X} + \mathbf{C}^T\mathbf{C})$ *is nonsingular.*

(*ii*) $(\mathbf{X}^T\mathbf{X} + \mathbf{C}^T\mathbf{C})^{-1}\mathbf{X}^T\mathbf{y}$ *is the unique solution to* $\mathbf{X}^T\mathbf{Xb} = \mathbf{X}^T\mathbf{y}$ *and* $\mathbf{Cb} = \mathbf{0}$.

(*iii*) $(\mathbf{X}^T\mathbf{X} + \mathbf{C}^T\mathbf{C})^{-1}$ *is a generalized inverse of* $\mathbf{X}^T\mathbf{X}$.

(*iv*) $\mathbf{C}(\mathbf{X}^T\mathbf{X} + \mathbf{C}^T\mathbf{C})^{-1}\mathbf{X}^T = \mathbf{0}$.

(*v*) $\mathbf{C}(\mathbf{X}^T\mathbf{X} + \mathbf{C}^T\mathbf{C})^{-1}\mathbf{C}^T = \mathbf{I}$.

Proof: For (i), from (3.6) note that

$$\mathcal{C}([\mathbf{X}^T \quad \mathbf{C}^T]) = \mathcal{C}\left([\mathbf{X}^T\ \mathbf{C}^T]\begin{bmatrix}\mathbf{X}\\ \mathbf{C}\end{bmatrix}\right) = \mathcal{C}((\mathbf{X}^T\mathbf{X} + \mathbf{C}^T\mathbf{C})) = R^p,$$

so that $(\mathbf{X}^T\mathbf{X} + \mathbf{C}^T\mathbf{C})$ is $p \times p$, with rank p, and hence nonsingular. For (ii), use (i) above to claim that the unique solution to $(\mathbf{X}^T\mathbf{X} + \mathbf{C}^T\mathbf{C})\mathbf{b} = \mathbf{X}^T\mathbf{y}$ is $(\mathbf{X}^T\mathbf{X} + \mathbf{C}^T\mathbf{C})^{-1}\mathbf{X}^T\mathbf{y}$, and from Lemma 3.1, also solves (3.2) above, as well as the normal equations. For (iii), using (ii) above, $\mathbf{X}^T\mathbf{X}(\mathbf{X}^T\mathbf{X} + \mathbf{C}^T\mathbf{C})^{-1}\mathbf{X}^T\mathbf{y} = \mathbf{X}^T\mathbf{y}$ for all $\mathbf{y}$ (a solution of the normal equations), so

$$\mathbf{X}^T\mathbf{X}(\mathbf{X}^T\mathbf{X} + \mathbf{C}^T\mathbf{C})^{-1}\mathbf{X}^T = \mathbf{X}^T.$$

Now just postmultiply by $\mathbf{X}$ to get (iii). From (ii) we know $\mathbf{C}(\mathbf{X}^T\mathbf{X} + \mathbf{C}^T\mathbf{C})^{-1}\mathbf{X}^T\mathbf{y} = \mathbf{0}$ for all $\mathbf{y}$ since the solution solves $\mathbf{Cb} = \mathbf{0}$; then use Result A.8 to get (iv). For (v), see Exercise 3.22. □

Example 3.7: continued

Consider the balanced case where $n_i = n$; then $\mathbf{X}^T\mathbf{X} + \mathbf{C}^T\mathbf{C}$ is

$$\begin{bmatrix} 3n & n & n & n \\ n & n & 0 & 0 \\ n & 0 & n & 0 \\ n & 0 & 0 & n \end{bmatrix} + \begin{bmatrix} 0 & 0 & 0 & 0 \\ 0 & 1 & 1 & 1 \\ 0 & 1 & 1 & 1 \\ 0 & 1 & 1 & 1 \end{bmatrix} = \begin{bmatrix} 3n & n & n & n \\ n & n+1 & 1 & 1 \\ n & 1 & n+1 & 1 \\ n & 1 & 1 & n+1 \end{bmatrix}$$

whose inverse is

$$\frac{1}{9n}\begin{bmatrix} 3+n & -n & -n & -n \\ -n & n+6 & n-3 & n-3 \\ -n & n-3 & n+6 & n-3 \\ -n & n-3 & n-3 & n+6 \end{bmatrix}$$

and so

$$\mathbf{C}(\mathbf{X}^T\mathbf{X} + \mathbf{C}^T\mathbf{C})^{-1} = \begin{bmatrix} -1/3 & 1/3 & 1/3 & 1/3 \end{bmatrix}.$$

From here we can see the consequences of Results 3.6(iv) and (v).

Recall that we began this discussion with a more direct approach to showing that imposing such a condition will lead to a unique solution. Since all solutions to the normal equations can be written as $\hat{\mathbf{b}}(\mathbf{z}) = (\mathbf{X}^T\mathbf{X})^g\mathbf{X}^T\mathbf{y} + (\mathbf{I} - (\mathbf{X}^T\mathbf{X})^g\mathbf{X}^T\mathbf{X})\mathbf{z}$, consider the effect of multiplying by $\mathbf{C}$ and solving for a unique $\mathbf{z}$. For a specific problem, such as the one-way ANOVA problem, this straighforward approach can work quite well. Adapting to the general case, however, requires parameterizing $\mathbf{C}$ in terms of $\mathbf{X}$ and basis vectors for $\mathcal{N}(\mathbf{X})$, and sheds little light on the geometry of the problem.

Example 3.7: continued

Consider again the balanced ANOVA where $n_i = n$; then the general solution to the normal equations takes the form

$$\hat{\mathbf{b}}(\mathbf{z}) = \begin{bmatrix} 0 \\ \overline{y}_{1.} \\ \overline{y}_{2.} \\ \overline{y}_{3.} \end{bmatrix} + z \begin{bmatrix} 1 \\ -1 \\ -1 \\ -1 \end{bmatrix}.$$

Satisfying the constraint $\sum_i \alpha_i = 0$ means $\sum_i \bar{y}_{i.} - 3z = 0$. Solving for z gives $z = \bar{y}_{..}$, yielding the familiar solution $(\bar{y}_{..}, \bar{y}_{1.} - \bar{y}_{..}, \bar{y}_{2.} - \bar{y}_{..}, \bar{y}_{3.} - \bar{y}_{..})^T$.

3.9 Constrained Parameter Space

We now consider the linear model with a parameter space restricted by a system of linear equations $\mathbf{P}^T\mathbf{b} = \delta$. We differentiate this discussion from the previous one by considering both estimable and nonestimable constraints. We are interested in how constraints affect estimability and how to minimize the error sum of squares subject to this constraint. We will return to this subject twice: in Chapter 4 to construct best-constrained estimators and in Chapter 6 for testing hypotheses.

We observe $\mathbf{y}$ whose expectation is $\mathbf{Xb}$ where the parameter space for $\mathbf{b}$ is restricted to a subset $\mathcal{T}$ of R^p where $\mathbf{b}$ satisfies the consistent system of linear equations $\mathbf{P}^T\mathbf{b} = \delta$, that is,

$$\mathcal{T} = \{\mathbf{b} : \mathbf{P}^T\mathbf{b} = \delta\}.$$

Here we will insist that the matrix $\mathbf{P}$ is $p \times q$ with full-column rank; otherwise, we will have redundant equations, but sufficient to ensure that the equations are consistent: $\delta \in \mathcal{C}(\mathbf{P}^T)$.

Definition 3.5 *The function $\lambda^T\mathbf{b}$ is estimable in the restricted model iff there exists c (scalar) and $\mathbf{a}$ ($N \times 1$) such that $E(c + \mathbf{a}^T\mathbf{y}) = \lambda^T\mathbf{b}$ for all $\mathbf{b}$ that satisfy $\mathbf{P}^T\mathbf{b} = \delta$.*

Notice that if $\lambda^T\mathbf{b}$ was estimable in the unrestricted model, where $E(c + \mathbf{a}^T\mathbf{y}) = \lambda^T\mathbf{b}$ for all $\mathbf{b}$, it would be estimable now that $\mathbf{b}$ is restricted to the subset $\mathcal{T}$.

Result 3.7 *In the restricted model, $(c + \mathbf{a}^T\mathbf{y})$ is unbiased for $\lambda^T\mathbf{b}$ iff there exists $\mathbf{d}$ such that $\lambda = \mathbf{X}^T\mathbf{a} + \mathbf{Pd}$ and $c = \mathbf{d}^T\delta$.*

Proof: (If) $E(c+\mathbf{a}^T\mathbf{y}) = c+\mathbf{a}^T\mathbf{Xb} = \mathbf{d}^T\delta+(\lambda-\mathbf{Pd})^T\mathbf{b} = \lambda^T\mathbf{b}+\mathbf{d}^T(\delta-\mathbf{P}^T\mathbf{b}) = \lambda^T\mathbf{b}$ for all $\mathbf{b}$ that satisfy $\mathbf{P}^T\mathbf{b} = \delta$. (Only if) If $E(c+\mathbf{a}^T\mathbf{y}) = \lambda^T\mathbf{b}$ for all $\mathbf{b}$ that satisfy $\mathbf{P}^T\mathbf{b} = \delta$, then write all solutions to these equations—all points in $\mathcal{T}$—in the form $\mathbf{b}_* + \mathbf{Wz}$ where $\mathbf{b}_*$ is a particular solution, some point in $\mathcal{T}$, and $\mathcal{C}(\mathbf{W}) = \mathcal{N}(\mathbf{P}^T)$. Then we have

$$c + \mathbf{a}^T\mathbf{X}(\mathbf{b}_* + \mathbf{Wz}) = \lambda^T(\mathbf{b}_* + \mathbf{Wz})$$

for all $\mathbf{z}$. Using Result A.8, we have $(\lambda - \mathbf{X}^T\mathbf{a})^T\mathbf{W} = \mathbf{0}$ or $\mathbf{W}^T(\lambda - \mathbf{X}^T\mathbf{a}) = \mathbf{0}$ or $(\lambda - \mathbf{X}^T\mathbf{a}) \in \mathcal{N}(\mathbf{W}^T) = \mathcal{C}(\mathbf{P})$, or there exists a vector $\mathbf{d}$ such that $\lambda = \mathbf{X}^T\mathbf{a} + \mathbf{Pd}$. A little bit of algebra yields the expression for c. □

Viewing this result another way, if there exists **a** and **d** such that $\lambda = \mathbf{X}^T\mathbf{a} + \mathbf{Pd}$, then $\lambda^T\mathbf{b}$ is estimable in the restricted model and an unbiased estimator of $\lambda^T\mathbf{b}$ is $\mathbf{d}^T\delta + \mathbf{a}^T\mathbf{y}$. From this viewpoint, the space of λ leading to estimable $\lambda^T\mathbf{b}$ has expanded to $\mathcal{C}([\mathbf{X}^T\ \mathbf{P}])$, where in the unrestricted model that space was just $\mathcal{C}(\mathbf{X}^T)$. If the constraints arise from something estimable, so the columns of **P** are in $\mathcal{C}(\mathbf{X}^T)$, then nothing changes in terms of estimability. If the columns of **P** arise from completely nonestimable functions, then new functions may be estimable in the constrained model. With enough nonestimable columns in **P**, everything will be estimable.

Example 3.8: One-Way ANOVA
Consider once more the usual one-way ANOVA model $y_{ij} = \mu + \alpha_i + e_{ij}$, but add the nonestimable constraint $\sum n_i\alpha_i = 0$. Here we can show that μ as well as α_i are all estimable. See Exercise 3.25.

Minimizing the error sum of squares function $Q(\mathbf{y}) = (\mathbf{y} - \mathbf{Xb})^T(\mathbf{y} - \mathbf{Xb})$ subject to the constraints $\mathbf{Pb} = \delta$ suggests the method of Langrange multipliers discussed in appendix B. This tool states that a constrained optimum is the stationary point of a function that includes the constraint by adding a linear combination of it, called the Lagrangian. In this case, the Langrangian will be a function of **b** and θ (Lagrange multipliers):

$$L(\mathbf{b}, \theta) = (\mathbf{y} - \mathbf{Xb})^T(\mathbf{y} - \mathbf{Xb}) + 2\theta^T(\mathbf{P}^T\mathbf{b} - \delta). \tag{3.9}$$

(N.B.: The constraint can be written as $\mathbf{P}^T\mathbf{b} = \delta$ or $\mathbf{P}^T\mathbf{b} - \delta = \mathbf{0}$ or its negative, or any constant multiple of it, like 2 here.) Finding the stationary point of $L(\mathbf{b}, \theta)$ means merely taking the derivatives and setting them to zero:

$$\frac{\partial L(\mathbf{b}, \theta)}{\partial \mathbf{b}} = -2\mathbf{X}^T(\mathbf{y} - \mathbf{Xb}) + 2\mathbf{P}\theta$$

$$\frac{\partial L(\mathbf{b}, \theta)}{\partial \theta} = 2(\mathbf{P}^T\mathbf{b} - \delta),$$

and setting these to zero leads to the restricted normal equations (RNEs):

$$\begin{bmatrix} \mathbf{X}^T\mathbf{X} & \mathbf{P} \\ \mathbf{P}^T & \mathbf{0} \end{bmatrix} \begin{bmatrix} \mathbf{b} \\ \theta \end{bmatrix} = \begin{bmatrix} \mathbf{X}^T\mathbf{y} \\ \delta \end{bmatrix}. \tag{3.10}$$

Example 3.8: continued
For the usual one-way ANOVA model $y_{ij} = \mu + \alpha_i + e_{ij}$, with the nonestimable constraint $\sum \alpha_i = 0$, and for simplicity $a = 3$, and balanced $n_i = n$, the RNEs and their solution take the form

$$\begin{bmatrix} 3n & n & n & n & 0 \\ n & n & 0 & 0 & 1 \\ n & 0 & n & 0 & 1 \\ n & 0 & 0 & n & 1 \\ 0 & 1 & 1 & 1 & 0 \end{bmatrix} \begin{bmatrix} \overline{y}_{..} \\ \overline{y}_{1.} - \overline{y}_{..} \\ \overline{y}_{2.} - \overline{y}_{..} \\ \overline{y}_{3.} - \overline{y}_{..} \\ 0 \end{bmatrix} = \begin{bmatrix} y_{..} \\ y_{1.} \\ y_{2.} \\ y_{3.} \\ 0 \end{bmatrix}.$$

Example 3.8: continued

For the usual one-way ANOVA model $y_{ij} = \mu + \alpha_i + e_{ij}$, with the estimable constraint $\alpha_1 - 2\alpha_2 + \alpha_3 = 0$, and again $a = 3$, the RNEs and their solution (a family of solutions this time) take the form

$$\begin{bmatrix} 3n & n & n & n & 0 \\ n & n & 0 & 0 & 1 \\ n & 0 & n & 0 & -2 \\ n & 0 & 0 & n & 1 \\ 0 & 1 & -2 & 1 & 0 \end{bmatrix} \left\{ \begin{bmatrix} \overline{y}_{..} \\ \frac{1}{2}\overline{y}_{1.} - \frac{1}{2}\overline{y}_{3.} \\ 0 \\ \frac{1}{2}\overline{y}_{3.} - \frac{1}{2}\overline{y}_{1.} \\ \frac{1}{6}\overline{y}_{1.} - \frac{1}{3}\overline{y}_{2.} + \frac{1}{6}\overline{y}_{3.} \end{bmatrix} + z \begin{bmatrix} -1 \\ 1 \\ 1 \\ 1 \\ 0 \end{bmatrix} \right\} = \begin{bmatrix} y_{..} \\ y_{1.} \\ y_{2.} \\ y_{3.} \\ 0 \end{bmatrix}.$$

We cannot accomplish very much unless we can show that these restricted normal equations are consistent. Recall that in Chapter 2 we needed to show that the normal equations $\mathbf{X}^T\mathbf{X}\mathbf{b} = \mathbf{X}^T\mathbf{y}$ are always consistent. We did this by showing $\mathcal{N}(\mathbf{X}^T\mathbf{X}) = \mathcal{N}(\mathbf{X})$, and hence their respective orthogonal complements were also equal: $\mathcal{C}(\mathbf{X}^T\mathbf{X}) = \mathcal{C}(\mathbf{X}^T)$. So since $\mathbf{X}^T\mathbf{y} \in \mathcal{C}(\mathbf{X}^T)$, then the normal equations are always consistent. Now it would have been sufficient to show that $\mathcal{N}(\mathbf{X}^T\mathbf{X}) \subseteq \mathcal{N}(\mathbf{X})$, so that $\mathcal{C}(\mathbf{X}^T) \subseteq \mathcal{C}(\mathbf{X}^T\mathbf{X})$, and since $\mathbf{X}^T\mathbf{y} \in \mathcal{C}(\mathbf{X}^T)$ gives $\mathbf{X}^T\mathbf{y} \in \mathcal{C}(\mathbf{X}^T\mathbf{X})$. Here, we will follow the same strategy.

Result 3.8 *The restricted normal Equations (3.10) are consistent.*

Proof: Here we have the right-hand-side vector in the column space of a $(p+q)$ by $(N+p)$ matrix, as follows:

$$\begin{bmatrix} \mathbf{X}^T\mathbf{y} \\ \delta \end{bmatrix} \in \mathcal{C}\left(\begin{bmatrix} \mathbf{X}^T & \mathbf{0} \\ \mathbf{0} & \mathbf{P}^T \end{bmatrix} \right)$$

since the equations $\mathbf{P}^T\mathbf{b} = \delta$ are consistent, that is, $\delta \in \mathcal{C}(\mathbf{P}^T)$. Consider a vector $\mathbf{v}$ ($p+q$ dimensions) in the nullspace of the matrix of the RNEs, that is,

$$\begin{bmatrix} \mathbf{X}^T\mathbf{X} & \mathbf{P} \\ \mathbf{P}^T & \mathbf{0} \end{bmatrix} \begin{bmatrix} \mathbf{v}_1 \\ \mathbf{v}_2 \end{bmatrix} = \begin{bmatrix} \mathbf{0} \\ \mathbf{0} \end{bmatrix}.$$

Then we have the following:

$$(i)\ \mathbf{X}^T\mathbf{X}\mathbf{v}_1 + \mathbf{P}\mathbf{v}_2 = \mathbf{0},$$

and

$$(ii)\ \mathbf{P}^T\mathbf{v}_1 = \mathbf{0}.$$

Premultiplying Equation (i) by $\mathbf{v}_1^T$ gives

$$(iii)\ \mathbf{v}_1^T\mathbf{X}^T\mathbf{X}\mathbf{v}_1 + \mathbf{v}_1^T\mathbf{P}\mathbf{v}_2 = \mathbf{0}.$$

Equation (ii) says that the second piece in (iii) is zero, leading to $\mathbf{v}_1^T\mathbf{X}^T\mathbf{X}\mathbf{v}_1 = \mathbf{0}$, and hence $\mathbf{X}\mathbf{v}_1 = \mathbf{0}$. Putting this back into (i) says that the second piece, $\mathbf{P}\mathbf{v}_2$, is also $\mathbf{0}$. Together, these say that the vectors in nullspace of the matrix of the RNEs have their first p components in $\mathcal{N}(\mathbf{X})$, and the last q components in $\mathcal{N}(\mathbf{P})$. We have shown

$$\mathcal{N}\left(\begin{bmatrix}\mathbf{X}^T\mathbf{X} & \mathbf{P}\\ \mathbf{P}^T & \mathbf{0}\end{bmatrix}\right) \subseteq \mathcal{N}\left(\begin{bmatrix}\mathbf{X} & \mathbf{0}\\ \mathbf{0} & \mathbf{P}\end{bmatrix}\right), \quad \text{hence } \mathcal{C}\left(\begin{bmatrix}\mathbf{X}^T & \mathbf{0}\\ \mathbf{0} & \mathbf{P}^T\end{bmatrix}\right) \subseteq \mathcal{C}\left(\begin{bmatrix}\mathbf{X}^T\mathbf{X} & \mathbf{P}\\ \mathbf{P}^T & \mathbf{0}\end{bmatrix}\right)$$

(via Result A.5), and we have shown that the restricted normal equations are consistent. □

Letting $Q(\mathbf{b}) = (\mathbf{y} - \mathbf{X}\mathbf{b})^T(\mathbf{y} - \mathbf{X}\mathbf{b})$, we have some results regarding solutions to the RNE, in particular the first component of the solution vector $\hat{\mathbf{b}}_H$, and the minimum of $Q(\mathbf{b})$.

Result 3.9 *If $\hat{\mathbf{b}}_H$ is the first component of a solution to the RNE, then $\hat{\mathbf{b}}_H$ minimizes $Q(\mathbf{b})$ over the restricted parameter space $\mathcal{T}$.*

Proof: Let $\tilde{\mathbf{b}}$ be any other vector satisfying $\mathbf{P}^T\mathbf{b} = \delta$; then write

$$\begin{aligned} Q(\tilde{\mathbf{b}}) &= (\mathbf{y} - \mathbf{X}\tilde{\mathbf{b}})^T(\mathbf{y} - \mathbf{X}\tilde{\mathbf{b}}) = (\mathbf{y} - \mathbf{X}\hat{\mathbf{b}}_H + \mathbf{X}(\hat{\mathbf{b}}_H - \tilde{\mathbf{b}}))^T(\mathbf{y} - \mathbf{X}\hat{\mathbf{b}}_H + \mathbf{X}(\hat{\mathbf{b}}_H - \tilde{\mathbf{b}}))\\ &= Q(\hat{\mathbf{b}}_H) + (\hat{\mathbf{b}}_H - \tilde{\mathbf{b}})^T\mathbf{X}^T\mathbf{X}(\hat{\mathbf{b}}_H - \tilde{\mathbf{b}})\end{aligned}$$

since the cross-product term is $2(\hat{\mathbf{b}}_H - \tilde{\mathbf{b}})^T\mathbf{X}^T(\mathbf{y} - \mathbf{X}\hat{\mathbf{b}}_H) = 2(\hat{\mathbf{b}}_H - \tilde{\mathbf{b}})^T\mathbf{P}\hat{\theta}_H$ where $\hat{\theta}_H$ is the second component of the solution. Notice, however, that $\mathbf{P}^T(\hat{\mathbf{b}}_H - \tilde{\mathbf{b}}) = \mathbf{0}$ since both $\tilde{\mathbf{b}}$ and $\hat{\mathbf{b}}_H$ satisfy the constraint, and so the cross-product term is zero. Hence $Q(\tilde{\mathbf{b}}) \geq Q(\hat{\mathbf{b}}_H)$ for all $\tilde{\mathbf{b}}$ satisfying $\mathbf{P}^T\tilde{\mathbf{b}} = \delta$, with equality iff $\mathbf{X}\hat{\mathbf{b}}_H = \mathbf{X}\tilde{\mathbf{b}}$. □

Result 3.10 *Let $\hat{\mathbf{b}}_H$ be the first component of a solution to the RNE; for $\tilde{\mathbf{b}}$ satisfying $\mathbf{P}^T\tilde{\mathbf{b}} = \delta$, $Q(\tilde{\mathbf{b}}) = Q(\hat{\mathbf{b}}_H)$ iff $\tilde{\mathbf{b}}$ is also part of a solution to the RNE (almost a converse to Result 3.9).*

Proof: (If) Following the analysis in the proof to the previous result, $Q(\tilde{\mathbf{b}}) = Q(\hat{\mathbf{b}}_H)$ for $\tilde{\mathbf{b}}$ satisfying $\mathbf{P}^T\tilde{\mathbf{b}} = \delta$ iff $\mathbf{X}\hat{\mathbf{b}}_H = \mathbf{X}\tilde{\mathbf{b}}$; hence $\mathbf{X}^T\mathbf{X}\hat{\mathbf{b}}_H = \mathbf{X}^T\mathbf{X}\tilde{\mathbf{b}}$, and since

$$\mathbf{X}^T\mathbf{X}\hat{\mathbf{b}}_H + \mathbf{P}\hat{\theta}_H = \mathbf{X}^T\mathbf{y} = \mathbf{X}^T\mathbf{X}\tilde{\mathbf{b}} + \mathbf{P}\hat{\theta}_H,$$

so also $(\tilde{\mathbf{b}}, \hat{\theta}_H)$ satisfies the RNE.

(Only if) $(\tilde{\mathbf{b}}, \tilde{\theta})$ satisfies the RNE, then $\mathbf{X}^T\mathbf{X}\hat{\mathbf{b}}_H + \mathbf{P}\hat{\theta}_H = \mathbf{X}^T\mathbf{y} = \mathbf{X}^T\mathbf{X}\tilde{\mathbf{b}} + \mathbf{P}\tilde{\theta}$ and so

$$\mathbf{X}^T\mathbf{X}(\hat{\mathbf{b}}_H - \tilde{\mathbf{b}}) = \mathbf{P}(\hat{\theta}_H - \tilde{\theta}),$$

which then gives

$$Q(\tilde{\mathbf{b}}) - Q(\hat{\mathbf{b}}_H) = (\hat{\mathbf{b}}_H - \tilde{\mathbf{b}})^T\mathbf{X}^T\mathbf{X}(\hat{\mathbf{b}}_H - \tilde{\mathbf{b}}) = (\hat{\mathbf{b}}_H - \tilde{\mathbf{b}})^T\mathbf{P}(\hat{\theta}_H - \tilde{\theta}) = 0$$

since $\mathbf{P}^T(\hat{\mathbf{b}}_H - \tilde{\mathbf{b}}) = \delta - \delta = \mathbf{0}$. □

3.10 Summary

1. We now add a mean assumption to the linear model $\mathbf{y} = \mathbf{Xb} + \mathbf{e}$ and $E(\mathbf{e}) = \mathbf{0}$, or $E(\mathbf{y}) = \mathbf{Xb}$.

2. A function $\lambda^T\mathbf{b}$ is estimable iff an unbiased estimator $\mathbf{a}^T\mathbf{y}$ can be constructed for it, that is, $E(\mathbf{a}^T\mathbf{y}) = \lambda^T\mathbf{b}$ for all $\mathbf{b}$.

3. In terms of the design matrix $\mathbf{X}$, $\lambda^T\mathbf{b}$ is estimable if $\lambda \in \mathcal{C}(\mathbf{X}^T)$ or, equivalently, $\lambda \perp \mathcal{N}(\mathbf{X})$.

4. The least squares estimator for an estimable function $\lambda^T\mathbf{b}$ is defined as $\lambda^T\hat{\mathbf{b}}$ where $\hat{\mathbf{b}}$ solves the normal equations.

5. Estimability and least squares estimators for the one-way and two-way crossed ANOVA models are discussed in detail.

6. Conditions are given for transferring estimability to a reparameterized model.

7. Commonly used full-rank reparameterizations for the one-way ANOVA model are given.

8. Imposing sufficient constraints on a set of jointly nonestimable functions can lead to a unique solution to the normal equations.

9. Estimability and least squares are discussed in the case of constrained parameter space.

3.11 Exercises

3.1. * Prove that if $\lambda^{(j)T}\mathbf{b}$, $j = 1, \ldots, k$, are estimable, then $\sum_j d_j\lambda^{(j)T}\mathbf{b}$ is also estimable, for any constants $d_1, \ldots, d_k$.

3.2. * Prove the converse to Result 3.2: if the least squares estimator $\lambda^T\hat{\mathbf{b}}$ is the same for all solutions $\hat{\mathbf{b}}$ to the normal equations, then $\lambda^T\mathbf{b}$ is estimable.

3.3. In the proof of Result 3.2, show that if $\mathbf{a}^{(1)}$ and $\mathbf{a}^{(2)}$ are solutions to $\mathbf{X}^T\mathbf{a} = \lambda$, then $\mathbf{P_X a}^{(1)} = \mathbf{P_X a}^{(2)}$.

3.4. Give an example using the one-way ANOVA model from Section 3.4 to show that if $\lambda^T\mathbf{b}$ is not an estimable function, then $\lambda^T\hat{\mathbf{b}}$ is not unbiased.

3.5. For the one-way ANOVA in Section 3.4

a. Find a third, asymmetric generalized inverse for $\mathbf{X}^T\mathbf{X}$.
b. Find all solutions to the normal equations for each generalized inverse.

c. Show that all three families of solutions are the same.
d. Find $(\mathbf{I} - \mathbf{P_X})\mathbf{y}$.

3.6. Recall the two-way crossed problem in Section 3.5.

a. Show that $\mathbf{G}^T$ is a generalized inverse for $\mathbf{X}^T\mathbf{X}$, and then compute both $(\mathbf{X}^T\mathbf{X})\mathbf{G}^T$ and $\mathbf{I} - \mathbf{G}(\mathbf{X}^T\mathbf{X})$.
b. Find the solution $\mathbf{G}\mathbf{X}^T\mathbf{y}$ to the normal equations. Also compute the solution $\mathbf{G}^T\mathbf{X}^T\mathbf{y}$.
c. Find the least squares estimators for the following estimable functions:
   - $\mu + \alpha_i + \beta_j$
   - $\alpha_i - \alpha_k$
   - $\beta_j - \beta_k$
   - $\sum d_i\alpha_i$ with $\sum d_i = 0$
   - $\sum f_j\beta_j$ with $\sum f_j = 0$

3.7. Consider the following cross-classified model without replicates:

$$y_{ij} = \mu + \alpha_i + \beta_j + e_{ij}, \quad E(\mathbf{e}) = \mathbf{0},$$

where $\mathbf{y}$, $\mathbf{X}$, and $\mathbf{b}$ are as follows:

$$\mathbf{y} = \begin{bmatrix} y_{11} \\ y_{12} \\ y_{13} \\ y_{21} \\ y_{22} \\ y_{23} \end{bmatrix} \quad \mathbf{X} = \begin{bmatrix} 1 & 1 & 0 & 1 & 0 & 0 \\ 1 & 1 & 0 & 0 & 1 & 0 \\ 1 & 1 & 0 & 0 & 0 & 1 \\ 1 & 0 & 1 & 1 & 0 & 0 \\ 1 & 0 & 1 & 0 & 1 & 0 \\ 1 & 0 & 1 & 0 & 0 & 1 \end{bmatrix} \quad \mathbf{b} = \begin{bmatrix} \mu \\ \alpha_1 \\ \alpha_2 \\ \beta_1 \\ \beta_2 \\ \beta_3 \end{bmatrix}.$$

a. What is $r = rank(X)$?
b. Write out the normal equations and find all solutions.
c. Give a set of basis vectors for $\mathcal{N}(\mathbf{X})$.
d. Give a list of r linearly independent estimable functions $\lambda^T\mathbf{b}$.
e. Show that $\alpha_1 - \alpha_2$ is estimable and give its least squares estimator.
f. Show that $\beta_1 - 2\beta_2 + \beta_3$ is estimable and give its least squares estimator.
g. Find all parameter vectors $\mathbf{b}$ so that $\mathbf{Xb} = (8, 7, 6, 6, 5, 4)^T$.

3.8. Suppose the experiment in Exercise 3.7 was not performed as planned. Two of the experimental units were not run properly—cells $(2, 2)$ and $(1, 3)$—those two observations became replicates for cells $(2, 3)$. These mistakes lead to the following relabeling of the observations, a slight change in the model and the design matrix $\mathbf{X}$, but the *same* $\mathbf{b}$:

$$y_{ijk} = \mu + \alpha_i + \beta_j + e_{ijk}, \quad E(\mathbf{e}) = \mathbf{0},$$

for $i = 1, 2$; $j = 1, 2, 3$; $k = 1, \ldots, n_{ij}$, where $n_{11} = n_{12} = n_{21} = 1, n_{23} = 3$, but $n_{22} = n_{13} = 0$.

$$\mathbf{y} = \begin{bmatrix} y_{111} \\ y_{121} \\ y_{211} \\ y_{231} \\ y_{232} \\ y_{233} \end{bmatrix} \quad \mathbf{X} = \begin{bmatrix} 1 & 1 & 0 & 1 & 0 & 0 \\ 1 & 1 & 0 & 0 & 1 & 0 \\ 1 & 0 & 1 & 1 & 0 & 0 \\ 1 & 0 & 1 & 0 & 0 & 1 \\ 1 & 0 & 1 & 0 & 0 & 1 \\ 1 & 0 & 1 & 0 & 0 & 1 \end{bmatrix} \quad \mathbf{b} = \begin{bmatrix} \mu \\ \alpha_1 \\ \alpha_2 \\ \beta_1 \\ \beta_2 \\ \beta_3 \end{bmatrix}$$

a. Find the rank of $\mathbf{X}$ and a set of basis vectors for $\mathcal{N}(\mathbf{X})$.
b. Show that $\alpha_1 - \alpha_2$ is still estimable by constructing an unbiased estimator for it.
c. For all of the functions that you listed in Exercise 3.7d, determine whether they are still estimable in this situation.

3.9. Consider the following design matrix for ANACOVA with two groups and one covariate:

$$y_{ij} = \mu + \alpha_i + \beta x_{ij} + e_{ij} \text{ where } \mathbf{X} = \begin{bmatrix} 1 & 1 & 0 & 1 \\ 1 & 1 & 0 & 2 \\ 1 & 1 & 0 & 3 \\ 1 & 1 & 0 & 4 \\ 1 & 0 & 1 & 1 \\ 1 & 0 & 1 & 2 \\ 1 & 0 & 1 & 3 \\ 1 & 0 & 1 & 4 \end{bmatrix} \quad \text{and} \quad \mathbf{b} = \begin{bmatrix} \mu \\ \alpha_1 \\ \alpha_2 \\ \beta \end{bmatrix}$$

so that $n_1 = 4, n_2 = 4$.

a. What is the rank of $\mathbf{X}$?
b. Find a generalized inverse for $\mathbf{X}^T\mathbf{X}$ and call it $\mathbf{G}$.
c. Compute $\mathbf{H} = \mathbf{G}\mathbf{X}^T\mathbf{X}$ and recall $\mathbf{H}^T$ is a projection onto $\mathcal{C}(\mathbf{X}^T) = \mathcal{C}(\mathbf{X}^T\mathbf{X})$.
d. Compute $(\mathbf{I} - \mathbf{H})$ and recall that it is a projection onto $\mathcal{N}(\mathbf{X}) = \mathcal{N}(\mathbf{X}^T\mathbf{X})$.
e. Find a basis for $\mathcal{C}(\mathbf{X}^T)$.
f. Find a basis for $\mathcal{N}(\mathbf{X})$.
g. Compute $\mathbf{P_X}$ and its trace.

(More practice) Construct a second generalized inverse in (b) and repeat (c)–(f).

3.10. To evaluate a new curriculum in biology, two teachers each taught two classes using the old curriculum, and three teachers taught two classes with the new. The response y_{ijk} is the average score for the class on the end-of-semester test.

The data are:

			n_{ij}	y_{ij1}	y_{ij2}
$i = 1$ (old)	$j = 1$	Dr. Able	2	100	80
	$j = 2$	Dr. Baker	2	80	80
$i = 2$ (new)	$j = 1$	Dr. Adams	2	110	90
	$j = 2$	Dr. Brown	2	100	140
	$j = 3$	Dr. Charles	2	110	150

A simple model for this experiment (all effects fixed) is the nested model

$$y_{ijk} = \mu + \alpha_i + \beta_{ij} + e_{ijk}$$

and assume $E(e_{ijk}) = 0$.

a. Write this as a linear model by writing $\mathbf{Xb}$.
b. What is $r = rank(\mathbf{X})$?
c. Write out the normal equations and find all solutions.
d. Give a set of basis vectors for $\mathcal{N}(\mathbf{X})$.
e. Give a list of r linearly independent estimable functions $\lambda^T\mathbf{b}$, and give the least squares estimator for each one.
f. Show that $\alpha_1 - \alpha_2$ is not estimable.
g. For which of the following sets of parameter values

$$\mathbf{b} = (\mu, \alpha_1, \alpha_2, \beta_{11}, \beta_{12}, \beta_{21}, \beta_{22}, \beta_{23})^T$$

is the mean vector $\mathbf{Xb}$ the same?

$$\mathbf{b}_1 = (100, 0, 0, 0, 0, 0, 0, 0)^T$$
$$\mathbf{b}_2 = (90, 0, 10, 10, 0, 10, 20, 20)^T$$
$$\mathbf{b}_3 = (50, 40, 30, 30, 10, 20, 20, 20)^T$$
$$\mathbf{b}_4 = (80, 20, 10, 10, 0, 10, 20, 20)^T$$
$$\mathbf{b}_5 = (90, 0, 20, 10, 0, 0, 10, 10)^T$$

h. For the parameter vectors in (g), which give the same $\mathbf{Xb}$, show that the estimable functions you gave in (e), the values of $\lambda^T\mathbf{b}$, are the same.

3.11. Consider the following two-way crossed ANOVA model with interaction and replication:

$$y_{ijk} = \mu + \alpha_i + \beta_j + \gamma_{ij} + e_{ijk}$$

for $i = 1, \ldots, a$, $j = 1, \ldots, b$, and $k = 1, \ldots, n_{ij} \geq 1$. First simplify matters with $a = 2, b = 3$, and $n_{ij} = 2$.

a. Write this as a linear model by writing $\mathbf{Xb}$.
b. Find $rank(\mathbf{X}) = r$.
c. Find a set of r linearly independent estimable functions.
d. Give a set of $p - r$ jointly nonestimable functions.

e. Redo parts (a)–(d) for general $n_{ij} = n$.
f. Redo parts (a)–(d) for the pattern of observations n_{ij} as in Exercise 3.10.

3.12. Find $\mathbf{S}_2$ and $\mathbf{S}_3$ in Example 3.3, so that $\mathbf{X} = \mathbf{ZS}_2 = \mathbf{US}_3$.

3.13. In Example 3.6, show that $\mathcal{C}(\mathbf{X}) = \mathcal{C}(\mathbf{W})$ by finding $\mathbf{S}$ and $\mathbf{T}$.

3.14. In Example 2.8, find $\mathbf{S}$ and $\mathbf{T}$ so that $\mathbf{X} = \mathbf{ZS}$ and $\mathbf{Z} = \mathbf{XT}$.

3.15. Find a second matrix $\mathbf{T}$ for Example 2.8.

3.16. Suppose $\mathcal{C}(\mathbf{X}) = \mathcal{C}(\mathbf{W})$, where $\mathbf{W}$ is full-column rank, and express the relationship between the two matrices as $\mathbf{W} = \mathbf{XT}$ and $\mathbf{X} = \mathbf{WS}$.

a. Suppose we have $\mathbf{S}$, give an easy way to find $\mathbf{T}$.
b. Suppose we have $\mathbf{T}$, give an easy way to find $\mathbf{S}$.
c. Suppose we have neither $\mathbf{S}$ nor $\mathbf{T}$, find an easy way to get one of them.

3.17. In Example 3.5, find $\mathbf{T}$.

3.18. For the three parameterizations of the one-way ANOVA model presented in Example 3.3, give both $\mathbf{T}$ and $\mathbf{S}$.

3.19. Can you construct an example for Remark 3.1 using your matrix $\mathbf{S}$ from Exercise 3.14?

3.20. Consider the general unbalanced one-way ANOVA model with the (usual) (nonestimable) constraint $\sum_i n_i\alpha_i = 0$. Write the constraint $\mathbf{Cb} = \mathbf{0}$ as $\mathbf{C} = (0, d^T)$ where $d_i = n_i, i = 1, \ldots, a$. Show that

$$(\mathbf{X}^T\mathbf{X} + \mathbf{C}^T\mathbf{C})^{-1} = \begin{bmatrix} \frac{N+1}{N^2} & -\frac{1}{N^2}\mathbf{1}_a^T \\ -\frac{1}{N^2}\mathbf{1}_a & \mathbf{D}^{-1} - \frac{N-1}{N^2}\mathbf{1}_a\mathbf{1}_a^T \end{bmatrix}$$

where $\mathbf{D} = diag(n_i)$. Also find the solution to the normal equations using this result and show that it satisfies the constraint.

3.21. Solve Exercise 3.20 directly by using the general solution to the normal equations from Section 3.3.

3.22. Prove Result 3.6(v). Hint: Let $\mathbf{Z} = \begin{bmatrix} \mathbf{X} \\ \mathbf{C} \end{bmatrix}$ and find idempotent $\mathbf{P}_Z$; then employ Results 3.6(iii and iv).

3.23. A quadratic regression model fits the U.S. decennial census population (y_t for years $t = 1790$ to 2000 in steps of 10) very well.

a. First fit the model $y_t = \beta_0 + \beta_1 t + \beta_2 t^2 + e_t$.
b. Then fit the model with time centered at c: $y_t = \alpha_0 + \alpha_1(t - c) + \alpha_2(t - c)^2 + e_t$ using centering values such as 1790, 1880, 2000.
c. For a given value of c, find the α's in terms of the β's and vice versa.

d. With the usual regression results from (a) or (b), suppose the task was to forecast the population for the year 2010—without a calculator.
e. Suppose the task was to estimate the (annual) rate of change of the population in the year 2000, how could you do that?

3.24. Consider the general balanced cross-classified model from Section 3.5, $y_{ij} = \mu + \alpha_i + \beta_j + e_{ij}$, and partition the design matrix as below:

$$\mathbf{Xb} = \begin{bmatrix} \mathbf{1}_b & \mathbf{1}_b & \mathbf{0} & \dots & \mathbf{0} & \mathbf{I}_b \\ \mathbf{1}_b & \mathbf{0} & \mathbf{1}_b & \dots & \mathbf{0} & \mathbf{I}_b \\ \dots & \dots & \dots & \dots & & \\ \mathbf{1}_b & \mathbf{0} & \mathbf{0} & \dots & \mathbf{0} & \mathbf{I}_b \\ \mathbf{1}_b & \mathbf{0} & \mathbf{0} & \dots & \mathbf{1}_b & \mathbf{I}_b \end{bmatrix} \begin{bmatrix} \mu \\ \alpha_1 \\ \alpha_2 \\ \dots \\ \alpha_a \\ \beta_1 \\ \beta_2 \\ \dots \\ \beta_b \end{bmatrix}$$

$$= \begin{bmatrix} \mathbf{1}_{ab} & \mathbf{X}_A & \mathbf{X}_B \end{bmatrix} \begin{bmatrix} \mu \\ \alpha's \\ \beta's \end{bmatrix}.$$

a. Show $\mathbf{P}_{\mathbf{X}_A}\mathbf{X}_B = \mathbf{P}_1\mathbf{X}_B$.
b. Show that $\mathbf{P}_{\mathbf{X}_A}\mathbf{P}_{\mathbf{X}_B} = \mathbf{P}_{\mathbf{1}_{ab}}$.
c. Show that $\mathbf{P}_{\mathbf{X}} - \mathbf{P}_{\mathbf{X}_A} = \mathbf{P}_{\mathbf{X}_B} - \mathbf{P}_1$.

3.25. In the one-way ANOVA model in Example 3.8 with the constraint $\sum n_i\alpha_i = 0$, construct unbiased estimators for both μ and α_i. Keep it simple with $a = 3$ and $n_i = 2$.

3.26. Prove the result: let $\begin{bmatrix} \hat{\mathbf{b}}_H \\ \hat{\theta}_H \end{bmatrix}$ and $\begin{bmatrix} \tilde{\mathbf{b}}_H \\ \tilde{\theta}_H \end{bmatrix}$ be two solutions to the RNE (3.10), then

a. $\mathbf{X}\hat{\mathbf{b}}_H = \mathbf{X}\tilde{\mathbf{b}}_H$, and
b. if $\lambda^T\mathbf{b}$ is estimable in the restricted model, then $\lambda^T\hat{\mathbf{b}}_H = \lambda^T\tilde{\mathbf{b}}_H$.

Chapter 4

Gauss–Markov Model

4.1 Model Assumptions

So far we have approached the linear model mostly as a method of mathematical approximation. In this chapter, we pose the Gauss–Markov model, which embodies the most common assumptions for the statistical approach to the linear model, leading to the Gauss–Markov Theorem. The Gauss–Markov model takes the form

$$\mathbf{y} = \mathbf{X}\mathbf{b} + \mathbf{e} \tag{4.1}$$

where $\mathbf{y}$ is the $(N \times 1)$ vector of observed responses and $\mathbf{X}$ is the $(N \times p)$ known design matrix. As before, the coefficient vector $\mathbf{b}$ is unknown and to be determined or estimated. The main features of the Gauss–Markov model are the assumptions on the error $\mathbf{e}$:

$$E(\mathbf{e}) = \mathbf{0} \quad \text{and} \quad Cov(\mathbf{e}) = \sigma^2 \mathbf{I}_N. \tag{4.2}$$

The notation for expectation and covariances above can be rewritten component by component:

$$(E(\mathbf{e}))_i = i^{th} \text{ component of } E(\mathbf{e}) = E(e_i)$$
$$(Cov(\mathbf{e}))_{ij} = i, j^{th} \text{ element of covariance matrix } = Cov(e_i, e_j)$$

so that the Gauss–Markov assumptions can be rewritten as

$$E(e_i) = 0, i = 1, \ldots, N$$
$$Cov(e_i, e_j) = \begin{cases} \sigma^2 & \text{for } i = j, \\ 0 & \text{for } i \neq j \end{cases}$$

that is, the errors in the model have a zero mean, constant variance, and are uncorrelated. An alternative view of the Gauss–Markov model does not employ the error vector $\mathbf{e}$:

$$E(\mathbf{y}) = \mathbf{X}\mathbf{b}, Cov(\mathbf{y}) = \sigma^2 \mathbf{I}_N.$$

The assumptions in the Gauss–Markov model are easily acceptable for many practical problems, and deviations from these assumptions will be considered in more detail later.

Before we get to the Gauss–Markov Theorem, we will need some simple tools for conveniently working with means and variances of vector random variables, in particular, a linear combination of variables $\mathbf{a}^T\mathbf{y}$ where $\mathbf{a}$ is a fixed vector. The rules are:

1. $E(\mathbf{a}^T\mathbf{y}) = \mathbf{a}^T E(\mathbf{y})$.
2. $Var(\mathbf{a}^T\mathbf{y}) = \mathbf{a}^T Cov(\mathbf{y})\mathbf{a}$.
3. $Cov(\mathbf{a}^T\mathbf{y}, \mathbf{c}^T\mathbf{y}) = \mathbf{a}^T Cov(\mathbf{y})\mathbf{c}$, for fixed $\mathbf{a}, \mathbf{c}$.
4. $Cov(\mathbf{A}^T\mathbf{y}) = \mathbf{A}^T Cov(\mathbf{y})\mathbf{A}$, for fixed matrix $\mathbf{A}$.

The key result is (3), from which (2) and (4) follow easily, and just algebraic book-keeping is necessary:

$$Cov(\mathbf{a}^T\mathbf{y}, \mathbf{c}^T\mathbf{y}) = Cov\left(\sum_i a_i y_i, \sum_j c_j y_j\right) = \sum_i a_i Cov\left(y_i, \sum_j c_j y_j\right)$$
$$= \sum_i \sum_j a_i c_j Cov(y_i, y_j) = \mathbf{a}^T Cov(\mathbf{y})\mathbf{c}.$$

Example 4.1: Variance and Covariance Calculations
Let

$$Cov(\mathbf{y}) = \begin{bmatrix} 4 & 2 & 4 \\ 2 & 5 & 0 \\ 4 & 0 & 25 \end{bmatrix}, \quad \mathbf{c} = \begin{bmatrix} 1 \\ -1 \\ 2 \end{bmatrix}, \quad \mathbf{a} = \begin{bmatrix} 2 \\ 0 \\ 1 \end{bmatrix},$$

then we have

$$Var(y_1 - y_2 + 2y_3) = Var(\mathbf{c}^T\mathbf{y}) = \mathbf{c}^T Cov(\mathbf{y})\mathbf{c}$$
$$= \begin{bmatrix} 1 & -1 & 2 \end{bmatrix} \begin{bmatrix} 4 & 2 & 4 \\ 2 & 5 & 0 \\ 4 & 0 & 25 \end{bmatrix} \begin{bmatrix} 1 \\ -1 \\ 2 \end{bmatrix} = \begin{bmatrix} 1 & -1 & 2 \end{bmatrix} \begin{bmatrix} 10 \\ -3 \\ 54 \end{bmatrix} = 121.$$

Also we have

$$Cov(2y_1 + y_3, y_1 - y_2 + 2y_3) = Cov(\mathbf{a}^T\mathbf{y}, \mathbf{c}^T\mathbf{y}) = \mathbf{a}^T Cov(y)\mathbf{c}$$
$$= \begin{bmatrix} 2 & 0 & 1 \end{bmatrix} \begin{bmatrix} 4 & 2 & 4 \\ 2 & 5 & 0 \\ 4 & 0 & 25 \end{bmatrix} \begin{bmatrix} 1 \\ -1 \\ 2 \end{bmatrix} = \begin{bmatrix} 2 & 0 & 1 \end{bmatrix} \begin{bmatrix} 10 \\ -3 \\ 54 \end{bmatrix} = 74.$$

Example 4.2: Variance of a Least Squares Estimator
Let $\mathbf{y} = \mathbf{Xb} + \mathbf{e}$, with the Gauss–Markov assumptions on $\mathbf{e}$, so that $Cov(\mathbf{y}) = \sigma^2\mathbf{I}_N$, and let $\lambda^T\mathbf{b}$ be an estimable function. Then the variance of the least squares estimator follows the calculation (see Exercise 4.2)

$$Var(\lambda^T\hat{\mathbf{b}}) = Var(\lambda^T(\mathbf{X}^T\mathbf{X})^g\mathbf{X}^T\mathbf{y}) = \lambda^T(\mathbf{X}^T\mathbf{X})^g\mathbf{X}^T Cov(\mathbf{y})\mathbf{X}(\mathbf{X}^T\mathbf{X})^{gT}\lambda$$
$$= \lambda^T(\mathbf{X}^T\mathbf{X})^g\mathbf{X}^T(\sigma^2\mathbf{I}_N)\mathbf{X}(\mathbf{X}^T\mathbf{X})^{gT}\lambda = \sigma^2\lambda^T(\mathbf{X}^T\mathbf{X})^g\lambda.$$

Example 4.3: Variance in Simple Linear Regression

In the simple linear regression case in Example 2.1, we solved the normal equations and found the usual slope estimate

$$\hat{\beta}_1 = \sum_{i=1}^{N} (x_i - \overline{x}) y_i / \sum_{i=1}^{N} (x_i - \overline{x})^2.$$

To compute the mean and variance of $\hat{\beta}_1$, we have two routes. One route is to treat $\hat{\beta}_1$ as a linear combination of the y_i's:

$$\hat{\beta}_1 = \sum_{i=1}^{N} (x_i - \overline{x}) y_i / \sum_{i=1}^{N} (x_i - \overline{x})^2 = \sum_{i=1}^{N} [(x_i - \overline{x})/S_{xx}]\, y_i$$

where

$$S_{xx} = \sum_{i=1}^{N} (x_i - \overline{x})^2.$$

Then

$$\begin{aligned} E(\hat{\beta}_1) &= \sum_{i=1}^{N} [(x_i - \overline{x})/S_{xx}]\, E(y_i) = \sum_{i=1}^{N} [(x_i - \overline{x})/S_{xx}]\, (\beta_0 + \beta_1 x_i) \\ &= \sum_{i=1}^{N} (x_i - \overline{x}) x_i \beta_1 / S_{xx} = \beta_1 \end{aligned}$$

and the usual slope estimate is unbiased. Its variance can be found from the following algebra:

$$\begin{aligned} Var(\hat{\beta}_1) &= \sum_{i=1}^{N} [(x_i - \overline{x})/S_{xx}] \sum_{j=1}^{N} \left[(x_j - \overline{x})/S_{xx}\right] Cov(y_i, y_j) \\ &= \sum_{i=1}^{N} [(x_i - \overline{x})/S_{xx}]^2\, Var(y_i) = \sigma^2 \sum_{i=1}^{N} (x_i - \overline{x})^2 / S_{xx}^2 = \sigma^2 / S_{xx} \end{aligned}$$

employing the usual assumptions of constant variance and uncorrelated observations. The other route would be to follow the algebra in Example 4.2 above and find the $(2, 2)$ element of

$$\sigma^2 (\mathbf{X}^T \mathbf{X})^{-1} = \sigma^2 \begin{bmatrix} N & \sum x_i \\ \sum x_i & \sum x_i^2 \end{bmatrix}^{-1} = \frac{\sigma^2}{N S_{xx}} \begin{bmatrix} \sum x_i^2 & -\sum x_i \\ -\sum x_i & N \end{bmatrix}.$$

4.2 The Gauss–Markov Theorem

The goal throughout this chapter is to show that the least squares estimators derived back in Section 2.2 are the best estimators in some sense. The Gauss–Markov model that we have been talking about consists of just those sets of assumptions that are sufficient. Recall that a linear estimator takes the form $c + \mathbf{a}^T\mathbf{y}$ and will be unbiased for estimable $\lambda^T\mathbf{b}$ when

$$E(c + \mathbf{a}^T\mathbf{y}) = \lambda^T\mathbf{b}$$

for all $\mathbf{b}$, which leads to $c = 0$, and $\lambda = \mathbf{X}^T\mathbf{a}$. In our one-way ANOVA model, $y_{ij} = \mu + \alpha_i + e_{ij}$, then y_{11}, $(y_{11} + 2y_{12})/3$, and $\overline{y}_{1.}$ are all unbiased estimators of $\mu + \alpha_1$. In simple linear regression, $y_i = \beta_0 + \beta_1 x_i + e_i$, with $x_i = i$, say, then $y_2 - y_1$ or $y_3 - y_2$ are both unbiased estimators of β_1. Which is a better estimator and how shall we measure? When two estimators are both unbiased, then the estimator with smaller variances is better, since variance is a measure of variability around its mean. We will show that the best estimator of $\lambda^T\mathbf{b}$ is our least squares estimator $\lambda^T\hat{\mathbf{b}}$ where $\hat{\mathbf{b}}$ is a solution to the normal equations. Recall that if we construct all solutions to the normal equations,

$$\hat{\mathbf{b}}(\mathbf{z}) = (\mathbf{X}^T\mathbf{X})^g\mathbf{X}^T\mathbf{y} + (\mathbf{I} - (\mathbf{X}^T\mathbf{X})^g\mathbf{X}^T\mathbf{X})\mathbf{z},$$

then for estimable $\lambda^T\mathbf{b}$, $\lambda^T\hat{\mathbf{b}}(\mathbf{z})$ is constant for all values of $\mathbf{z}$—all solutions to the normal equations lead to the same least squares estimator.

Theorem 4.1 (Gauss–Markov Theorem) *Under the assumptions of the Gauss–Markov model,*

$$\mathbf{y} = \mathbf{X}\mathbf{b} + \mathbf{e}, \quad \textit{where} \quad E(\mathbf{e}) = \mathbf{0} \quad \textit{and} \quad Cov(\mathbf{e}) = \sigma^2\mathbf{I}_N,$$

if $\lambda^T\mathbf{b}$ is estimable, then $\lambda^T\hat{\mathbf{b}}$ is the best (minimum variance) linear unbiased estimator (BLUE) of $\lambda^T\mathbf{b}$, where $\hat{\mathbf{b}}$ solves the normal equations $\mathbf{X}^T\mathbf{X}\mathbf{b} = \mathbf{X}^T\mathbf{y}$.

Proof: Suppose $c + \mathbf{d}^T\mathbf{y}$ is another unbiased estimator of $\lambda^T\mathbf{b}$. Then $c = 0$ and $\mathbf{d}^T\mathbf{X} = \lambda^T$ since $E(c + \mathbf{d}^T\mathbf{y}) = c + \mathbf{d}^T\mathbf{X}\mathbf{b} = \lambda^T\mathbf{b}$ for all $\mathbf{b}$. Now,

$$\begin{aligned}
&Var\,(c + \mathbf{d}^T\mathbf{y}) \\
&\quad = Var\,(\mathbf{d}^T\mathbf{y}) = Var\,(\lambda^T\hat{\mathbf{b}} + \mathbf{d}^T\mathbf{y} - \lambda^T\hat{\mathbf{b}}) \\
&\quad = Var\,(\lambda^T\hat{\mathbf{b}}) + Var\,(\mathbf{d}^T\mathbf{y} - \lambda^T\hat{\mathbf{b}}) + 2Cov\,(\lambda^T\hat{\mathbf{b}}, \mathbf{d}^T\mathbf{y} - \lambda^T\hat{\mathbf{b}}) \\
&\quad = Var\,(\lambda^T\hat{\mathbf{b}}) + Var\,(\mathbf{d}^T\mathbf{y} - \lambda^T\hat{\mathbf{b}}) + 2\lambda^T(\mathbf{X}^T\mathbf{X})^g\mathbf{X}^T(\sigma^2\mathbf{I}_N)(\mathbf{d} - \mathbf{X}(\mathbf{X}^T\mathbf{X})^{gT}\lambda) \\
&\quad = Var\,(\lambda^T\hat{\mathbf{b}}) + Var\,(\mathbf{d}^T\mathbf{y} - \lambda^T\hat{\mathbf{b}}) + 2\sigma^2\lambda^T(\mathbf{X}^T\mathbf{X})^g(\mathbf{X}^T\mathbf{d} - \mathbf{X}^T\mathbf{X}(\mathbf{X}^T\mathbf{X})^{gT}\lambda) \\
&\quad = Var\,(\lambda^T\hat{\mathbf{b}}) + Var\,(\mathbf{d}^T\mathbf{y} - \lambda^T\hat{\mathbf{b}}) + 2\sigma^2\lambda^T(\mathbf{X}^T\mathbf{X})^g(\lambda - \lambda) \\
&\quad = Var\,(\lambda^T\hat{\mathbf{b}}) + Var\,(\mathbf{d}^T\mathbf{y} - \lambda^T\hat{\mathbf{b}}).
\end{aligned}$$

Since $\lambda^T\mathbf{b}$ is estimable, we have $\lambda \in \mathcal{C}(\mathbf{X}^T)$ and a projection $\mathbf{X}^T\mathbf{X}(\mathbf{X}^T\mathbf{X})^{gT}$ onto it; hence $\mathbf{X}^T\mathbf{X}(\mathbf{X}^T\mathbf{X})^{gT}\lambda = \lambda$. So $Var\,(\mathbf{d}^T\mathbf{y}) \geq Var(\lambda^T\hat{\mathbf{b}})$, with equality iff

$$Var\,(\mathbf{d}^T\mathbf{y} - \lambda^T\hat{\mathbf{b}}) = Var\,((\mathbf{d} - \mathbf{X}(\mathbf{X}^T\mathbf{X})^g\lambda)^T\mathbf{y}) = \sigma^2\|\mathbf{d} - \mathbf{X}(\mathbf{X}^T\mathbf{X})^{gT}\lambda\|^2 = 0.$$

So equality occurs (and an estimator with equal variance) iff $\mathbf{d} = \mathbf{X}(\mathbf{X}^T\mathbf{X})^{gT}\lambda$ or $\mathbf{d}^T\mathbf{y} = \lambda^T\hat{\mathbf{b}}$. In other words, the best linear unbiased estimator is unique. □

Note that in the proof above, the crucial step is showing $Cov\,(\lambda^T\hat{\mathbf{b}}, \mathbf{d}^T\mathbf{y} - \lambda^T\hat{\mathbf{b}}) = 0$, where $\mathbf{a}^T\mathbf{X} = \lambda^T$. Now what is $\mathbf{d}^T\mathbf{y} - \lambda^T\hat{\mathbf{b}}$ estimating? Notice that

$$E(\mathbf{d}^T\mathbf{y} - \lambda^T\hat{\mathbf{b}}) = \mathbf{d}^T\mathbf{X}\mathbf{b} - \lambda^T(\mathbf{X}^T\mathbf{X})^g\mathbf{X}^T\mathbf{X}\mathbf{b} = \lambda^T\mathbf{b} - \lambda^T\mathbf{b} = 0$$

so that $\mathbf{d}^T\mathbf{y} - \lambda^T\hat{\mathbf{b}}$ is an unbiased estimator of zero, and the best linear unbiased estimator $\lambda^T\hat{\mathbf{b}}$ is uncorrelated with it.

Result 4.1 *The BLUE $\lambda^T\hat{\mathbf{b}}$ of estimable $\lambda^T\mathbf{b}$ is uncorrelated with all unbiased estimators of zero.*

Proof: First, characterize unbiased estimators of zero as $c + \mathbf{a}^T\mathbf{y}$ such that $E(c + \mathbf{a}^T\mathbf{y}) = c + \mathbf{a}^T\mathbf{X}\mathbf{b} = 0$ for all $\mathbf{b}$, or $c = 0$ and $\mathbf{X}^T\mathbf{a} = \mathbf{0}$, or $\mathbf{a} \in \mathcal{N}(\mathbf{X}^T)$. Computing the covariance between $\lambda^T\hat{\mathbf{b}}$ and $\mathbf{a}^T\mathbf{y}$ we have

$$Cov\,(\lambda^T\hat{\mathbf{b}}, \mathbf{a}^T\mathbf{y}) = \lambda^T(\mathbf{X}^T\mathbf{X})^g\mathbf{X}^T Cov\,(\mathbf{y})\mathbf{a} = \sigma^2\lambda^T(\mathbf{X}^T\mathbf{X})^g\mathbf{X}^T\mathbf{a} = 0.$$

□

The Gauss–Markov Theorem can be extended to the vector case. Let the columns $\lambda^{(j)}$ of the matrix $\mathbf{\Lambda}$ be linear independent such that $\lambda^{(j)T}\mathbf{b}$ are linearly independent estimable functions; then

$$\begin{aligned} Cov\,(\mathbf{\Lambda}^T\hat{\mathbf{b}}) &= Cov\,(\mathbf{\Lambda}^T(\mathbf{X}^T\mathbf{X})^g\mathbf{X}^T\mathbf{y}) = \sigma^2\mathbf{\Lambda}^T(\mathbf{X}^T\mathbf{X})^g\mathbf{X}^T\mathbf{X}(\mathbf{X}^T\mathbf{X})^{gT}\mathbf{\Lambda} \\ &= \sigma^2\mathbf{\Lambda}^T(\mathbf{X}^T\mathbf{X})^g\mathbf{\Lambda}. \end{aligned}$$

If we have any set of unbiased estimators $\mathbf{\Lambda}^T\mathbf{b}$, say $\mathbf{C}^T\mathbf{y}$, that is, $E(\mathbf{C}^T\mathbf{y}) = \mathbf{\Lambda}^T\mathbf{b}$ for all $\mathbf{b}$, then the difference in their covariance matrices

$$Cov\left(\mathbf{C}^T\mathbf{y}\right) - Cov\left(\mathbf{\Lambda}^T\hat{\mathbf{b}}\right)$$

is nonnegative definite. If $\mathbf{X}$ has full-column rank and applying $\mathbf{\Lambda} = \mathbf{I}$, we have $Cov\,(\hat{\mathbf{b}}) = \sigma^2(\mathbf{X}^T\mathbf{X})^{-1}$. Moreover, in the full-column rank model, the Gauss–Markov Theorem says that if $\tilde{\mathbf{b}}$ is any other unbiased estimator of $\mathbf{b}$, then $Cov\,(\tilde{\mathbf{b}}) - Cov\,(\hat{\mathbf{b}})$ is nonnegative definite. See also Exercise 4.9.

4.3 Variance Estimation

Throughout this discussion, our focus has been on estimating linear functions of the coefficients $\mathbf{b}$ and no attention has been paid to the other unknown parameter, σ^2. So far, we have used $\mathbf{P_X}\mathbf{y}$ to estimate $\mathbf{b}$, as another version of the normal equations is $\mathbf{X}\mathbf{b} = \mathbf{P_X}\mathbf{y}$. As the reader might guess, we will use $(\mathbf{I} - \mathbf{P_X})\mathbf{y}$, or, more specifically, its sum of squares $SSE = \|(\mathbf{I} - \mathbf{P_X})\mathbf{y}\|^2$ to estimate σ^2. To construct an unbiased estimator for σ^2, we need the following lemma.

Lemma 4.1 *Let* $\mathbf{Z}$ *be a vector random variable with* $E(\mathbf{Z}) = \mu$ *and* $Cov(\mathbf{Z}) = \mathbf{\Sigma}$. *Then* $E(\mathbf{Z}^T\mathbf{A}\mathbf{Z}) = \mu^T\mathbf{A}\mu + tr(\mathbf{A}\mathbf{\Sigma})$.

Proof: Note that

$$\begin{aligned} E(Z_i - \mu_i)(Z_j - \mu_j) = \Sigma_{ij} &= E(Z_i Z_j) - \mu_i E(Z_j) - \mu_j E(Z_i) + \mu_i\mu_j \\ &= E(Z_i Z_j) - \mu_i\mu_j \end{aligned}$$

so that

$$\left[E(\mathbf{Z}\mathbf{Z}^T)\right]_{ij} = E(Z_i Z_j) = \Sigma_{ij} + \mu_i\mu_j,$$

or $E(\mathbf{Z}\mathbf{Z}^T) = \mu\mu^T + \mathbf{\Sigma}$. Now using the linearity of trace and expectation operations, we have

$$\begin{aligned} E(\mathbf{Z}^T\mathbf{A}\mathbf{Z}) &= E(tr(\mathbf{Z}^T\mathbf{A}\mathbf{Z})) = E(tr(\mathbf{A}\mathbf{Z}\mathbf{Z}^T)) = tr(\mathbf{A}E(\mathbf{Z}\mathbf{Z}^T)) = tr(\mathbf{A}(\mu\mu^T + \mathbf{\Sigma})) \\ &= tr(\mathbf{A}\mu\mu^T) + tr(\mathbf{A}\mathbf{\Sigma}) = \mu^T\mathbf{A}\mu + tr(\mathbf{A}\mathbf{\Sigma}). \end{aligned}$$ □

Result 4.2 *Consider the Gauss–Markov model given in (4.1) and (4.2). An unbiased estimator of* σ^2 *is* $\hat{\sigma}^2 = SSE/(N - r)$, *where* $SSE = \hat{\mathbf{e}}^T\hat{\mathbf{e}} = \mathbf{y}^T(\mathbf{I} - \mathbf{P_X})\mathbf{y}$ *and* $r = rank(\mathbf{X})$.

Proof: Since $E(\mathbf{y}) = \mathbf{Xb}$, $Cov(\mathbf{y}) = \sigma^2\mathbf{I}_N$, applying Lemma 4.1 above yields

$$E(\mathbf{y}^T(\mathbf{I} - \mathbf{P_X})\mathbf{y}) = (\mathbf{Xb})^T(\mathbf{I} - \mathbf{P_X})\mathbf{Xb} + tr((\mathbf{I} - \mathbf{P_X})(\sigma^2\mathbf{I}_N)) = \sigma^2(N - r).$$

Dividing by the degrees of freedom parameter $tr(\mathbf{I} - \mathbf{P_X}) = (N - r)$ associated with $SSE = \mathbf{y}^T(\mathbf{I} - \mathbf{P_X})\mathbf{y}$ gives the desired result. □

If we examine the regression sum of squares, $SSR = \hat{\mathbf{y}}^T\hat{\mathbf{y}} = \mathbf{y}^T\mathbf{P_X}\mathbf{y}$, we find $E(SSR) = E(\mathbf{y}^T\mathbf{P_X}\mathbf{y}) = \mathbf{b}^T\mathbf{X}^T\mathbf{P_X}\mathbf{Xb} + tr(\mathbf{P_X}\sigma^2\mathbf{I}_N) = \|\mathbf{Xb}\|^2 + r\sigma^2$.

Little else can be said about $\hat{\sigma}^2$ without further distributional assumptions, even when e_i are iid (independent, identically distributed) since its variance depends on the third and fourth moments of the underlying distribution. The algebra for computing $Var(\hat{\sigma}^2)$ in terms of the third and fourth moments is given in an addendum to this chapter.

4.4 Implications of Model Selection

In the application of linear models, sometimes the appropriate statistical model is not obvious. Even in designed experiments, researchers will debate whether to include certain interactions into the model. In the case of observational studies with continuous covariates the problem of model selection becomes quite difficult and leads to many diagnostics and proposed methodologies (see Miller [30], Rawlings et al. [39]). In the context of this book, however, we are concerned with the theoretical aspects that drive the practical model selection. At this point an important distinction must be made between the underlying true model for the data and the model that the researcher is employing. In practice, of course, we never know the true model.

The issue of model selection in multiple regression is often described as steering between two unpleasant situations: *overfitting*, that is, including explanatory variables that are not needed, and *underfitting*, not including an important explanatory variable. We will focus on the implications of these two situations.

4.4.1 Underfitting or Misspecification

In the case of underfitting, also referred to as *misspecification*, we can write the model for the truth as

$$\mathbf{y} = \mathbf{X}\mathbf{b} + \eta + \mathbf{e} \tag{4.3}$$

where $\mathbf{X}$ is the design matrix for the model that the researcher is using and η includes the omitted variables and their coefficients. Assume, as usual, $E(\mathbf{e}) = \mathbf{0}$ and $Cov(\mathbf{e}) = \sigma^2\mathbf{I}_N$, and notice that if $E(\mathbf{e})$ were not zero, η would capture its effect. Consider first the least squares estimators:

$$E(\lambda^T\hat{\mathbf{b}}) = \lambda^T(\mathbf{X}^T\mathbf{X})^g\mathbf{X}^T E(\mathbf{y}) = \lambda^T(\mathbf{X}^T\mathbf{X})^g\mathbf{X}^T(\mathbf{X}\mathbf{b} + \eta) = \lambda^T\mathbf{b} + \mathbf{a}^T\mathbf{P}_\mathbf{X}\eta$$

where $\lambda = \mathbf{X}^T\mathbf{a}$. When the model is misspecified, the least squares estimators are biased:

$$E(\lambda^T\hat{\mathbf{b}}) - \lambda^T\mathbf{b} = \mathbf{a}^T\mathbf{P}_\mathbf{X}\eta = \lambda^T(\mathbf{X}^T\mathbf{X})^g\mathbf{X}^T\eta$$

where the bias above depends on:

- the magnitude of the misspecified effect η;
- how much of that effect lies in $\mathcal{C}(\mathbf{X})$, $\mathbf{P}_\mathbf{X}\eta$; and
- how much it relates to the function at hand, $\lambda^T(\mathbf{X}^T\mathbf{X})^g\mathbf{X}^T\eta$.

If the missing signal η is orthogonal to $\mathcal{C}(\mathbf{X})$, then $\mathbf{P}_\mathbf{X}\eta = \mathbf{0}$ and the estimation of coefficients will be unaffected. If the misspecification is due to omitted explanatory variables that are uncorrelated with those included in $\mathbf{X}$, that is, $\mathbf{X}^T\eta = \mathbf{0}$, the estimates are not biased.

The estimation of the variance is also affected by misspecification:

$$\begin{aligned} E(\mathbf{y}^T(\mathbf{I} - \mathbf{P}_\mathbf{X})\mathbf{y}) &= (\mathbf{X}\mathbf{b} + \eta)^T(\mathbf{I} - \mathbf{P}_\mathbf{X})(\mathbf{X}\mathbf{b} + \eta) + tr((\mathbf{I} - \mathbf{P}_\mathbf{X})(\sigma^2\mathbf{I}_N)) \\ &= \eta^T(\mathbf{I} - \mathbf{P}_\mathbf{X})\eta + \sigma^2(N - r). \end{aligned} \tag{4.4}$$

Therefore, the bias in the variance estimate disappears only when the misspecification disappears, or if the misspecification lies only in $\mathcal{C}(\mathbf{X})$. Note that the bias in $\hat{\sigma}^2$ is zero if and only if $\eta^T(\mathbf{I} - \mathbf{P_X})\eta = 0$, or $\eta \in \mathcal{C}(\mathbf{X})$, or $\eta = \mathbf{Xd}$ for some $\mathbf{d}$. In that case, the model is then

$$\mathbf{y} = \mathbf{Xb} + \mathbf{Xd} + \mathbf{e} = \mathbf{X}(\mathbf{b} + \mathbf{d}) + \mathbf{e}.$$

One view of misspecification is that the signal that is not accounted for η is partitioned into two pieces:

$$\mathbf{y} = \mathbf{Xb} + \eta + \mathbf{e} = \mathbf{P_X y} + (\mathbf{I} - \mathbf{P_X})\mathbf{y} = (\mathbf{Xb} + \mathbf{P_X}\eta + \mathbf{P_X e}) + (\mathbf{I} - \mathbf{P_X})(\eta + \mathbf{e}).$$

One piece $\mathbf{P_X}\eta$ in $\mathcal{C}(\mathbf{X})$ affects the estimation of $\lambda^T\mathbf{b}$; the other part $(\mathbf{I} - \mathbf{P_X})\eta$ affects the estimation of the variance.

Example 4.4: Simple Misspecification
Suppose $y_i = \beta_0 + \beta_1 x_i + e_i$, but the covariate x_i is ignored and we just estimate the mean and variance. Here we have $\eta_i = \beta_1 x_i$ and $\mathbf{X} = \mathbf{1}$ so that $\hat{\beta}_0 = \overline{y}$ and $E(\overline{y}) = \beta_0 + \beta_1\overline{x}$. As for the variance, our usual variance estimate is $\hat{\sigma}^2 = \sum_i (y_i - \overline{y})^2/(N-1)$. Applying (4.4), the size of the bias, $E(\hat{\sigma}^2) - \sigma^2 = \beta_1^2 \sum_i (x_i - \overline{x})^2/(N-1)$ depends on the size of the departure of the mean response from the model, whose mean is constant.

Example 4.5: Electricity Problem
Consider the analysis of electricity consumption, using the following model as the true model for households:

$$bill_i = \beta_0 + \beta_1\, income_i + \beta_2\, persons_i + \beta_3\, area_i + e_i \tag{4.5}$$

where $bill_i$ = monthly electric bill for household i, $income_i$ = monthly disposable income, $persons_i$ = number in household, and $area_i$ = heating living area of home or apartment. In such analyses, often income may not be available, so consider the consequences in estimating the regression coefficients β_2 and β_3 when income is dropped from the model:

$$E(bill_i) = \beta_0 + \beta_2\, persons_i + \beta_3\, area_i. \tag{4.6}$$

In this situation, rows of $\mathbf{X}$ contain $[1\ persons_i\ area_i]$, $\mathbf{b}^T = [\beta_0\ \beta_2\ \beta_3]$, and $\eta_i = \beta_1\, income_i$. Another approach is to construct a regression model for the missing variable $income$, using the remaining variables as explanatory variables:

$$income_i = \gamma_0 + \gamma_1\, persons_i + \gamma_2\, area_i + f_i. \tag{4.7}$$

Combining the true model (4.5) and the expression above for the missing variable (4.7), the misspecified model we are fitting (4.6) really now becomes

$$bill_i = (\beta_0 + \beta_1\gamma_0) + (\beta_2 + \beta_1\gamma_1)\, persons_i + (\beta_3 + \beta_1\gamma_2)\, area_i + (e_i + \beta_1 f_i). \tag{4.8}$$

If income is not related to persons, then $\gamma_1 = 0$ and the estimate of β_2 remains unbiased; however, if income is related to area, then $E(\hat{\beta}_3) = \beta_3 + \beta_1\gamma_2$, with the size of the bias depending on the importance of the missing variable β_1 and the strength of the relationship with the variable of interest γ_2. Also note the expression for the error in the misspecified model (4.8), $e_i + \beta_1 f_i$, contains both the original error e_i and the effect of the misspecification.

4.4.2 Overfitting and Multicollinearity

The case of overfitting can be viewed in the following context, by partitioning the explanatory variables into two groups or blocks:

$$\mathbf{y} = \mathbf{X}_1\mathbf{b}_1 + \mathbf{X}_2\mathbf{b}_2 + \mathbf{e}$$

where the second group of explanatory variables is not needed, since $\mathbf{b}_2 = \mathbf{0}$. To make comparisons, let us simplify matters and assume the Gauss–Markov model and that both $\mathbf{X}_1$ and $\mathbf{X} = [\mathbf{X}_1\ \mathbf{X}_2]$ have full-column rank. Using the smallest model with a design matrix of $\mathbf{X}_1$ leads to the least squares estimator we will denote as $\tilde{\mathbf{b}}_1$, constructed as

$$\tilde{\mathbf{b}}_1 = (\mathbf{X}_1^T\mathbf{X}_1)^{-1}\mathbf{X}_1^T\mathbf{y}.$$

Using previous results, we can show that this estimator is unbiased, $E(\tilde{\mathbf{b}}_1) = \mathbf{b}_1$, and its covariance matrix is

$$Cov\,(\tilde{\mathbf{b}}_1) = \sigma^2(\mathbf{X}_1^T\mathbf{X}_1)^{-1}.$$

Including the second block of explanatory variables into the model that the researcher fits leads to the least squares estimators

$$\begin{bmatrix}\hat{\mathbf{b}}_1\\ \hat{\mathbf{b}}_2\end{bmatrix} = \begin{bmatrix}\mathbf{X}_1^T\mathbf{X}_1 & \mathbf{X}_1^T\mathbf{X}_2\\ \mathbf{X}_2^T\mathbf{X}_1 & \mathbf{X}_2^T\mathbf{X}_2\end{bmatrix}^{-1}\begin{bmatrix}\mathbf{X}_1^T\mathbf{y}\\ \mathbf{X}_2^T\mathbf{y}\end{bmatrix}.$$

Again, it is easy to show similar results for this estimator:

$$E\left(\begin{bmatrix}\hat{\mathbf{b}}_1\\ \hat{\mathbf{b}}_2\end{bmatrix}\right) = \begin{bmatrix}\mathbf{b}_1\\ \mathbf{0}\end{bmatrix} \quad \text{and} \quad Cov\left(\begin{bmatrix}\hat{\mathbf{b}}_1\\ \hat{\mathbf{b}}_2\end{bmatrix}\right) = \sigma^2\begin{bmatrix}\mathbf{X}_1^T\mathbf{X}_1 & \mathbf{X}_1^T\mathbf{X}_2\\ \mathbf{X}_2^T\mathbf{X}_1 & \mathbf{X}_2^T\mathbf{X}_2\end{bmatrix}^{-1}.$$

Using Exercise A.72 (partitioned inverse) we can show

$$Cov\,(\hat{\mathbf{b}}_1) = \sigma^2(\mathbf{X}_1^T\mathbf{X}_1)^{-1} + \sigma^2(\mathbf{X}_1^T\mathbf{X}_1)^{-1}\mathbf{X}_1^T\mathbf{X}_2\left[\mathbf{X}_2^T(\mathbf{I}-\mathbf{P}_{\mathbf{X}_1})\mathbf{X}_2\right]^{-1}\mathbf{X}_2^T\mathbf{X}_1(\mathbf{X}_1^T\mathbf{X}_1)^{-1}, \tag{4.9}$$

and so the penalty for including the second block of variables $\mathbf{X}_2$ is increased variance in the coefficient estimators.

Denoting $rank(\mathbf{X}_1) = r_1$ and $rank(\mathbf{X}) = r$, the variance estimators are both unbiased:

$$E\left[\mathbf{y}^T(\mathbf{I}-\mathbf{P}_{\mathbf{X}_1})\mathbf{y}\right]/(N-r_1) = \sigma^2$$
$$E\left[\mathbf{y}^T(\mathbf{I}-\mathbf{P}_{\mathbf{X}})\mathbf{y}\right]/(N-r) = \sigma^2$$

since there is no misspecification, and the only difference between the two estimators is in the degrees of freedom. Except for some applications where error degrees of freedom are few and precious, there is little lost in variance estimation by overfitting.

Example 4.6: No Intercept
Suppose we have the simple linear regression problem, with $y_i = \beta_0 + \beta_1 x_i + e_i$, but $\beta_0 = 0$. The variance of the least squares estimate of the slope, where we include the intercept, is $\sigma^2 / \sum_i (x_i - \overline{x})^2$. If we drop the intercept, the slope estimator is simply $\sum_i x_i y_i / \sum_i x_i^2$ and its variance is $\sigma^2 / \sum_i x_i^2$, which is smaller, since $\sum_i x_i^2 > \sum_i (x_i - \overline{x})^2$. See also Exercise 4.7.

Returning to the coefficient estimation, examination of the difference of the two covariance matrices shows the effect of including the second block of variables:

$$Cov(\hat{\mathbf{b}}_1) - Cov(\tilde{\mathbf{b}}_1) = \sigma^2 (\mathbf{X}_1^T\mathbf{X}_1)^{-1}\mathbf{X}_1^T\mathbf{X}_2\left[\mathbf{X}_2^T(\mathbf{I}-\mathbf{P}_{\mathbf{X}_1})\mathbf{X}_2\right]^{-1}\mathbf{X}_2^T\mathbf{X}_1(\mathbf{X}_1^T\mathbf{X}_1)^{-1}.$$

If the second block of explanatory variables is orthogonal to the first, that is, $\mathbf{X}_1^T\mathbf{X}_2 = \mathbf{0}$, then the estimators not only have the same variance, but they are the same estimators, as $\mathbf{X}^T\mathbf{X}$ becomes block diagonal. As $\mathbf{X}_2$ gets closer to $\mathcal{C}(\mathbf{X}_1)$, then $\mathbf{X}_2^T(\mathbf{I}-\mathbf{P}_{\mathbf{X}_1})\mathbf{X}_2$ gets smaller and, when inverted, causes the $Cov(\hat{\mathbf{b}}_1)$ to explode. This condition, known as *multicollinearity*, is the other feared consequence in model selection. One signal for severe multicollinearity is the appearance of unexpected signs of coefficient estimators due to their wild inaccuracy. Without multicollinearity, and with enough degrees of freedom to estimate the variance, there is little to be lost in overfitting. With multicollinearity, overfitting can be catastrophic.

One measure of the effect of multicollinearity is called the *variance inflation factor* (VIF). If the explanatory variables were mutually orthogonal, then following, say, Example 4.3, $Var(\hat{\mathbf{b}}_j) = \sigma^2/S_{xx}$ where $S_{xx} = \sum_i (X_{ij} - \overline{x}_{.j})^2$. In practice, when the explanatory variables are not orthogonal, the defining expression is

$$Var(\hat{\mathbf{b}}_j) = (VIF) \times \sigma^2/S_{xx}. \tag{4.10}$$

This relationship arises from the other form of the partitioned inverse result (Exercise A.72). Without loss of generality, consider $j = 1$ and partition the design matrix $\mathbf{X}$ as above, with just the first column in $\mathbf{X}_1$ and all of the other columns in $\mathbf{X}_2$. Then employing the other form of the partitioned inverse result, we have

$$Cov(\hat{\mathbf{b}}_1) = \sigma^2\left[\mathbf{X}_1^T(\mathbf{I}-\mathbf{P}_{\mathbf{X}_2})\mathbf{X}_1\right]^{-1}. \tag{4.11}$$

Putting (4.10) and (4.11) together, we find that

$$VIF = \frac{S_{xx}}{\mathbf{X}_1^T(\mathbf{I}-\mathbf{P}_{\mathbf{X}_2})\mathbf{X}_1} = \frac{1}{1-R_j^2} \tag{4.12}$$

where R_j^2 is what we would have for R^2 if we took column j for the response vector and employed the remaining explanatory variables. Clearly, when an explanatory variable can itself be closely approximated by a linear combination of other variables, R_j^2 will be close to one and VIF very large, indicating a serious multicollinearity problem.

Another view of multicollinearity arises from the examination of a scalar measure of the accuracy of the estimator of a vector quantity, mean squared error.

Definition 4.1 *The* mean squared error *of an estimator* $\hat{\theta}$ *of a parameter vector* θ *is* $E\{\|\hat{\theta} - \theta\|^2\}$.

For an unbiased estimator, such as $\hat{\mathbf{b}}$, we have the following simplifying algebra:

$$E\{\|\hat{\mathbf{b}} - \mathbf{b}\|^2\} = E\{(\hat{\mathbf{b}} - \mathbf{b})^T(\hat{\mathbf{b}} - \mathbf{b})\} = tr\, E\{(\hat{\mathbf{b}} - \mathbf{b})(\hat{\mathbf{b}} - \mathbf{b})^T\}$$
$$= trace\ Cov(\hat{\mathbf{b}}) = \sigma^2 tr(\mathbf{X}^T\mathbf{X})^{-1}.$$

The singular value decomposition (Exercise A.50) of the design matrix

$$\mathbf{X} = \mathbf{U}\begin{bmatrix}\mathbf{\Lambda} \\ \mathbf{0}\end{bmatrix}\mathbf{V}^T$$

(for simplicity we are assuming $\mathbf{X}$ is full-column rank) leads to the expression for the mean squared error as

$$E\{\|\hat{\mathbf{b}} - \mathbf{b}\|^2\} = \sigma^2 \sum_i \lambda_i^{-2}.$$

Now if multicollinearity is serious, the design matrix $\mathbf{X}$ is nearly singular, the smallest singular value λ_p is very small, and its effect on the mean squared error is to make it very large. Another measure of multicollinearity is the *condition number*, which is the ratio of the largest to smallest singular value λ_1/λ_p.

The quantity $\mathbf{X}_2^T(\mathbf{I} - \mathbf{P}_{X_1})\mathbf{X}_2$ is employed by some algorithms, for example, the sweep operator in SAS (see, e.g., Goodnight [14] or Monahan [32]), to determine rank of $\mathbf{X}$ or $\mathbf{X}^T\mathbf{X}$ in regression problems. In practice, this works quite well, although the finite precision of floating-point arithmetic limits its effectiveness. When most of the columns of the design matrix $\mathbf{X}$ are composed of 0, 1, or some small integer, the computed elements of $\mathbf{X}_2^T(\mathbf{I} - \mathbf{P}_{\mathbf{X}_1})\mathbf{X}_2$ can be zero or nearly so. However, in the case of continuous covariates, $\mathbf{X}_2^T(\mathbf{I} - \mathbf{P}_{\mathbf{X}_1})\mathbf{X}_2$ will rarely be zero even in the case of perfect collinearity due to the effects of rounding error. In practice, most computer software for regression determine rank by testing whether $\mathbf{X}_2^T(\mathbf{I} - \mathbf{P}_{\mathbf{X}_1})\mathbf{X}_2$, or something similar, is close to zero. While the mathematics presented here appears to proceed smoothly in the case of dependence in the columns of $\mathbf{X}$—by using the generalized inverse when the inverse does not exist—the reality is that the effect is dramatic: certain components or functions of $\mathbf{b}$ are no longer estimable. The practical matter is that the rank of the design matrix should be known in advance by the researcher. If the computer software determines a smaller rank than expected, then either an unexpected dependency or catastrophic multicollinearity problem exists. Finding a larger rank than expected indicates the inability of the software to detect dependence. While numerical analysts (e.g., Stewart [44] or Golub and van Loan [13]) consider the singular value decomposition to be the most reliable method for determining rank, here it is best employed to measure multicollinearity (see Exercises 4.17 and 4.18).

4.5 The Aitken Model and Generalized Least Squares

The Aitken model is a slight extension of the Gauss–Markov model in that only different moment assumptions are made on the errors. The Aitken model takes the form $\mathbf{y} = \mathbf{Xb} + \mathbf{e}$, where $E(\mathbf{e}) = \mathbf{0}$, but $Cov\,(e) = \sigma^2\mathbf{V}$ where the matrix $\mathbf{V}$ is a *known* positive definite matrix. In this way, it is similar to the Gauss–Markov model in that the covariance is known up to a scalar σ^2. If $\mathbf{V} = \mathbf{I}$, then we have the Gauss–Markov model. In practice, however, usually $\mathbf{V}$ is unknown, or has just a few parameters; this case will be addressed later. In an Aitken model, the least squares estimator $\lambda^T\hat{\mathbf{b}}$ of an estimable function $\lambda^T\mathbf{b}$ may no longer be the BLUE for $\lambda^T\mathbf{b}$. We will now construct a generalized least squares (GLS) estimator of $\lambda^T\mathbf{b}$ and show that it is the BLUE for $\lambda^T\mathbf{b}$.

The crucial step in GLS is the construction of a square root of the unscaled covariance matrix, that is, find $\mathbf{R}$ such that $\mathbf{RVR}^T = \mathbf{I}_N$. As discussed in appendix A, there are two approaches for constructing a square root of a positive definite matrix: Cholesky factorization and spectral decomposition. The Cholesky factorization writes $\mathbf{V}$ as the product of a lower triangular matrix $\mathbf{L}$ and its transpose, $\mathbf{V} = \mathbf{LL}^T$. Taking this route, use $\mathbf{R} = \mathbf{L}^{-1}$. The spectral decomposition uses the eigenvector-eigenvalue decomposition $\mathbf{V} = \mathbf{Q\Lambda Q}^T$ where $\mathbf{\Lambda}$ is the diagonal matrix of eigenvalues and $\mathbf{Q}$ is the (orthogonal) matrix of eigenvectors stacked as columns. This route suggests taking $\mathbf{R} = \mathbf{Q\Lambda}^{-1/2}\mathbf{Q}^T$, so that the square root matrix is symmetric in this case. Note that we will insist that $\mathbf{V}$ be positive definite, so that $\mathbf{L}$ or $\mathbf{\Lambda}$ is nonsingular. If $\mathbf{V}$ were singular, then there would be a linear combination of observations with zero variance, and this case would more properly be treated as a linear constraint.

Using the matrix $\mathbf{R}$, we can reformulate the Aitken model using a transformed response variable

$$\mathbf{z} = \mathbf{Ry} = \mathbf{RXb} + \mathbf{Re}, \text{ or } \mathbf{z} = \mathbf{Ub} + \mathbf{f}, \text{ where } E(\mathbf{f}) = \mathbf{0} \quad \text{and} \quad Cov(\mathbf{f}) = \sigma^2\mathbf{I}_N \tag{4.13}$$

with $\mathbf{U} = \mathbf{RX}$ which looks just like the Gauss–Markov model. Now we can tackle all of the same issues as before, and then transform back to the Aitken model.

4.5.1 Estimability

The linear function $\lambda^T\mathbf{b}$ is estimable if λ is in the column space of the transpose of the design matrix. Here this means $\lambda \in \mathcal{C}(\mathbf{U}^T) = \mathcal{C}(\mathbf{X}^T\mathbf{R}^T) = \mathcal{C}(\mathbf{X}^T)$ since $\mathbf{R}$ is nonsingular. From another viewpoint, estimability did not involve the second moment anyway, so that estimability should not be affected by the fact that $\mathbf{V}$ is not a constant diagonal matrix.

4.5.2 Linear Estimator

Note that any linear estimator $\mathbf{g} + \mathbf{h}^T\mathbf{z}$ that is linear in $\mathbf{z}$ is also a linear estimator $\mathbf{g} + \mathbf{h}^T\mathbf{R}\mathbf{y} = \mathbf{g} + \mathbf{a}^T\mathbf{y}$, and vice versa. In other words, the class of linear estimators in $\mathbf{z}$ is the same as the class of linear estimators in $\mathbf{y}$.

4.5.3 Generalized Least Squares Estimators

In the transformed model, the normal equations are

$$\mathbf{U}^T\mathbf{U}\mathbf{b} = \mathbf{U}^T\mathbf{z}, \tag{4.14}$$

and so the least squares estimator from (4.13) solves (4.14) above. However, these normal equations can be easily rewritten as

$$\begin{aligned} (\mathbf{RX})^T(\mathbf{RX})\mathbf{b} &= (\mathbf{RX})^T(\mathbf{Ry}) \\ \text{or} \quad \mathbf{X}^T\mathbf{V}^{-1}\mathbf{X}\mathbf{b} &= \mathbf{X}^T\mathbf{V}^{-1}\mathbf{y} \end{aligned} \tag{4.15}$$

which are known as the *Aitken equations*, and the solution to (4.15) will be denoted as $\hat{\mathbf{b}}_{GLS}$, a generalized least squares estimator of $\mathbf{b}$. When needed for clarity, the solution to the usual normal equations $\mathbf{X}^T\mathbf{X}\mathbf{b} = \mathbf{X}^T\mathbf{y}$ will be denoted by $\hat{\mathbf{b}}_{OLS}$, for *ordinary least squares*. From Section 4.1, we should expect that $\lambda^T\hat{\mathbf{b}}_{GLS}$ is BLUE for $\lambda^T\mathbf{b}$, as stated in the following theorem. Its proof is left as exercises; one uses some later results.

Theorem 4.2 (Aitken's Theorem) *Consider the Aitken model given by* $\mathbf{y} = \mathbf{X}\mathbf{b} + \mathbf{e}$, *where* $E(\mathbf{e}) = \mathbf{0}$, *and* $Cov(\mathbf{e}) = \sigma^2\mathbf{V}$, *where* $\mathbf{V}$ *is a known positive definite matrix. If* $\lambda^T\mathbf{b}$ *is estimable, then* $\lambda^T\hat{\mathbf{b}}_{GLS}$ *is the BLUE for* $\lambda^T\mathbf{b}$.

Proof: See Exercises 4.22 and 4.23.

From Chapter 2, we know that $\hat{\mathbf{b}}_{GLS}$ minimizes a sum of squares, in particular,

$$\|\mathbf{z} - \mathbf{U}\mathbf{b}\|^2 = \|\mathbf{R}(\mathbf{y} - \mathbf{X}\mathbf{b})\|^2 = (\mathbf{y} - \mathbf{X}\mathbf{b})\mathbf{V}^{-1}(\mathbf{y} - \mathbf{X}\mathbf{b}),$$

so that this sum of squares is often called *weighted least squares* or *generalized least squares*. In the simplest case, say, simple linear regression and $\mathbf{V}$ diagonal, we have

$$(\mathbf{y} - \mathbf{X}\mathbf{b})\mathbf{V}^{-1}(\mathbf{y} - \mathbf{X}\mathbf{b}) = \sum_i \frac{1}{V_{ii}}(y_i - \beta_0 - \beta_1 x_i)^2,$$

and the name should be apparent. □

4.5.4 Estimation of σ^2

Since the transformed model follows the Gauss–Markov assumptions, the estimator constructed in Section 4.3 is the natural unbiased estimator for σ^2:

$$\begin{aligned} \hat{\sigma}^2_{GLS} &= (\mathbf{z} - \mathbf{U}\hat{\mathbf{b}}_{GLS})^T(\mathbf{z} - \mathbf{U}\hat{\mathbf{b}}_{GLS})/(N - r) \\ &= (\mathbf{y} - \mathbf{X}\hat{\mathbf{b}}_{GLS})\mathbf{V}^{-1}(\mathbf{y} - \mathbf{X}\hat{\mathbf{b}}_{GLS})/(N - r). \end{aligned}$$

Example 4.7: Simplest Linear Model
In the case of the one sample problem, we have $E(\mathbf{y}) = \mu\mathbf{1}$, $Cov(\mathbf{y}) = \sigma^2\mathbf{V}$, and so the Aitken equations take the simple form $\mathbf{1}^T\mathbf{V}^{-1}\mathbf{1}\mu = \mathbf{1}^T\mathbf{V}^{-1}\mathbf{y}$. See Exercise 4.16.

Example 4.8: Heteroskedasticity
Consider the simple linear regression model through the origin with heteroskedastic (different variances) errors, with the variances proportional to the squares of the explanatory variables: $y_i = \beta x_i + e_i$ where $E(e_i) = 0$, $Var(e_i) = \sigma^2 x_i^2$, e_i uncorrelated and $x_i \neq 0$. Notice that $Var(e_i/x_i) = \sigma^2$ so that the obvious step is to transform by dividing by x_i:

$$z_i = y_i/x_i = \beta + e_i/x_i.$$

The BLUE of β, then, is $\hat{\beta}_{GLS} = \bar{z} = \frac{1}{N}\sum_i(y_i/x_i)$, and $Var(\hat{\beta}_{GLS}) = \sigma^2/N$. For comparison, $\hat{\beta}_{OLS} = \sum_i x_i y_i / \sum_i x_i^2$, and see Exercise 4.4.

Example 4.9: Autoregressive Errors
Suppose we have the usual multiple regression model $y_i = \mathbf{x}_i^T\mathbf{b} + e_i$ where the errors have the usual zero mean $E(e_i) = 0$, but the covariance structure is induced by the model $e_i = \rho e_{i-1} + a_i$ where the a_i's are uncorrelated with zero mean and variance σ_a^2. Then it can be shown that $Var(e_i) = \sigma^2/(1-\rho^2)$ and the covariance matrix of the original errors e_i is given by

$$Cov(\mathbf{y}) = Cov(\mathbf{e}) = \sigma^2\mathbf{V} = \frac{\sigma_a^2}{1-\rho^2}\begin{bmatrix} 1 & \rho & \rho^2 & \dots & \rho^{N-1} \\ \rho & 1 & \rho & \dots & \rho^{N-2} \\ \rho^2 & \rho & 1 & \dots & \\ \dots & & & \dots & \rho \\ \rho^{N-1} & \dots & & \rho & 1 \end{bmatrix}$$

so that $V_{ij} = \rho^{|i-j|}/(1-\rho^2)$. This error structure is known as a first-order autoregressive model. The following transformation, known as the Cochrane–Orcutt [8] transformation, restores the usual Gauss–Markov assumptions:

$$\begin{aligned} z_1 &= \sqrt{1-\rho^2}y_1 = \sqrt{1-\rho^2}\mathbf{x}_1^T\mathbf{b} + \sqrt{1-\rho^2}e_1 \\ z_i &= y_i - \rho y_{i-1} = \mathbf{x}_i^T\mathbf{b} - \rho\mathbf{x}_{i-1}^T\mathbf{b} + e_i - \rho e_{i-1} \\ &= (\mathbf{x}_i - \rho\mathbf{x}_{i-1})^T\mathbf{b} + a_i = \mathbf{u}_i^T\mathbf{b} + a_i \text{ for } i = 2, \dots, N. \end{aligned}$$

The single parameter ρ of this model is not usually known, and the usual approach is to begin with ordinary least squares to estimate $\mathbf{b}$, estimate ρ from the residuals, do GLS with estimated ρ to reestimate $\mathbf{b}$, reestimate ρ, and iterate until convergence. This procedure is known as *estimated generalized least squares* (EGLS) since ρ is estimated. See also Exercise 4.5.

Finding the BLUE estimator under the Aitken model really is not difficult, as you have seen; all that was necessary was to transform $\mathbf{y}$ and $\mathbf{X}$ to $\mathbf{z}$ and $\mathbf{U}$ so that the

Gauss–Markov assumptions held. A more interesting pursuit, however, is the set of conditions that make the usual $\hat{b}_{OLS}$ to be BLUE under the Aitken model assumptions.

Result 4.3 (Generalization of Result 4.1) *The estimator* $\mathbf{t}^T\mathbf{y}$ *is the BLUE for* $E(\mathbf{t}^T\mathbf{y})$ *iff* $\mathbf{t}^T\mathbf{y}$ *is uncorrelated with all unbiased estimators of zero.*

Proof: (If) Let $\mathbf{a}^T\mathbf{y}$ be another unbiased estimator of $E(\mathbf{t}^T\mathbf{y})$, that is, $E(\mathbf{a}^T\mathbf{y}) = E(\mathbf{t}^T\mathbf{y})$. Then

$$\begin{aligned} Var(\mathbf{a}^T\mathbf{y}) &= Var(\mathbf{t}^T\mathbf{y} + \mathbf{a}^T\mathbf{y} - \mathbf{t}^T\mathbf{y}) \\ &= Var(\mathbf{t}^T\mathbf{y}) + Var(\mathbf{a}^T\mathbf{y} - \mathbf{t}^T\mathbf{y}) + 2\,Cov(\mathbf{t}^T\mathbf{y}, \mathbf{a}^T\mathbf{y} - \mathbf{t}^T\mathbf{y}) \end{aligned}$$

Since $\mathbf{a}^T\mathbf{y} - \mathbf{t}^T\mathbf{y}$ is an unbiased estimator of zero, the covariance term drops out of the equation above, leading to

$$Var(\mathbf{a}^T\mathbf{y}) = Var(\mathbf{t}^T\mathbf{y}) + Var(\mathbf{a}^T\mathbf{y} - \mathbf{t}^T\mathbf{y}) \geq Var(\mathbf{t}^T\mathbf{y}).$$

(Only if) Suppose there is an unbiased estimator of zero $\mathbf{h}^T\mathbf{y}$, so that $E(\mathbf{h}^T\mathbf{y}) = 0$ where $\mathbf{h} \neq \mathbf{0}$, and let $Cov(\mathbf{t}^T\mathbf{y}, \mathbf{h}^T\mathbf{y}) = c$ and $Var(\mathbf{h}^T\mathbf{y}) = d$. Then consider the estimator of $E(\mathbf{t}^T\mathbf{y})$ given by

$$\mathbf{a}^T\mathbf{y} = \mathbf{t}^T\mathbf{y} - (c/d)\mathbf{h}^T\mathbf{y}.$$

This estimator is also unbiased for $E(\mathbf{t}^T\mathbf{y})$. Its variance is

$$\begin{aligned} Var(\mathbf{a}^T\mathbf{y}) &= Var(\mathbf{t}^T\mathbf{y}) + (c/d)^2 Var(\mathbf{h}^T\mathbf{y}) - 2(c/d)Cov(\mathbf{t}^T\mathbf{y}, \mathbf{h}^T\mathbf{y}) \\ &= Var(\mathbf{t}^T\mathbf{y}) - c^2/d \leq Var(\mathbf{t}^T\mathbf{y}). \end{aligned}$$

Thus if $\mathbf{t}^T\mathbf{y}$ is the BLUE, then $c = 0$; otherwise the estimator $\mathbf{a}^T\mathbf{y}$ constructed above will also be unbiased and have smaller variance. □

Corollary 4.1 *Under the Aitken model, the estimator* $\mathbf{t}^T\mathbf{y}$ *is the BLUE for* $E(\mathbf{t}^T\mathbf{y})$ *iff* $\mathbf{Vt} \in \mathcal{C}(\mathbf{X})$.

Proof: From result 4.3, $\mathbf{t}^T\mathbf{y}$ is the BLUE for $E(\mathbf{t}^T\mathbf{y})$ iff $Cov(\mathbf{t}^T\mathbf{y}, \mathbf{h}^T\mathbf{y}) = 0$ for all $\mathbf{h}$ such that $E(\mathbf{h}^T\mathbf{y}) = 0$. Note that if $\mathbf{h}^T\mathbf{y}$ is an unbiased estimator of zero, then we have $E(\mathbf{h}^T\mathbf{y}) = \mathbf{h}^T\mathbf{X}\mathbf{b} = 0$ for all $\mathbf{b}$. This means $\mathbf{h}^T\mathbf{X} = \mathbf{0}$ or $\mathbf{h} \in \mathcal{N}(\mathbf{X}^T)$. Now $Cov(\mathbf{t}^T\mathbf{y}, \mathbf{h}^T\mathbf{y}) = \sigma^2\mathbf{t}^T\mathbf{Vh}$, and this is zero iff $\mathbf{h}$ is orthogonal to $\mathbf{Vt}$, or, since $\mathbf{h} \in \mathcal{N}(\mathbf{X}^T)$, iff $\mathbf{Vt} \in \mathcal{C}(\mathbf{X})$. □

Result 4.4 *Under the Aitken model, all OLS estimators are BLUE (that is, each* $\lambda^T\hat{\mathbf{b}}_{OLS}$ *is the BLUE for the corresponding estimable* $\lambda^T\mathbf{b}$*) iff there exists a matrix* $\mathbf{Q}$ *such that* $\mathbf{VX} = \mathbf{XQ}$.

Proof: (If $\mathbf{VX} = \mathbf{XQ}$) First write $\lambda^T\hat{\mathbf{b}}_{OLS} = \lambda^T(\mathbf{X}^T\mathbf{X})^g\mathbf{X}^T\mathbf{y} = \mathbf{t}^T\mathbf{y}$ for $\mathbf{t} = \mathbf{X}(\mathbf{X}^T\mathbf{X})^{gT}\lambda$ so that $\lambda^T\hat{\mathbf{b}}_{OLS}$ is the BLUE for $\lambda^T\mathbf{b}$ iff $\mathbf{Vt} \in \mathcal{C}(\mathbf{X})$. Now if $\mathbf{VX} = \mathbf{XQ}$, then

$$\mathbf{Vt} = \mathbf{VX}(\mathbf{X}^T\mathbf{X})^g\lambda = \mathbf{XQ}(\mathbf{X}^T\mathbf{X})^g\lambda \in \mathcal{C}(\mathbf{X})$$

and employ Corollary 4.1.
(Only if) Now if $\lambda^T\hat{\mathbf{b}}_{OLS}$ is the BLUE for $\lambda^T\mathbf{b}$, take $\lambda^{(j)}$ as column j of $\mathbf{X}^T\mathbf{X}$, so that $\mathbf{t}^{(j)} = \mathbf{X}(\mathbf{X}^T\mathbf{X})^{gT}\lambda^{(j)}$; then from Corollary 4.1, $\mathbf{Vt}^{(j)} \in \mathcal{C}(\mathbf{X})$, or there exists a vector $\mathbf{q}^{(j)}$ such that $\mathbf{Vt}^{(j)} = \mathbf{Xq}^{(j)}$. Stacking these columns side by side to form matrices $\mathbf{T}$ and $\mathbf{Q}$, we have

$$\mathbf{VT} = \mathbf{VX}(\mathbf{X}^T\mathbf{X})^g\mathbf{X}^T\mathbf{X} = \mathbf{VX} = \mathbf{XQ}.$$

□

Example 4.10: Equicorrelation
Consider the regression problem $\mathbf{y} = \mathbf{Xb} + \mathbf{e}$ with an equicorrelated covariance structure, that is, $Cov\,(\mathbf{e}) = \sigma^2\mathbf{V} = \sigma^2\mathbf{I}_N + \tau^2\mathbf{1}_N\mathbf{1}_N^T$, and with an intercept and the following partitioning:

$$\mathbf{Xb} = [\mathbf{1} \quad \mathbf{X}^*]\begin{bmatrix} b_1 \\ \mathbf{b}_2 \end{bmatrix}$$

In this case, is $\lambda^T\hat{\mathbf{b}}_{OLS}$ the BLUE for estimable $\lambda^T\mathbf{b}$? It will be if we can find $\mathbf{Q}$ such that $\mathbf{VX} = \mathbf{XQ}$.

$$\mathbf{VX} = \left(\sigma^2\mathbf{I}_N + \tau^2\mathbf{1}_N\mathbf{1}_N^T\right), [\mathbf{1} \quad \mathbf{X}^*] = \left[\sigma^2\mathbf{1}_N + N\tau^2\mathbf{1}_N \quad \sigma^2\mathbf{X}^* + \tau^2\mathbf{1}_N\mathbf{1}_N^T\mathbf{X}^*\right]$$

$$\mathbf{XQ} = [\mathbf{1} \quad \mathbf{X}^*]\begin{bmatrix} Q_{11} & \mathbf{Q}_{12} \\ \mathbf{Q}_{21} & \mathbf{Q}_{22} \end{bmatrix} = \left[\mathbf{1}Q_{11} + \mathbf{X}^*\mathbf{Q}_{21} \quad \mathbf{1}\mathbf{Q}_{12} + \mathbf{X}^*\mathbf{Q}_{22}\right]$$

Matching the first entries suggests choosing $Q_{11} = \sigma^2 + N\tau^2$ and $\mathbf{Q}_{21} = \mathbf{0}$, and matching the second yields $\mathbf{Q}_{12} = \tau^2\mathbf{1}^T\mathbf{X}^*$ and $\mathbf{Q}_{22} = \sigma^2\mathbf{I}_{p-1}$. See also Exercise 4.8.

Example 4.11: Multivariate and Seemingly Unrelated Regressions
Suppose we have m individuals, each with n responses following regression models:

$$\mathbf{y}^{(i)} = \mathbf{X}^{(i)}\mathbf{b}^{(i)} + \mathbf{e}^{(i)}, \quad i = 1, \ldots m,$$

where $\mathbf{y}^{(i)}$ and $\mathbf{e}^{(i)}$ are $n \times 1$, $\mathbf{X}^{(i)}$ is $n \times p$, and $\mathbf{b}^{(i)}$ is $p \times 1$. The covariances in the errors $\mathbf{e}^{(i)}$ tie these regressions together:

$$Cov\,(\mathbf{e}^{(i)}, \mathbf{e}^{(j)}) = \sigma_{ij}\mathbf{I}_n.$$

For example, the individuals may be companies, and the responses are quarterly sales that would be contemporaneously correlated. We can write this as one large linear model by combining these pieces:

$$\mathbf{y} = \begin{bmatrix} \mathbf{y}^{(1)} \\ \mathbf{y}^{(2)} \\ \cdots \\ \mathbf{y}^{(m)} \end{bmatrix}, \quad \mathbf{X} = \begin{bmatrix} \mathbf{X}^{(1)} & \mathbf{0} & \cdots & \mathbf{0} \\ \mathbf{0} & \mathbf{X}^{(2)} & & \mathbf{0} \\ \cdots & & \cdots & \\ \mathbf{0} & \mathbf{0} & & \mathbf{X}^{(m)} \end{bmatrix}, \quad \mathbf{b} = \begin{bmatrix} \mathbf{b}^{(1)} \\ \mathbf{b}^{(2)} \\ \cdots \\ \mathbf{b}^{(m)} \end{bmatrix},$$

$$\sigma^2 \mathbf{V} = \begin{bmatrix} \sigma_{11}\mathbf{I}_n & \sigma_{12}\mathbf{I}_n & \dots & \sigma_{1m}\mathbf{I}_n \\ \sigma_{21}\mathbf{I}_n & \sigma_{22}\mathbf{I}_n & & \sigma_{2m}\mathbf{I}_n \\ & \dots & & \dots \\ \sigma_{m1}\mathbf{I}_n & \sigma_{m2}\mathbf{I}_n & \dots & \sigma_{mm}\mathbf{I}_n \end{bmatrix}.$$

What are the best estimators in this case? Or, more specifically, when are the least squares estimators BLUE? In general, the least squares estimators are not always the best. However, some specific cases are interesting.

If $\sigma_{ij} = 0$ for $i \neq j$, then the problem completely decouples into m individual least squares problems and, not surprisingly, the least squares estimators are BLUE. The other interesting case has $\mathbf{X}^{(i)} = \mathbf{X}^{(1)}$, that is, the design matrices are the same for each company. In this case, we can show $\mathbf{VX} = \mathbf{XV}$, so that taking $\mathbf{Q} = \mathbf{V}$, the least squares estimators are BLUE. This latter situation is known as *multivariate regression*, which will be discussed in detail (and written differently) in Chapter 9.

4.6 Application: Aggregation Bias

In many situations, the theory from either economics or physical sciences may dictate that the response follow a linear model at the level of the individual, but that data may only be available aggregated over time or individuals. For example, ice cream sales may depend linearly on temperature, but data may not be available each day at each location, but only weekly averages or averages for a county or state. Let us begin with a simple linear regression model,

$$y_{ij} = \beta_0 + \beta_1 x_{ij} + e_{ij}, \quad \text{for} \quad i = 1, \dots, a; \; j = 1, \dots, n_i \tag{4.16}$$

where the errors e_{ij} follow the usual Gauss–Markov assumptions with variance σ^2. But now suppose that the only data available were averages:

$$\overline{y}_{i.} = n_i^{-1} \sum_{j=1}^{n_i} y_{ij}, \quad \text{and} \quad \overline{x}_{i.} = n_i^{-1} \sum_{j=1}^{n_i} x_{ij}.$$

Then, looking at the available responses, we find

$$E(\overline{y}_{i.}) = \beta_0 + \beta_1 \overline{x}_{i.} \quad \text{and} \quad Var(\overline{y}_{i.}) = \sigma^2 / n_i. \tag{4.17}$$

As a result, using only $(\overline{x}_{i.}, \overline{y}_{i.}, i = 1, \dots, a)$ we can find unbiased estimators for the unknown coefficients β_0 and β_1, as well as σ^2 using the Aitken model with $\mathbf{V} = diag\,(n_i^{-1})$. There may be some loss of information (Exercise 4.28), but misspecification is not a problem.

Let's extend this situation one step to a quadratic response at the individual level:

$$y_{ij} = \beta_0 + \beta_1 x_{ij} + \beta_2 x_{ij}^2 + e_{ij}, \quad \text{for} \quad i = 1, \dots, a; \; j = 1, \dots, n_i, \tag{4.18}$$

again with the usual Gauss–Markov assumptions. If only $(\overline{x}_{i.}, \overline{y}_{i.}, i = 1, \ldots, a)$ are observed, we have the model for mean reponses as

$$\begin{aligned} E(\overline{y}_{i.}) &= \beta_0 + \beta_1 \overline{x}_{i.} + \beta_2 n_i^{-1} \sum x_{ij}^2 \\ &= \beta_0 + \beta_1 \overline{x}_{i.} + \beta_2 \overline{x}_{ij}^2 + \beta_2 n_i^{-1} \sum (x_{ij} - \overline{x}_{i.})^2 \end{aligned} \tag{4.19}$$

by reworking the familiar formula $\sum_j (x_{ij} - \overline{x}_{i.})^2 = \sum_j x_{ij}^2 - n_i \overline{x}_{i.}^2$. While the heteroskedasticity can be easily accommodated in the same way as before, the model is misspecified, with $\eta_i = \beta_2 n_i^{-1} \sum (x_{ij} - \overline{x}_{i.})^2$. Also see Exercise 4.29.

In a different light, suppose the aggregation is over a heterogeneous population with different regression lines, say

$$y_{ij} = \beta_{0j} + \beta_{1j} x_{ij} + e_{ij}, \quad \text{for } i = 1, \ldots, a; \; j = 1, \ldots, n_i. \tag{4.20}$$

For simplicity, assume that we have the same individuals $j = 1, \ldots, n$ at each time point i. Again, the expectation of the mean response takes an unwelcome form:

$$\begin{aligned} E(\overline{y}_{i.}) &= \overline{\beta}_{0.} + n^{-1} \sum_j \beta_{1j} x_{ij} \\ &= \overline{\beta}_{0.} + n^{-1} \sum_j (\beta_{1j} - \overline{\beta}_{1.} + \overline{\beta}_{1.}) x_{ij} \\ &= \overline{\beta}_{0.} + \overline{\beta}_{1.} \overline{x}_{i.} + n^{-1} \sum_j (\beta_{1j} - \overline{\beta}_{1.}) x_{ij}, \end{aligned} \tag{4.21}$$

and the last term expresses the misspecification.

4.7 Best Estimation in a Constrained Parameter Space

In Section 3.9, we discussed estimability and least squares for a constrained parameter space $\mathcal{T} = \{\mathbf{b} : \mathbf{P}^T \mathbf{b} = \delta\}$. Now with the Gauss–Markov assumptions, we would like to show that using the solution to the restricted normal Equations (3.10) leads to best estimators, just as the solution to the normal equations leads to BLUE. Some of the following steps appear cryptic in their intent, so the reader is advised to be patient.

Lemma 4.2 *If $\lambda^T \mathbf{b}$ is estimable under the constrained model, then the following Equations (4.22) have a solution:*

$$\begin{bmatrix} \mathbf{X}^T\mathbf{X} & \mathbf{P} \\ \mathbf{P}^T & \mathbf{0} \end{bmatrix} \begin{bmatrix} \mathbf{v}_1 \\ \mathbf{v}_2 \end{bmatrix} = \begin{bmatrix} \lambda \\ \mathbf{0} \end{bmatrix} \tag{4.22}$$

Proof: From Result 3.7, we can write $\lambda = \mathbf{X}^T \mathbf{a} + \mathbf{P}\mathbf{d}$. From Result 3.8, we know that the restricted normal Equations (3.10) are consistent, so take $\mathbf{y} = \mathbf{a}$ and $\delta = \mathbf{0}$, so that the following equations

$$\begin{bmatrix} \mathbf{X}^T\mathbf{X} & \mathbf{P} \\ \mathbf{P}^T & \mathbf{0} \end{bmatrix} \begin{bmatrix} \mathbf{w}_1 \\ \mathbf{w}_2 \end{bmatrix} = \begin{bmatrix} \mathbf{X}^T\mathbf{a} \\ \mathbf{0} \end{bmatrix}$$

have a solution, call it $\mathbf{w}^*$. Then the following vector,

$$\mathbf{w}^* + \begin{bmatrix} \mathbf{0} \\ \mathbf{d} \end{bmatrix} = \begin{bmatrix} \mathbf{w}_1^* \\ \mathbf{w}_2^* + \mathbf{d} \end{bmatrix},$$

is a solution to Equation (4.22). □

Lemma 4.3 *If $\hat{\mathbf{b}}_H$ is the first component of a solution to the RNEs (3.10), and if $\lambda^T\mathbf{b}$ is estimable in the constrained model, then $\lambda^T\hat{\mathbf{b}}_H$ is an unbiased estimator of $\lambda^T\mathbf{b}$.*

Proof: The first step is to show that $\lambda^T\hat{\mathbf{b}}_H$ is a linear estimator, taking the form $c + \mathbf{a}^T\mathbf{y}$. Substitute the solution $\mathbf{v}$ of the Equations above (4.22) in the expression

$$\begin{aligned} \lambda^T\hat{\mathbf{b}}_H &= [\lambda^T \quad \mathbf{0}] \begin{bmatrix} \hat{\mathbf{b}}_H \\ \hat{\theta}_H \end{bmatrix} = [\mathbf{v}_1 \quad \mathbf{v}_2]^T \begin{bmatrix} \mathbf{X}^T\mathbf{X} & \mathbf{P} \\ \mathbf{P}^T & \mathbf{0} \end{bmatrix} \begin{bmatrix} \hat{\mathbf{b}}_H \\ \hat{\theta}_H \end{bmatrix} \\ &= [\mathbf{v}_1 \quad \mathbf{v}_2]^T \begin{bmatrix} \mathbf{X}^T\mathbf{y} \\ \delta \end{bmatrix} \end{aligned}$$

so that $\lambda^T\hat{\mathbf{b}}_H = \mathbf{v}_1^T\mathbf{X}^T\mathbf{y} + \mathbf{v}_2^T\delta$, which is linear in $\mathbf{y}$. As for unbiasedness, we have

$$E(\lambda^T\hat{\mathbf{b}}_H) = E\left(\mathbf{v}_1^T\mathbf{X}^T\mathbf{y} + \mathbf{v}_2^T\delta\right) = \mathbf{v}_1^T\mathbf{X}^T\mathbf{X}\mathbf{b} + \mathbf{v}_2^T\delta = (\lambda - \mathbf{P}\mathbf{v}_2)^T\mathbf{b} + \mathbf{v}_2^T\delta = \lambda^T\mathbf{b}$$

for all $\mathbf{b}$ in the constrained parameter space $\mathcal{T}$. The substitution above, $\mathbf{X}^T\mathbf{X}\mathbf{v}_1 = \lambda - \mathbf{P}\mathbf{v}_2$, follows from the first part of Equation (4.22). □

Result 4.5 *Under the Gauss–Markov assumptions, if $\hat{\mathbf{b}}_H$ is the first component of a solution to the RNEs (3.10), and if $\lambda^T\mathbf{b}$ is estimable in the constrained model, then $\lambda^T\hat{\mathbf{b}}_H$ is the BLUE of $\lambda^T\mathbf{b}$ in the constrained model.*

Proof: Following the same strategy for proving the Gauss–Markov Theorem, we wish to show that any other unbiased estimator has larger variance. Let $\mathbf{a}^T\mathbf{y} + \mathbf{d}^T\delta$ be another unbiased estimator of $\lambda^T\mathbf{b}$ and follow the same steps:

$$\begin{aligned} Var\,(\mathbf{a}^T\mathbf{y} + \mathbf{d}^T\delta) = Var\left(\lambda^T\hat{\mathbf{b}}_H\right) &+ Var\left(\lambda^T\hat{\mathbf{b}}_H - \mathbf{a}^T\mathbf{y} - \mathbf{d}^T\delta\right) \\ &+ 2Cov\left(\lambda^T\hat{\mathbf{b}}_H, \lambda^T\hat{\mathbf{b}}_H - \mathbf{a}^T\mathbf{y} - \mathbf{d}^T\delta\right). \end{aligned}$$

Again, using the solution to Equation (4.22) and Lemma 4.2, we have

$$
\begin{aligned}
Cov(\lambda^T\hat{\mathbf{b}}_H, \lambda^T\hat{\mathbf{b}}_H - \mathbf{a}^T\mathbf{y} - \mathbf{d}^T\delta) \\
&= Cov\left(\mathbf{v}_1^T\mathbf{X}^T\mathbf{y} + \mathbf{v}_2^T\delta, \mathbf{v}_1^T\mathbf{X}^T\mathbf{y} + \mathbf{v}_2^T\delta - \mathbf{a}^T\mathbf{y} - \mathbf{d}^T\delta\right) \\
&= Cov\left(\mathbf{v}_1^T\mathbf{X}^T\mathbf{y}, \mathbf{v}_1^T\mathbf{X}^T\mathbf{y} - \mathbf{a}^T\mathbf{y}\right) \\
&= \sigma^2\mathbf{v}_1^T\mathbf{X}^T(\mathbf{X}\mathbf{v}_1 - \mathbf{a}) \\
&= \sigma^2\mathbf{v}_1^T\left(\mathbf{X}^T\mathbf{X}\mathbf{v}_1 - \mathbf{X}^T\mathbf{a}\right) \\
&= \sigma^2\mathbf{v}_1^T(\lambda - \mathbf{P}\mathbf{v}_2 - \lambda + \mathbf{P}\mathbf{d}) = \sigma^2\mathbf{v}_1^T\mathbf{P}(\mathbf{d} - \mathbf{v}_2) = 0
\end{aligned}
$$

since $\mathbf{P}^T\mathbf{v}_1 = \mathbf{0}$ from Equation (4.22). □

4.8 Summary

1. The Gauss–Markov assumptions on the errors in a linear model are introduced. They specify that the errors have zero mean, are uncorrelated, and have constant variance.
2. The Gauss–Markov Theorem says that the least squares estimator $\lambda^T\hat{\mathbf{b}}$ has the smallest variance of all linear unbiased estimators of an estimable function $\lambda^T\mathbf{b}$ when the Gauss–Markov assumptions hold.
3. The estimator $\hat{\sigma}^2 = SSE/(N-r) = \mathbf{y}^T(\mathbf{I} - \mathbf{P_X})\mathbf{y}/(N-r)$ is an unbiased estimator of the variance parameter σ^2.
4. The consequences of underfitting or misspecification and overfitting are evaluated.
5. Generalized least squares estimators are introduced for cases where the Gauss–Markov assumptions on the errors may not hold and optimal estimates constructed.
6. The problem of aggregation illustrates the issues of misspecification and heteroskedasticity.
7. The solution to the restricted normal equations leads to the BLUE in the constrained model under Gauss–Markov assumptions.

4.9 Notes

- See Milliken and Albohali [31] on conditions where OLS estimators are BLUE.
- As noted, except in some simple cases of weighting such as the aggregation example, generalized least squares is rarely used in its pure form—most of the

time some parameter of the covariance matrix $\mathbf{V}$ must be estimated. This will be the case later in Chapter 8.

- The bias/variance trade-offs from underfitting and overfitting form the basis of the burgeoning field of model selection; see, for example, Miller [30].
- Finding the BLUE in the case of the constrained parameter space is difficult to prove; Rao's [36] proof is remarkable.

4.10 Exercises

4.1. * Suppose the random variable Y_i represents the number of votes that a candidate receives in county i. A reasonable model would be that $Y_i, i = 1, \ldots, N$ would be independent binomial random variables with parameters n_i = number registered of voters in county i and

$$p = Pr \text{ (voter correctly votes for candidate)}.$$

a. What are $E(Y_i)$ and $Var\,(Y_i)$?
b. Write this as a linear model.
c. Find the BLUE of p in this situation.

4.2. * Under the Gauss–Markov model, show that for λ such that $\lambda^T\mathbf{b}$ is estimable, $Var\,(\lambda^T\hat{\mathbf{b}}) = \sigma^2\lambda^T(\mathbf{X}^T\mathbf{X})^g\lambda$ does not depend on the choice of generalized inverse $(\mathbf{X}^T\mathbf{X})^g$.

4.3. In Example 4.8 (heteroskedasticity) find $Var(\hat{b}_{OLS})$ and compare it to $Var\,(\hat{b}_{GLS})$.

4.4. Consider the heteroskedasticity situation in Example 4.7, but suppose $Var(e_i) = \sigma^2 x_i$ where $x_i > 0$. Find $Var\,(\hat{b}_{OLS})$ and compare it to $Var\,(\hat{b}_{GLS})$.

4.5. Show that the response $\mathbf{z}$ arising from the Cochrane–Orcutt transformation in Example 4.9 satisfies the Gauss–Markov assumptions.

4.6. Suppose we have the simple linear regression model $y_i = \beta_0 + \beta_1 x_i + e_i, i = 1, \ldots, N$ where e_i are uncorrelated and $Var(e_i) = \sigma^2$. Consider the instrumental variables estimator of the slope $\tilde{b}_1 = \sum_{i=1}^N (z_i - \bar{z})(y_i - \bar{y}) / \sum_{i=1}^N (z_i - \bar{z})(x_i - \bar{x})$ where $z_1, \ldots, z_N$ are known constants.

a. Is $\tilde{b}_1$ an unbiased estimator of the slope parameter β_1?
b. Find the variance of $\tilde{b}_1$.
c. We know that taking $z_i = x_i$ gives our familiar least squares estimator $\hat{b}_1$. Find the ratio of the two variances, $Var\,(\hat{b}_1)/Var\,(\tilde{b}_1)$, and show that it is less than or equal to 1 ($\hat{b}_1$ is BLUE).

You may find the following results useful: $\sum_{i=1}(x_i - \bar{x}) = \sum_{i=1}(z_i - \bar{z}) = 0$, and $\sum_{i=1}(x_i - \bar{x})(y_i - \bar{y}) = \sum_{i=1}(x_i - \bar{x})y_i = \sum_{i=1} x_i(y_i - \bar{y})$.

4.7. In a reversal of example 4.6, suppose we have the simple linear regression problem, with $y_i = \beta_0 + \beta_1 x_i + e_i$, and the usual Gauss–Markov assumptions. Compute the bias in the slope estimator $\sum_i x_i y_i / \sum_i x_i^2$ when $\beta_0 \neq 0$.

4.8. * (Compare with Example 4.10 (equicorrelation).) Consider the linear model $\mathbf{y} = \mathbf{X}\mathbf{b} + \mathbf{e}$ where $\mathbf{e} = \mathbf{u} + Z\mathbf{1}$ (Z is a scalar random variable), where $Cov(\mathbf{u}) = \sigma^2 \mathbf{I}_N$, $Var(Z) = \tau^2$, and Z and $\mathbf{u}$ are uncorrelated. Find $\mathbf{V}$ and derive conditions under which the OLS estimator of every estimable function is BLUE.

4.9. If $\mathbf{X}$ has full-column rank and $\mathbf{C}^T\mathbf{y}$ is also unbiased for estimable $\mathbf{\Lambda}^T\mathbf{b}$, show that $Cov(\mathbf{C}^T\mathbf{y}) - Cov(\mathbf{\Lambda}^T\hat{\mathbf{b}})$ is nonnegative definite.

4.10. Suppose $y_i = \beta_0 + \beta_1 x_i + \beta_2 x_i^2 + e_i$, where $E(e_i) = 0$, $Var(e_i) = \sigma^2$, and the e_i are independent. But suppose we fit a simple linear regression model: $E(y_i) = \beta_0 + \beta_1 x_i$. To illustrate, consider the following simple situation: $x_i = i, i = 1, 2, \ldots, n = 8$, for $\beta_0 = 2, \beta_1 = 3$. Consider also various values of β_2, say, -2 to 4 by ones. (Hint: I suggest using a regression program, such as SAS's PROC REG for computations.)

a. Compute the bias in the least squares estimators, $\hat{\beta}_0, \hat{\beta}_1$.
b. Compute the bias in our usual variance estimate $\hat{\sigma}^2$.
c. Would your results change if the values of β_0 and β_1 were changed? Explain.

4.11. Consider the simple linear regression problem, $y_i = \beta_0 + \beta_1 x_i + e_i$, for $i = 1, \ldots, 4 = n$ with $x_i = i$ and the Gauss–Markov assumptions on e_i. A direct route for getting the BLUE would be to construct all linear unbiased estimators and then directly minimize the variance. Write the unbiased estimators as $\sum_{i=1}^n a_i y_i$ and focus on estimating the slope β_1.

a. Write the two linear equations in the a_i's that express constraints so that $\sum a_i y_i$ is an unbiased estimator of β_1.
b. Construct the family of solutions to the equations in (a). (Hint: You'll need two z's.) This will parameterize all unbiased estimators with just two parameters.
c. Compute the variance of the estimators in (b) and minimize the variance. You should get a familiar solution.

4.12. Prove the Gauss–Markov Theorem directly, that is, by constructing all linear estimators $\mathbf{a}^T\mathbf{y}$ that are unbiased for $\lambda^T\mathbf{b}$ (find a family of solutions $\mathbf{a}(\mathbf{z})$), and then minimizing the variance $\sigma^2\mathbf{a}^T\mathbf{a}$.

4.13. Show that if $\mathbf{R}$ is square and nonsingular and $\mathbf{R}^T\mathbf{V}\mathbf{R} = \mathbf{I}$, then $\mathbf{V}^{-1} = \mathbf{R}^T\mathbf{R}$. Do you need the assumption that $\mathbf{R}$ is square? Nonsingular?

4.14. Let $\mathbf{Q} = \mathbf{X}(\mathbf{X}^T\mathbf{V}^{-1}\mathbf{X})^g\mathbf{X}^T\mathbf{V}^{-1}$. Show that $\mathbf{Q}$ is a projection onto $\mathcal{C}(\mathbf{X})$. (This is difficult without the hint: first factor $\mathbf{V} = \mathbf{L}\mathbf{L}^T$ with Cholesky and work with a symmetric version $\mathbf{Q}* = \mathbf{L}^{-1}\mathbf{Q}\mathbf{L}$.)

4.15. Let $\sigma^2\mathbf{V}$ be the $N \times N$ covariance matrix for a first-order moving average process:

$$\begin{aligned} V_{ij} &= 1+\alpha^2 && \text{if} \quad i = j \\ V_{ij} &= \alpha && \text{if} \quad |i-j| = 1 \\ V_{ij} &= 0 && \text{if} \quad |i-j| > 1 \end{aligned}$$

Notice that $\mathbf{V}$ is banded with zeros outside of three bands, on the diagonal and above and below the diagonal.

a. Show that the Cholesky factor of $\mathbf{V}$ is also banded (lower triangular, with nonzeros on the diagonal and below the diagonal).
b. Find the limit of the two nonzero elements in row N as $N \to \infty$.

4.16. Consider the simplest linear model in Example 4.7, where $\mathbf{V}$ follows the covariance structure in Exercise 4.15. For the case $\sigma^2 = 1, \alpha = .5$, and $N = 5$, find $var(\overline{y})$ numerically and compare it to the variance of $\hat{\mu}_{GLS}$.

4.17. Consider the multiple regression problem including an intercept with the following list of explanatory variables:

$$\begin{aligned} c1 &= cos\,(2\pi i/7) & s1 &= sin\,(2\pi i/7) \\ c2 &= cos\,(2\pi 2i/7) & s2 &= sin\,(2\pi 2i/7) \\ c3 &= cos\,(2\pi 3i/7) & s3 &= sin\,(2\pi 3i/7) \\ c4 &= cos\,(2\pi 4i/7) & s4 &= sin\,(2\pi 4i/7) \\ c5 &= cos\,(2\pi 5i/7) & s5 &= sin\,(2\pi 5i/7) \\ c6 &= cos\,(2\pi 6i/7) & s6 &= sin\,(2\pi 6i/7) \end{aligned}$$

for $i = 1, \ldots, N$.

a. Show that the last six variables (c4, s4, ..., s6) are linearly dependent on the first six (c1, s1, ..., s3) and an intercept.
b. Test whether the regression software that you commonly use can detect dependencies among the explanatory variables, using 3.1416 as your approximation for π, and various values of N.
c. Repeat this exercise with a cruder approximation of 3.14 for π.
d. Repeat this exercise with 4 in place of 7 (that is, $2\pi i/4, 4\pi i/4$, etc.).

4.18. Using the variables (1, c1, s1, c2, s2, c3, s3) from Exercise 4.17, is there any multicollinearity problem?

4.19. Consider the usual simple linear regression situation $E(y_i) = \alpha + \beta x_i$ for $i = 1, \ldots, 5$ with $x_i = i$, $Var(y_i) = \sigma^2$ and y_i being independent. Note the simple form of x_i and that we have only five observations.

a. Find the least squares estimator $\hat{\beta}$ and express it in terms of $\mathbf{t}^T\mathbf{y}$ by explicitly giving $\mathbf{t}$.
b. Show that $\hat{\beta}$ is unbiased and find its variance.

c. Show that $\hat{\gamma} = (y_4 - y_2)/2$ and $\hat{\eta} = (y_5 - y_1)/4$ are also unbiased for β and also find the variance of each.

Consider now another estimator of the slope parameter β, the estimator $\hat{\delta} = c\hat{\gamma} + (1 - c)\hat{\eta}$.

d. Find the variance of $\hat{\delta}$ in terms of c and σ^2.
e. Find the value of c that minimizes the variance found in part (d).
f. Suppose we knew that $\alpha = 0$, would $\hat{\beta}$ still be BLUE?

4.20. Under Gauss–Markov assumptions, show that if $Cov(\mathbf{a}^T\mathbf{y}, \mathbf{d}^T\hat{\mathbf{e}}) = 0$ for all $\mathbf{d}$, then $\mathbf{a}^T\mathbf{y}$ is the BLUE for its expectation.

4.21. Under the Aitken model, with $Cov(\mathbf{e}) = \sigma^2\mathbf{V}$, if $cov\,(\mathbf{a}^T\mathbf{y}\,, \mathbf{d}^T\hat{\mathbf{e}}) = 0$ for all $\mathbf{d}$, then is $\mathbf{a}^T\mathbf{y}$ still the BLUE for its expectation?

4.22. * Prove Aitken's Theorem (Theorem 4.2) indirectly using Corollary 4.1. Is this argument circular, that is, does the corollary use Aitken's Theorem?

4.23. Prove Aitken's Theorem (Theorem 4.2) directly. Consider the Aitken model, and let $\lambda^T\mathbf{b}$ be estimable. Let $\mathbf{a}^T\mathbf{y}$ be a linear unbiased estimator of $\lambda^T\mathbf{b}$. Show that

$$Var\,(\mathbf{a}^T\mathbf{y}) = Var\,(\lambda^T\hat{\mathbf{b}}_{GLS}) + Var\,(\mathbf{a}^T\mathbf{y} - \lambda^T\hat{\mathbf{b}}_{GLS}).$$

4.24. Recall the analysis of the U.S. population in Exercise 3.23. Find the VIFs for each of the coefficients in the uncentered model, using $t = 1790$ through 2000. If you center using $c = 1890$, what happens to the VIFs? Also examine the condition number of the design matrix.

4.25. Prove that if $\mathbf{V}$ is positive definite, then $\mathcal{C}(\mathbf{X}^T\mathbf{V}^{-1}\mathbf{X}) = \mathcal{C}(\mathbf{X}^T)$.

4.26. Ridge regression is a technique that has been recommended by some statisticians to address multicollinearity problems arising in multiple regression. In our usual linear models framework with $E(\mathbf{y}) = \mathbf{X}\mathbf{b}$, and $Cov\,(\mathbf{y}) = \sigma^2\mathbf{I}_N$, the ridge regression estimator takes the form

$$\tilde{\mathbf{b}} = (\mathbf{X}^T\mathbf{X} + k\mathbf{I}_P)^{-1}\mathbf{X}^T\mathbf{y}$$

where $k > 0$. Assume here that the $\mathbf{X}$ has full-column rank, that is, *rank* $(\mathbf{X}) = p$.

a. Find $E(\tilde{\mathbf{b}})$.
b. Is $\lambda^T\tilde{\mathbf{b}}$ an unbiased estimator of $\lambda^T\mathbf{b}$?
c. Find $Cov\,(\tilde{\mathbf{b}})$.
d. Mean squared error is commonly used to assess the quality of an estimator. For the ridge regression estimator $\tilde{\mathbf{b}}$, find its mean squared error $E(\|\tilde{\mathbf{b}} - \mathbf{b}\|^2) = E((\tilde{\mathbf{b}} - \mathbf{b})^T(\tilde{\mathbf{b}} - \mathbf{b}))$.

Consider applying ridge regression to a multivariate regression problem with two centered covariates ($\sum x_i = \sum z_i = 0$), taking the form

$$\mathbf{Xb} = \begin{bmatrix} 1 & x_1 & z_1 \\ 1 & x_2 & z_2 \\ \dots & \dots & \dots \\ 1 & x_N & z_N \end{bmatrix} \begin{bmatrix} \beta_0 \\ \beta_1 \\ \beta_2 \end{bmatrix}$$

yielding the inner product matrix

$$\mathbf{X}^T\mathbf{X} = \begin{bmatrix} N & 0 & 0 \\ 0 & \sum x_i^2 & \sum x_i z_i \\ 0 & \sum x_i z_i & \sum z_i^2 \end{bmatrix}.$$

For simplicity, suppose $N = 10$, $\sum x_i^2 = \sum z_i^2 = 5$, and $\sum x_i z_i = 4$.

e. Find the covariance matrix of our usual least squares estimator $\hat{\mathbf{b}} = (\mathbf{X}^T\mathbf{X})^{-1}\mathbf{X}^T\mathbf{y}$ in this situation.
f. Find a value of k such that one component of $\tilde{\mathbf{b}}$ (your choice of component) has smaller variance than the corresponding least squares estimator.
g. Part (f) looks like it violates the Gauss–Markov Theorem. Does it?

4.27. Principal components regression is considered another alternative to least squares estimators in the case of multicollinearity. Begin with the singular value decomposition of the design matrix $\mathbf{X} = \mathbf{U}[\begin{smallmatrix}\mathbf{\Lambda} \\ \mathbf{0}\end{smallmatrix}]\mathbf{V}^T = \mathbf{U}_1\mathbf{\Lambda}\mathbf{V}^T$, partitioning off the first p columns of $\mathbf{U}$, and express the least squares estimator as $\hat{\mathbf{b}} = \mathbf{V}\mathbf{\Lambda}^{-1}\mathbf{U}_1^T\mathbf{y}$. Setting some of the small singular values to zero, and using $\mathbf{\Lambda}_*^- = diag\{\lambda_1^{-1}, \dots, \lambda_k^{-1}, 0, \cdots, 0\}$, changing the rank from p to k, leads to the estimator $\tilde{\mathbf{b}} = \mathbf{V}\mathbf{\Lambda}_*^-\mathbf{U}_1^T\mathbf{y}$. Find its bias and mean squared error.

4.28. For the heteroskedastic regression model (4.17), denote the generalized least squares estimator of the slope as $\tilde{\beta}_1$. Find the variance of $\tilde{\beta}_1$ and show that it is greater than the variance of the usual estimator

$$Var\,(\hat{\beta}_1) = \sigma^2 / \left(\sum_i \sum_j (x_{ij} - \overline{x}_{..}) \right)^2.$$

4.29. In the aggregation problem with a quadratic response expressed in (4.18), if variation of the explanatory variable $\sum (x_{ij} - \overline{x}_{i.})^2$ were constant over i, can an unbiased estimate of slope be constructed?

4.11 Addendum: Variance of Variance Estimator

Computing $Var\,(\hat{\sigma}^2)$ requires considerable detail, although the result is none too deep. The main result will be given with some generality.

Result 4.6 *Let* $\mathbf{P}$ *be a symmetric matrix and* $\mathbf{e}$ *be a random vector. The components* e_i *are iid with the following four moments:* $E(e_i) = 0$, $Var(e_i) = E(e_i^2) = \sigma^2$, $E(e_i^3) = \gamma_3$, *and* $E(e_i^4) = \gamma_4$; *then*

$$\begin{aligned} Var((\mu+\mathbf{e})^T\mathbf{P}(\mu+\mathbf{e})) = & 4\sigma^2(\mu^T\mathbf{P}^2\mu) + 4\gamma_3 \sum_i \mu_i P_{ii} \sum_j P_{ji} \\ & + 2\sigma^4 \sum_{i \neq j} P_{ij}^2 + \sum_i (\gamma_4 - \sigma^4) P_{ii}^2 \end{aligned}$$

Proof: Begin with

$$\begin{aligned} Var((\mu+\mathbf{e})^T\mathbf{P}(\mu+\mathbf{e})) = & E[(\mu+\mathbf{e})^T\mathbf{P}(\mu+\mathbf{e})(\mu+\mathbf{e})^T\mathbf{P}(\mu+\mathbf{e})] \\ & - E[(\mu+\mathbf{e})^T\mathbf{P}(\mu+\mathbf{e})]^2 \end{aligned}$$

Note that $E[(\mu+\mathbf{e})^T\mathbf{P}(\mu+\mathbf{e})] = \mu^T\mathbf{P}\mu + \sigma^2 tr(\mathbf{P})$ from Lemma 4.1. Now write out all sixteen terms of $E[(\mu+\mathbf{e})^T\mathbf{P}(\mu+\mathbf{e})(\mu+\mathbf{e})^T\mathbf{P}(\mu+\mathbf{e})]$ as

$$E[\mu^T\mathbf{P}\mu\mu^T\mathbf{P}\mu] = (\mu^T\mathbf{P}\mu)^2$$

$$E[\mu^T\mathbf{P}\mu\mathbf{e}^T\mathbf{P}\mu] = 0$$

$$E[\mu^T\mathbf{P}\mu\mu^T\mathbf{P}\mathbf{e}] = 0$$

$$E[\mu^T\mathbf{P}\mu\mathbf{e}^T\mathbf{P}\mathbf{e}] = (\mu^T\mathbf{P}\mu)E[\mathbf{e}^T\mathbf{P}\mathbf{e}] = (\mu^T\mathbf{P}\mu)\sigma^2 trace(\mathbf{P})$$

$$E[\mathbf{e}^T\mathbf{P}\mu\mu^T\mathbf{P}\mu] = 0$$

$$E[\mathbf{e}^T\mathbf{P}\mu\mathbf{e}^T\mathbf{P}\mu] = Var(\mu^T\mathbf{P}\mathbf{e}) = E[\mathbf{e}^T\mathbf{P}\mu\mu^T\mathbf{P}\mathbf{e}] = \sigma^2(\mu^T\mathbf{P}^2\mu)$$

$$E[\mathbf{e}^T\mathbf{P}\mu\mu^T\mathbf{P}\mathbf{e}] = \sigma^2(\mu^T\mathbf{P}^2\mu)$$

$$E[\mathbf{e}^T\mathbf{P}\mu\mathbf{e}^T\mathbf{P}\mathbf{e}] = E[\mu^T\mathbf{P}\mathbf{e}\mathbf{e}^T\mathbf{P}\mathbf{e}] \text{ (see below)}$$

$$E[\mu^T\mathbf{P}\mathbf{e}\mu^T\mathbf{P}\mu] = 0$$

$$E[\mu^T\mathbf{P}\mathbf{e}\mathbf{e}^T\mathbf{P}\mu] = \sigma^2(\mu^T\mathbf{P}^2\mu)$$

$$E[\mu^T\mathbf{P}\mathbf{e}\mu^T\mathbf{P}\mathbf{e}] = \sigma^2(\mu^T\mathbf{P}^2\mu)$$

$$E[\mu^T\mathbf{P}\mathbf{e}\mathbf{e}^T\mathbf{P}\mathbf{e}] = \text{ (see below)}$$

$$E[\mathbf{e}^T\mathbf{P}\mathbf{e}\mu^T\mathbf{P}\mu] = (\mu^T\mathbf{P}\mu)E[\mathbf{e}^T\mathbf{P}\mathbf{e}] = (\mu^T\mathbf{P}\mu)\sigma^2 trace(\mathbf{P})$$

$$E[\mathbf{e}^T\mathbf{P}\mathbf{e}\mathbf{e}^T\mathbf{P}\mu] = E[\mu^T\mathbf{P}\mathbf{e}\mathbf{e}^T\mathbf{P}\mathbf{e}] \text{ (see below)}$$

$$E[\mathbf{e}^T\mathbf{P}\mathbf{e}\mu^T\mathbf{P}\mathbf{e}] = E[\mu^T\mathbf{P}\mathbf{e}\mathbf{e}^T\mathbf{P}\mathbf{e}] \text{ (see below)}$$

$$E[\mathbf{e}^T\mathbf{P}\mathbf{e}\mathbf{e}^T\mathbf{P}\mathbf{e}] = \text{ (see below)}$$

Only two difficult expressions remain: $E[\mu^T\mathbf{P}\mathbf{e}\mathbf{e}^T\mathbf{P}\mathbf{e}]$ and $E[\mathbf{e}^T\mathbf{P}\mathbf{e}\mathbf{e}^T\mathbf{P}\mathbf{e}]$. For both we need the following algebra:

$$E[\mathbf{a}^T\mathbf{P}\mathbf{b}\mathbf{c}^T\mathbf{P}\mathbf{d}] = \sum_i\sum_j\sum_k\sum_l E[a_i b_j c_k d_l P_{ij} P_{kl}]$$

We are interested in all cases where the indices are the same:

iiii	(all same)	n
iiij	(one differs)	n(n–1) of each of (iiij, iiji, ijii, jiii)
iijj	(two pair)	n(n–1) of each of (iijj, ijij, ijji)
iijk	(one pair)	n(n–1)(n-2) of each of (iijk, ijik, ijki, jiik, jiki, jkii)
ijkl	(all different)	n(n–1)(n–2)(n–3)

For $E[\mu^T \mathbf{P}\mathbf{e}\mathbf{e}^T \mathbf{P}\mathbf{e}] = \sum_i \sum_j \sum_k \sum_l E[\mu_i e_j e_k e_l P_{ij} P_{kl}]$ we have

iiii $\sum_i E\left[\mu_i e_i e_i e_i P_{ii} P_{ii}\right] = \sum_i \mu_i \gamma_3 P_{ii}^2$

iiij $\sum_{i \neq j} E\left[\mu_j e_i e_i e_i P_{ji} P_{ii}\right] = \sum_{i \neq j} \mu_i \gamma_3 P_{ji} P_{ii}$

Note that the other three cases are zero, for example, $E[\mu_i e_j e_i e_i P_{ij} P_{ii}] = 0$. This extends to the other three possible combinations (two pair, one pair, all differ). Adding these two pieces produces

$$E[\mu^T \mathbf{P}\mathbf{e}\mathbf{e}^T \mathbf{P}\mathbf{e}] = \sum_i \mu_i \gamma_3 P_{ii}^2 + \sum_{i \neq j} \mu_i \gamma_3 P_{ji} P_{ii} = \gamma_3 \sum_i \mu_i P_{ii} \sum_j P_{ji},$$

which does not appear to simplify further.

For $E[\mathbf{e}^T \mathbf{P}\mathbf{e}\mathbf{e}^T \mathbf{P}\mathbf{e}] = \sum_i \sum_j \sum_k \sum_l E[e_i e_j e_k e_l P_{ij} P_{kl}]$ we have

iiii $\sum_i E\left[e_i e_i e_i e_i P_{ii} P_{ii}\right] = \sum_i \gamma_4 P_{ii}^2$

iiij $\sum_{i \neq j} E\left[e_j e_i e_i e_i P_{ji} P_{ii}\right] = 0$ (all four cases: jiii, ijii, iiji, iiij)

iijj $\sum_{i \neq j} E\left[e_i e_i e_j e_j P_{ii} P_{jj}\right] = \sigma^4 \sum_{i \neq j} P_{ii} P_{jj}$

ijij $\sum_{i \neq j} E\left[e_i e_j e_i e_j P_{ij} P_{ij}\right] = \sigma^4 \sum_{i \neq j} P_{ij} P_{ij}$

ijji $\sum_{i \neq j} E\left[e_i e_j e_j e_i P_{ij} P_{ji}\right] = \sigma^4 \sum_{i \neq j} P_{ij} P_{ji}$

and the other cases are all zero. Gathering up the pieces produces the following:

$$\begin{aligned}
Var\,((\mu + \mathbf{e})^T \mathbf{P}(\mu + \mathbf{e})) &= E[(\mu + \mathbf{e})^T \mathbf{P}(\mu + \mathbf{e})(\mu + \mathbf{e})^T \mathbf{P}(\mu + \mathbf{e})] \\
&\quad - E[(\mu + \mathbf{e})^T \mathbf{P}(\mu + \mathbf{e})]^2 \\
&= (\mu^T \mathbf{P}\mu)^2 + 2(\mu^T \mathbf{P}\mu)\sigma^2 trace(\mathbf{P}) + 4\sigma^2(\mu^T \mathbf{P}^2 \mu) + 4\gamma_3 \sum_i \mu_i \mathbf{P}_{ii} \sum_j P_{ji} \\
&\quad + \sigma^4 \sum_{i \neq j} (P_{ii} P_{jj} + 2P_{ij}^2) + \sum_i \gamma_4 P_{ii}^2 - (\mu^T \mathbf{P}\mu + \sigma^2 tr(\mathbf{P}))(\mu^T \mathbf{P}\mu + \sigma^2 tr(\mathbf{P})) \\
&= 4\sigma^2(\mu^T \mathbf{P}^2 \mu) + 4\gamma_3 \sum_i \mu_i P_{ii} \sum_j P_{ji} + \sigma^4 \sum_{i \neq j} (P_{ii} P_{jj} + 2P_{ij}^2) \\
&\quad - \sigma^4 tr(\mathbf{P})^2 + \sum_i \gamma_4 P_{ii}^2 \\
&= 4\sigma^2(\mu^T \mathbf{P}^2 \mu) + 4\gamma_3 \sum_i \mu_i P_{ii} \sum_j P_{ji} + 2\sigma^4 \sum_{i \neq j} P_{ij}^2 + \sum_i (\gamma_4 - \sigma^4) P_{ii}^2
\end{aligned}$$

The last step employs

$$tr(\mathbf{P})^2 = \left(\sum_i P_{ii}\right)^2 = \sum_i P_{ii}^2 + \sum_{i \neq j} P_{ii} P_{jj}.$$

□

Corollary 4.2 *Let* **P** *be symmetric and idempotent, and the components of the random vector* **e** *iid with four moments existing, then*

$$Var\,(\mathbf{e}^T \mathbf{P} \mathbf{e}) = 2\sigma^4 \sum_{i \neq j} P_{ij}^2 + \sum_i (\gamma_4 - \sigma^4) P_{ii}^2$$

Corollary 4.3 *Let* **P** *be symmetric and* $\gamma_3 = 0$, $\gamma_4 = 3\sigma^4$ *(which hold if* e_i *are iid Normal*$(0, \sigma^2)$*), then*

$$Var\,((\mathbf{e} + \mu)^T \mathbf{P} (\mathbf{e} + \mu)) = 4\sigma^2 (\mu^T \mathbf{P}^2 \mu) + 2\sigma^4 tr\,(\mathbf{P}^2)$$

Proof: Here $\gamma_3 = 0$, $\gamma_4 = 3\sigma^4$ and starting with

$$Var\,(\mathbf{e}^T \mathbf{P} \mathbf{e}) = 4\sigma^2 (\mu^T \mathbf{P}^2 \mu) + 2\sigma^4 \sum_{i \neq j} \mathbf{P}_{ij}^2 + \sum_i (3\sigma^4 - \sigma^4) P_{ii}^2,$$

and now use

$$tr\,(\mathbf{P}^2) = \sum_i \sum_j P_{ij}^2 = \sum_i P_{ii}^2 + \sum_{i \neq j} P_{ij}^2.$$

□

Chapter 5

Distributional Theory

5.1 Introduction

Although the motivations of the linear least squares model and Gauss–Markov model were different, they both led in the same direction using minimal assumptions. From a statistical viewpoint, the Gauss–Markov model employed only moment assumptions on the errors: $E(\mathbf{e}) = \mathbf{0}$, $Cov(\mathbf{e}) = \sigma^2\mathbf{I}$. The goal of this chapter is to extend the assumptions on the errors $\mathbf{e}$ to specifying its joint distribution, so that in Chapter 6 we can then look for best estimators. The distributional assumptions will also permit construction of hypothesis tests and confidence regions.

Most of this exposition, of course, will follow the traditional route using the normal distribution. In applications, we often observe a process whose deviations from the mean can be thought of as the sum of a large number of independent random effects, with no few of them dominating. Consequently, the central limit theorem's conclusion makes the assumption that the errors are normally distributed to be the most reasonable distribution to assume.

5.2 Multivariate Normal Distribution

Let us begin by defining the (univariate) normal distribution.

Definition 5.1 *A random variable Y has the normal distribution with mean μ and variance σ^2, denoted $Y \sim N(\mu, \sigma^2)$ whose density is given by*

$$p_Y(y) = (2\pi\sigma^2)^{-\frac{1}{2}}\, exp\,\{-(y-\mu)^2/(2\sigma^2)\}. \tag{5.1}$$

Another route to this distribution, as we will take a parallel route to the multivariate normal, is to begin with the standard normal distribution, where $\mu = 0$ and $\sigma^2 = 1$, denoted $Z \sim N(0, 1)$, whose density is

$$p_z(z) = (2\pi)^{-\frac{1}{2}}\, exp\,\{-z^2/2\}.$$

The moment generating function (mgf) for the standard normal is easily obtained:

$$m_z(t) \equiv E[e^{tZ}] = \int_{-\infty}^{\infty} e^{tz} p_z(z)dz = \int_{-\infty}^{\infty} (2\pi)^{-\frac{1}{2}} exp\{tz - z^2/2\}dz$$
$$= \int_{-\infty}^{\infty} (2\pi)^{-\frac{1}{2}} exp\{-(z-t)^2/2 + t^2/2\}dz = exp\{t^2/2\}.$$

Now the more general distribution can be constructed from the standard normal using the transformation $Y = \mu + \sigma Z$. Using the rules for transformations, the density for Y can be derived as above (5.1), and the rules for moment generating functions can also be employed:

$$m_Y(t) \equiv E\{e^{tY}\} = E\{e^{t(\mu+\sigma Z)}\}$$
$$= e^{t\mu} \times E\{e^{t\sigma Z}\} = e^{t\mu} \times m_z(t\sigma) = exp\{t\mu + t^2\sigma^2/2\}. \quad (5.2)$$

The general case of the multivariate normal distribution will be defined as a transformation of this standard multivariate normal: let $\mathbf{Z} \sim N_p(0, \mathbf{I}_p)$, then $\mathbf{X} = \mathbf{AZ} + \mu$ will have a multivariate normal distribution. To find its properties, we will go a somewhat circuitous route by first constructing its moment generating function.

Definition 5.2 *Let* $\mathbf{Z}$ *be a* $p \times 1$ *vector with each component* $Z_i, i = 1, \ldots, p$ *independently distributed with* $Z_i \sim N(0, 1)$. *Then* $\mathbf{Z}$ *has the* standard multivariate normal distribution, *denoted* $\mathbf{Z} \sim N_p(0, \mathbf{I}_p)$, *in p dimensions. The joint density of the standard multivariate normal can be written then as*

$$p_{\mathbf{Z}}(\mathbf{z}) = (2\pi)^{-p/2} exp\left\{-\sum_{i=1}^{p} z_i^2/2\right\}.$$

Definition 5.3 *The moment generating function of a multivariate random variable* $\mathbf{X}$ *is given by*

$$m_{\mathbf{X}}(\mathbf{t}) = E\{e^{\mathbf{t}^T\mathbf{X}}\}$$

provided this expectation exists in a rectangle that includes the origin. More precisely, there exists $h_i > 0, i = 1, \ldots, p$, *so that the expectation exists for all* $\mathbf{t}$ *such that* $-h_i < t_i < h_i, i=1, \ldots, p$.

The following two results, which will not be proven, provide the rules for handling multivariate mgf's.

Result 5.1 *If moment generating functions for two random vectors* $\mathbf{X}_1$ *and* $\mathbf{X}_2$ *exist, then the cdf's for* $\mathbf{X}_1$ *and* $\mathbf{X}_2$ *are identical iff the mgf's are identical in an open rectangle that includes the origin.*

Result 5.2 *Assume the random vectors* $\mathbf{X}_1, \mathbf{X}_2, \ldots, \mathbf{X}_p$ *each have mgf's* $m_{\mathbf{X}_j}(\mathbf{t}_j)$, $j = 1, \ldots, p$, *and that* $\mathbf{X} = (\mathbf{X}_1^T, \mathbf{X}_2^T, \ldots, \mathbf{X}_p^T)^T$ *has mgf* $m_{\mathbf{X}}(\mathbf{t})$, *where* $\mathbf{t}$ *is partitioned* *similarly. Then* $\mathbf{X}_1, \mathbf{X}_2, \ldots, \mathbf{X}_p$ *are mutually independent iff*

$$m_{\mathbf{X}}(\mathbf{t}) = m_{\mathbf{X}_1}(\mathbf{t}_1) \times m_{\mathbf{X}_2}(\mathbf{t}_2) \times \ldots \times m_{\mathbf{X}_p}(\mathbf{t}_p)$$

for all $\mathbf{t}$ *in an open rectangle that includes the origin.*

The moment generating function for the standard multivariate normal distribution $\mathbf{Z} \sim N_p(0, \mathbf{I}_p)$ can be easily computed:

$$\begin{aligned} m_{\mathbf{z}}(\mathbf{t}) &= E\{exp\,(\mathbf{t}^T\mathbf{Z})\} = E\left\{exp\left(\sum_{i=1}^{p} t_i Z_i\right)\right\} = \prod_{i=1}^{p} m_{z_i}(t_i) \\ &= exp\left\{\sum_{i=1}^{p} t_i^2/2\right\} = exp\,\{\mathbf{t}^T\mathbf{t}/2\}. \end{aligned}$$

From this the moment generating function for $\mathbf{X} = \mu + \mathbf{AZ}$ can be constructed:

$$m_{\mathbf{X}}(\mathbf{t}) = E\left[e^{\mathbf{t}^T\mathbf{X}}\right] = E\left[e^{\mathbf{t}^T\mu + \mathbf{t}^T\mathbf{AZ}}\right] = e^{\mathbf{t}^T\mu} \times m_z(\mathbf{A}^T\mathbf{t}) = exp\,\{\mathbf{t}^T\mu + \mathbf{t}^T\mathbf{AA}^T\mathbf{t}/2\},$$

which is clearly a function of just μ and $\mathbf{AA}^T$. Now just from the rules of means and covariances in Section 4.1, we know that $E\,[\mathbf{X}] = \mu$ and $Cov(\mathbf{X}) = \mathbf{AA}^T$, and so we can see that the multivariate normal distribution is characterized by its mean vector and covariance matrix. From appendix A.4, for any nonnegative definite matrix $\mathbf{V}$, we can find a matrix $\mathbf{A}$ such that $\mathbf{V} = \mathbf{AA}^T$, so that for any vector μ and any nonnegative definite matrix $\mathbf{V}$, we can define the multivariate normal distribution with mean vector μ and covariance matrix $\mathbf{V}$ in terms of its moment generating function.

Definition 5.4 *The p-dimensional vector* $\mathbf{X}$ *has the multivariate normal distribution with mean* μ *and covariance matrix* $\mathbf{V}$*, denoted by* $\mathbf{X} \sim N_p(\mu, \mathbf{V})$*, if and only if its moment generating function takes the form*

$$m_{\mathbf{X}}(\mathbf{t}) = exp\,\{\mathbf{t}^T\mu + \mathbf{t}^T\mathbf{V}\mathbf{t}/2\}.$$

An important point to be emphasized here is that the covariance matrix may be singular, leading to the singular multivariate normal distribution. Nonetheless, the moment generating function can be constructed, although, as we will see later, the probability density function may not exist. In this singular normal distribution, the probability mass lies in a subspace, and the dimension of the subspace—the rank of the covariance matrix—will be important (see Exercise 5.1). Another point to be noted is that when defining the multivariate normal distribution with given covariance matrix $\mathbf{V}$, the choice of the square root matrix $\mathbf{A}$ does not matter, as long as $\mathbf{V} = \mathbf{AA}^T$ (see also Exercise 5.2). Although different choices of $\mathbf{A}$ may be made, they all will lead to the same moment generating function, and the probability of sets in $\mathbf{Y}$ will be the same since the moment generating functions will be the same.

Result 5.3 *If* $\mathbf{X} \sim N_p(\mu, \mathbf{V})$ *and* $\mathbf{Y} = \mathbf{a} + \mathbf{BX}$ *where* $\mathbf{a}$ *is* $q \times 1$*, and* $\mathbf{B}$ *is* $q \times p$*, then* $\mathbf{Y} \sim N_q(\mathbf{a} + \mathbf{B}\mu, \mathbf{BVB}^T)$.

Proof: Construct the moment generating function for $\mathbf{Y}$ as follows:

$$m_{\mathbf{Y}}(\mathbf{t}) = E\left[e^{\mathbf{t}^T\mathbf{Y}}\right] = E\left[\, e^{\mathbf{t}^T(\mathbf{a}+\mathbf{B}\mathbf{X})}\right] = e^{\mathbf{t}^T\mathbf{a}} \times m_{\mathbf{X}}(\mathbf{B}^T\mathbf{t})$$
$$= e^{\mathbf{t}^T\mathbf{a}} \times exp\{\mathbf{t}^T\mathbf{B}\mu + \mathbf{t}^T\mathbf{B}\mathbf{V}\mathbf{B}^T\mathbf{t}/2\},$$

which is the mgf for $N_q(\mathbf{a}+\mathbf{B}\mu, \mathbf{B}\mathbf{V}\mathbf{B}^T)$. ∎

The reader should notice that no provision was made for whether p or q was larger, or whether the distribution for $\mathbf{X}$ or $\mathbf{Y}$ was singular. Linear combinations of jointly normal random variables have a normal distribution.

Corollary 5.1 *If $\mathbf{X}$ is multivariate normal, then the joint distribution of any subset is multivariate normal.*

Proof: Without loss of generality, partition $\mathbf{X}$, μ, and $\mathbf{V}$ in the following fashion:

$$\mathbf{X} = \begin{bmatrix}\mathbf{X}_1\\ \mathbf{X}_2\end{bmatrix}\begin{matrix}p_1\\ p_2\end{matrix}, \quad \mu = \begin{bmatrix}\mu_1\\ \mu_2\end{bmatrix}\begin{matrix}p_1\\ p_2\end{matrix}, \quad \mathbf{V} = \begin{bmatrix}\mathbf{V}_{11} & \mathbf{V}_{12}\\ \mathbf{V}_{21} & \mathbf{V}_{22}\end{bmatrix}\begin{matrix}p_1\\ p_2\end{matrix},$$

and using $\mathbf{a} = \mathbf{0}$ and $\mathbf{B} = [\mathbf{I}_{p_1} \quad \mathbf{0}]$ with Result 5.3, we have $\mathbf{X}_1 \sim N(\mu_1, \mathbf{V}_{11})$. ∎

Corollary 5.2 *If $\mathbf{X} \sim N_p(\mu, \mathbf{V})$ and $\mathbf{V}$ is nonsingular, then*

a) a nonsingular matrix $\mathbf{A}$ exists such that $\mathbf{V} = \mathbf{A}\mathbf{A}^T$,

b) $\mathbf{A}^{-1}(\mathbf{X}-\mu) \sim N_p(\mathbf{0}, \mathbf{I}_p)$, and

c) the pdf is $(2\pi)^{-p/2}|\mathbf{V}|^{-\frac{1}{2}} exp\{-\frac{1}{2}(\mathbf{x}-\mu)^T\mathbf{V}^{-1}(\mathbf{x}-\mu)\}$.

Proof: See Exercise 5.4. ∎

Result 5.4 *If $\mathbf{X} \sim N_p(\mu, \mathbf{V})$, and we partition*

$$\mathbf{X} = \begin{bmatrix}\mathbf{X}_1\\ \mathbf{X}_2\\ \cdots\\ \mathbf{X}_m\end{bmatrix}\begin{matrix}p_1\\ p_2\\ \\ p_m\end{matrix}, \quad \mu = \begin{bmatrix}\mu_1\\ \mu_2\\ \cdots\\ \mu_m\end{bmatrix}\begin{matrix}p_1\\ p_2\\ \\ p_m\end{matrix}, \quad \mathbf{V} = \begin{bmatrix}\mathbf{V}_{11} & \mathbf{V}_{12} & \cdots & \mathbf{V}_{1m}\\ \mathbf{V}_{21} & \mathbf{V}_{22} & \cdots & \mathbf{V}_{2m}\\ \cdots & \cdots & \cdots & \cdots\\ \mathbf{V}_{m1} & \mathbf{V}_{m2} & \cdots & \mathbf{V}_{mm}\end{bmatrix}\begin{matrix}p_1\\ p_2\\ \\ p_m\end{matrix}$$

then $\mathbf{X}_1, \mathbf{X}_2, \ldots, \mathbf{X}_m$ are jointly independent iff $\mathbf{V}_{ij} = \mathbf{0}$ for all $i \neq j$.

Proof: If $\mathbf{X}_1, \mathbf{X}_2, \ldots, \mathbf{X}_m$ are jointly independent, then

$$\begin{aligned}\mathbf{V}_{ij} &= cov(\mathbf{X}_i, \mathbf{X}_j) = E\left[(\mathbf{X}_i - \mu_i)(\mathbf{X}_j - \mu_j)^T\right] \\ &= E\left[(\mathbf{X}_i - \mu_i)\right] E\left[(\mathbf{X}_j - \mu_j)\right]^T = \mathbf{0} \times \mathbf{0}.\end{aligned}$$

Conversely, if $\mathbf{V}_{ij} = \mathbf{0}$, then the mfg for $\mathbf{X}$ is

$$\begin{aligned}m_{\mathbf{X}}(\mathbf{t}) &= exp\{\mathbf{t}^T\mu + \mathbf{t}^T\mathbf{V}\mathbf{t}/2\} = exp\left\{\sum_{i=1}^m \mathbf{t}_i^T\mu_i + \sum_{i=1}^m \mathbf{t}_i^T\mathbf{V}_{ii}\mathbf{t}_i/2\right\} \\ &= \prod_{i=1}^m exp\{\mathbf{t}_i^T\mu_i + \mathbf{t}_i^T\mathbf{V}_{ii}\mathbf{t}_i/2\} = m_{\mathbf{X}_1}(\mathbf{t}_1) \times m_{\mathbf{X}_2}(\mathbf{t}_2) \times \ldots \times m_{\mathbf{X}_m}(\mathbf{t}_m),\end{aligned}$$

so, by Result 5.2, we have joint independence. □

Corollary 5.3 *Let* $\mathbf{X} \sim N_p(\mu, \mathbf{V})$, *and* $\mathbf{Y}_1 = \mathbf{a}_1 + \mathbf{B}_1\mathbf{X}$, $\mathbf{Y}_2 = \mathbf{a}_2 + \mathbf{B}_2\mathbf{X}$, *then* $\mathbf{Y}_1$ *and* $\mathbf{Y}_2$ *are independent iff* $\mathbf{B}_1\mathbf{V}\mathbf{B}_2^T = \mathbf{0}$.

Proof: Let

$$\mathbf{Y} = \begin{bmatrix}\mathbf{Y}_1 \\ \mathbf{Y}_2\end{bmatrix} = \begin{bmatrix}\mathbf{a}_1 \\ \mathbf{a}_2\end{bmatrix} + \begin{bmatrix}\mathbf{B}_1 \\ \mathbf{B}_2\end{bmatrix}\mathbf{X},$$

then since $cov(\mathbf{Y}_1, \mathbf{Y}_2) = \mathbf{B}_1\mathbf{V}\mathbf{B}_2^T$, apply Result 5.4. □

Some words of emphasis are in order before leaving this section. The assumption of joint normality is an extremely powerful assumption as a great deal is known about how to work with various functions of jointly normal random variables. The results included in this chapter represent only a small fraction of this literature.

5.3 Chi-Square and Related Distributions

The key to this chapter is the distribution of quadratic forms in multivariate normal vectors. With the multivariate normal distribution established, we need to lay the groundwork for sums of squares of normals, first with the (central) chi-square distribution, then the noncentral chi-square. The remainder of the journey proceeds with many small steps. Result 5.5 is the first strong step toward the goal; the remaining ones are small details.

Definition 5.5 *Let* $\mathbf{Z} \sim N_p(\mathbf{0}, \mathbf{I}_p)$, *then* $\mathbf{U} = \mathbf{Z}^T\mathbf{Z} = \sum_{i=1}^p \mathbf{Z}_i^2$ *has the* chi-square distribution with p degrees of freedom, *denoted by* $U \sim \chi_p^2$.

The moment generating function for U can be computed directly from the normal distribution as

$$m_U(t) = E\left[e^{tU}\right] = E\left[exp\left\{t\sum_{i=1}^{p} Z_i^2\right\}\right]$$
$$= \prod_{i=1}^{p}\int_{-\infty}^{\infty}(2\pi)^{-\frac{1}{2}}\, exp\left\{tz_i^2 - \frac{1}{2}z_i^2\right\}dz_i = (1-2t)^{-\frac{p}{2}} \tag{5.3}$$

since

$$\int_{-\infty}^{\infty}(2\pi)^{-\frac{1}{2}}\, exp\left\{tz^2 - \frac{1}{2}z^2\right\}dz = \int_{-\infty}^{\infty}(2\pi)^{-\frac{1}{2}}\, exp\left\{-\frac{1}{2}(1-2t)z^2\right\}dz = (1-2t)^{-\frac{1}{2}}.$$

See also Exercise 5.5. The density for U can be derived following two routes, either obtaining the density for χ_1^2, and then getting the density of a convolution (see Exercise 5.6), or, with a good guess, as below, constructing the mgf to match (5.3). The density for $U \sim \chi_p^2$ is given by

$$p_U(u) = \frac{u^{(p-2)/2}e^{-u/2}}{\Gamma(p/2)2^{p/2}}$$

for $u > 0$, and zero otherwise. Obtaining the mgf from the density we have

$$m_U(t) = \int_0^{\infty} e^{tu}\, p_U(u)\, du = \int_0^{\infty}\frac{u^{(p-2)/2}e^{-u(\frac{1}{2}-t)}}{\Gamma(p/2)2^{p/2}}\, du$$
$$= \frac{\Gamma(p/2)\left(\frac{1}{2}-t\right)^{-p/2}}{\Gamma(p/2)2^{p/2}} = (1-2t)^{-p/2}$$

which is the same as (5.3). We will follow a more direct route for the noncentral chi-square.

Definition 5.6 *Let $J \sim Poisson\,(\phi)$, and $(U|J=j) \sim \chi_{p+2j}^2$, then unconditionally, U has the noncentral chi-square distribution with noncentrality parameter ϕ, denoted by $U \sim \chi_p^2(\phi)$.*

Using the characterization above, the density of the noncentral χ^2 can be written as a Poisson-weighted mixture:

$$p_U(u) = \sum_{j=0}^{\infty}\left[\frac{e^{-\phi}\phi^j}{j!}\right] \times \frac{u^{(p+2j-2)/2}e^{-u/2}}{\Gamma\left(\frac{p+2j}{2}\right)2^{j+p/2}} \tag{5.4}$$

for $u > 0$ and zero otherwise. The reader should note that the definition for the noncentrality parameter varies across the statistical literature. Here we use the Poisson rate parameter as the noncentrality parameter, as does Searle [42]. However, Rao [37], among others, defines the noncentrality parameter as twice the rate parameter.

Result 5.5 *If $U \sim \chi_p^2(\phi)$, then its mgf is $m_U(t) = (1-2t)^{-p/2} exp\{2\phi t/(1-2t)\}$.*

Proof: Taking the conditional route rather than directly using the density and employing Result 5.8, we have

$$
\begin{aligned}
E\left[e^{tU}\right] = E\left[E\left[e^{tU}|J=j\right]\right] &= E\left[(1-2t)^{-(p+2J)/2}\right] \\
&= \sum_{j=0}^{\infty}(1-2t)^{-(p+2j)/2}\phi^j e^{-\phi}/j! \\
&= (1-2t)^{-p/2}e^{-\phi}\sum_{j=0}^{\infty}[\phi/(1-2t)]^j/j! \\
&= (1-2t)^{-p/2}e^{-\phi}e^{\phi/(1-2t)}.
\end{aligned}
$$

□

Result 5.6 *If $U \sim \chi_p^2(\phi)$, then $E(U) = p + 2\phi$ and $Var(U) = 2p + 8\phi$.*

Proof: Exercise 5.11. □

Result 5.7 *If $U_1, U_2, \ldots, U_m$ are jointly independent, and $U_i \sim \chi_{p_i}^2(\phi_i)$, then $U = \sum_{i=1}^m U_i \sim \chi_p^2(\phi)$ where $p = \sum_{i=1}^m p_i$ and $\phi = \sum_{i=1}^m \phi_i$.*

Proof: Obtaining the mgf for U we have

$$
\begin{aligned}
m_U(t) &= E\left[e^{t(\sum U_i)}\right] = \prod_{i=1}^m m_{U_i}(t) = \prod_{i=1}^m \left[(1-2t)^{-p_i/2} exp\{2t\phi_i/(1-2t)\}\right] \\
&= (1-2t)^{-p/2} exp\{2t\phi/(1-2t)\}.
\end{aligned}
$$

□

Result 5.8 *If $X \sim N(\mu, 1)$, then $U = X^2 \sim \chi_1^2(\mu^2/2)$.*

Proof: Finding the moment generating function for U, we have

$$
\begin{aligned}
m_U(t) = E\left[e^{tX^2}\right] &= \int_{-\infty}^{\infty}(2\pi)^{-\frac{1}{2}} exp\{tx^2 - (x-\mu)^2/2\}dx \\
&= \int_{-\infty}^{\infty}(2\pi)^{-\frac{1}{2}} exp\left\{-\frac{1}{2}\left[x^2 - 2x\mu + \mu^2 - 2tx^2\right]\right\}dx \\
&= \int_{-\infty}^{\infty}(2\pi)^{-\frac{1}{2}} exp\left\{-(1-2t)(x-\mu/(1-2t))^2/2\right\}dx \\
&\quad \times exp\left\{-\frac{1}{2}(\mu^2 - \mu^2/(1-2t)\right\} \\
&= (1-2t)^{-\frac{1}{2}} \times exp\left\{\left(\frac{1}{2}\mu^2\right)2t/(1-2t)\right\}.
\end{aligned}
$$

The missing algebra on the square brackets is

$$[.] = x^2(1-2t) - 2x\mu + \mu^2 = (1-2t)(x - \mu/(1-2t))^2 + \mu^2 - \mu^2/(1-2t).$$

□

Result 5.9 *If $X \sim N_p(\mu, \mathbf{I}_p)$, then $W = \mathbf{X}^T\mathbf{X} = \sum_{i=1}^p X_i^2 \sim \chi_p^2\left(\frac{1}{2}\mu^T\mu\right)$.*

Proof: Since $W = \sum_{i=1}^p U_i$ where U_i are independent (since $V_{ij} = 0$ for $i \neq j$), and $U_i \sim \chi_{p_i}^2(\phi_i)$ where $p_i = 1$, $\phi_i = \frac{1}{2}\mu_i^2$, Result 5.7 provides the result, since $\sum_{i=1}^p \phi_i = \frac{1}{2}\mu^T\mu$. □

Result 5.10 *If $\mathbf{X} \sim N_p(\mu, \mathbf{V})$ where $\mathbf{V}$ is nonsingular, then $W = \mathbf{X}^T\mathbf{V}^{-1}\mathbf{X} \sim \chi_p^2(\frac{1}{2}\mu^T\mathbf{V}^{-1}\mu)$.*

Proof: Since $\mathbf{V}$ is nonsingular, we can construct a nonsingular matrix $\mathbf{A}$ such that $\mathbf{A}^T\mathbf{A} = \mathbf{V}$. So define $\mathbf{Z} = \mathbf{A}^{-T}\mathbf{X}$, then $\mathbf{Z} \sim N_p(\mathbf{A}^{-T}\mu, \mathbf{A}^{-T}\mathbf{A}^T\mathbf{A}\mathbf{A}^{-1} = \mathbf{I}_p)$, and using Result 5.9 we have

$$\mathbf{Z}^T\mathbf{Z} = W = \mathbf{X}^T\mathbf{V}^{-1}\mathbf{X} \sim \chi_p^2\left(\frac{1}{2}\mu^T\mathbf{A}^{-1}\mathbf{A}^{-T}\mu = \frac{1}{2}\mu^T\mathbf{V}^{-1}\mu\right).$$

□

These two results (5.9 and 5.10) are key for finding the distribution of quadratic forms—sums of squares—in the linear model. The next result shows that an increasing noncentrality parameter leads to stochastically larger random variables, as Figures 5.1 and 5.2 illustrate. This property carries over to the F-distribution, whose definition follows, with the implication that the tests we will construct in Chapter 6 will be unbiased. This discourse on distributions concludes with the definition of the Student's t-distribution.

Result 5.11 *Let $U \sim \chi_p^2(\phi)$, then $Pr(U > c)$ is strictly increasing in ϕ for fixed p and $c > 0$.*

Proof: First define

$$v_k = Pr\left(U > c | U \sim \chi_k^2\right) = \int_c^\infty \frac{u^{(k-2)/2}e^{-u/2}}{\Gamma\left(\frac{k}{2}\right)2^{k/2}}\, du\,.$$

Take the derivative with respect to ϕ of

$$Pr(U > c) = \sum_{j=0}^{\infty}\left[\frac{e^{-\phi}\phi^j}{j!}\right] \times \int_c^\infty \frac{u^{(p+2j-2)/2}e^{-u/2}}{\Gamma\left(\frac{p+2j}{2}\right)2^{j+p/2}}\, du = \sum_{j=0}^{\infty}\left[\frac{e^{-\phi}\phi^j}{j!}\right] \times v_{p+2j}$$

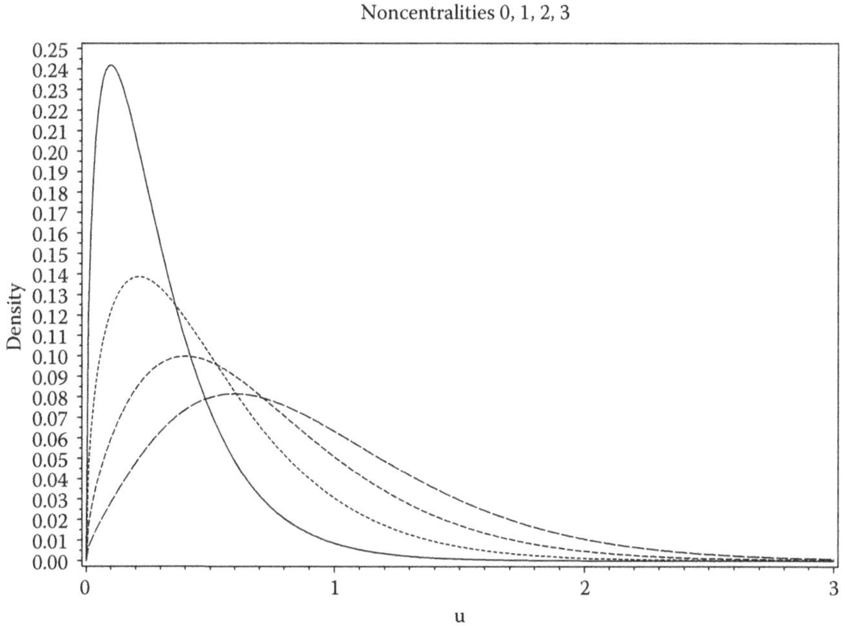

Figure 5.1: Noncentral chi-square densities with 3 degrees of freedom. Increasing the noncentrality parameter flattens the density and shifts the distribution to the right.

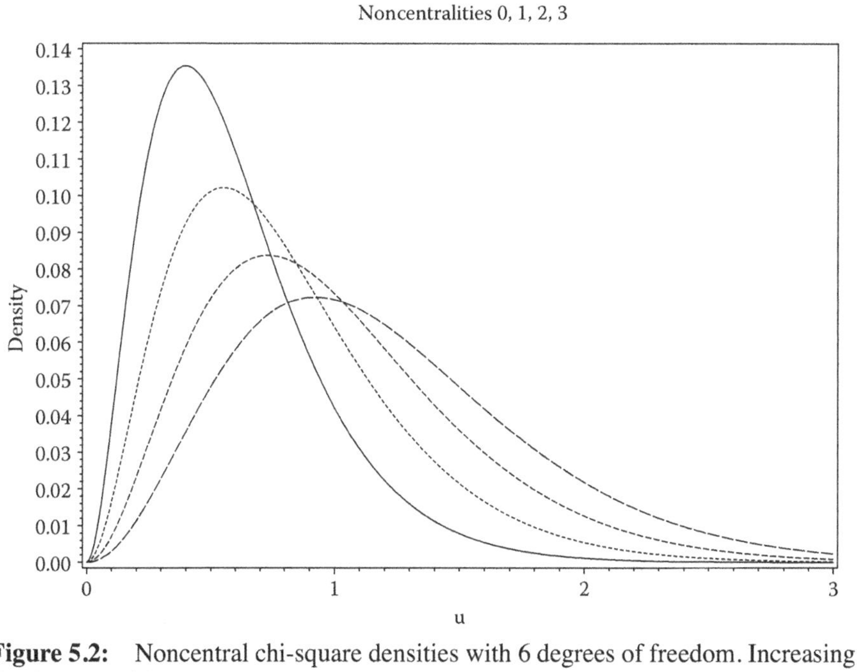

Figure 5.2: Noncentral chi-square densities with 6 degrees of freedom. Increasing the degrees of freedom parameter also flattens the density and shifts the distribution to the right.

$$dPr(U > c)/d\phi = \sum_{j=0}^{\infty} \left[\frac{j\phi^{j-1}e^{-\phi} - e^{-\phi}\phi^j}{j!} \right] \times v_{p+2j}$$

$$= \sum_{j=1}^{\infty} \left[\frac{e^{-\phi}\phi^{j-1}}{(j-1)!} \right] \times v_{p+2j} - \sum_{j=0}^{\infty} \left[\frac{e^{-\phi}\phi^j}{j!} \right] \times v_{p+2j}$$

$$= \sum_{k=0}^{\infty} \left[\frac{e^{-\phi}\phi^k}{k!} \right] \times v_{p+2k+2} - \sum_{j=0}^{\infty} \left[\frac{e^{-\phi}\phi^j}{j!} \right] \times v_{p+2j}$$

$$= \sum_{k=0}^{\infty} \left[\frac{e^{-\phi}\phi^k}{k!} \right] \times \{v_{p+2k+2} - v_{p+2k}\}.$$

The change in index from j to k uses $k = j - 1$; the last step reestablishes a common index. See Exercise 5.7 for a (positive) expression for the quantity in braces $\{.\}$ above.

$\square$

Definition 5.7   Let U_1 and U_2 be independent random variables, with $U_1 \sim \chi^2_{p_1}$ and $U_2 \sim \chi^2_{p_2}$; then $F = \frac{U_1/p_1}{U_2/p_2}$ has the F-distribution with p_1 and p_2 degrees of freedom, denoted as $F \sim F_{p_1, p_2}$.

Result 5.12   The density of the F-distribution with p_1 and p_2 degrees of freedom is

$$p_F(f) = \frac{\Gamma\left(\frac{p_1+p_2}{2}\right)(p_2/p_1)^{p_1/2}}{\Gamma\left(\frac{1}{2}p_1\right)\Gamma\left(\frac{1}{2}p_2\right)} f^{(p_1/2)-1} \left(1 + \frac{p_1}{p_2}f\right)^{-(p_1+p_2)/2}$$

for $f > 0$ and zero otherwise.

Proof: Exercise 5.9. $\square$

Definition 5.8   Let U_1 and U_2 be independent random variables, with $U_1 \sim \chi^2_{p_1}(\phi)$ and $U_2 \sim \chi^2_{p_2}$; then $F = \frac{U_1/p_1}{U_2/p_2}$ has the noncentral F-distribution with p_1 and p_2 degrees of freedom, noncentrality ϕ, denoted as $F \sim F_{p_1, p_2}(\phi)$.

Result 5.13   Let $W \sim F_{p_1, p_2}(\phi)$; then for fixed p_1, p_2 and $c > 0$, $Pr(W > c)$ is strictly increasing in ϕ.

Proof: Write $W = \frac{U_1/p_1}{U_2/p_2}$ where U_1 and U_2 are independent, with $U_1 \sim \chi^2_{p_1}(\phi)$ and $U_2 \sim \chi^2_{p_2}$, then we have

$$Pr(W > c) = Pr\left(\frac{U_1/p_1}{U_2/p_2} > c\right) = Pr\left(U_1 > \frac{p_1}{p_2}cU_2\right)$$

$$= \int_0^{\infty} Pr\left(U_1 > \frac{p_1}{p_2}cu_2 | U_2 = u_2\right) p_{U_2}(u_2)du_2.$$

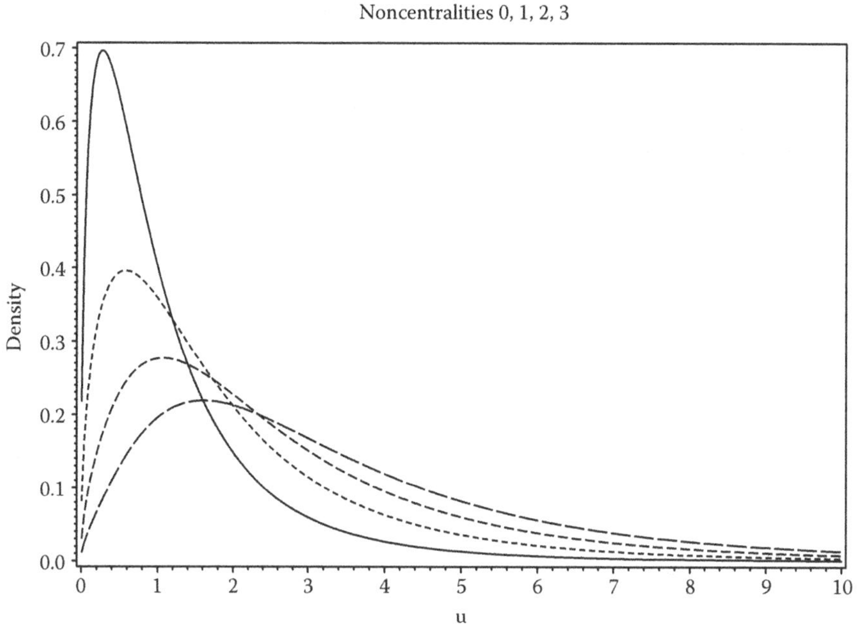

Figure 5.3: Noncentral F densities with 3 and 10 degrees of freedom. Increasing the noncentrality parameter flattens the density and shifts the distribution to the right.

From Result 5.11, the conditional probability is strictly increasing in ϕ for fixed p_1, p_2, c, and u_2. □

Figures 5.3 and 5.4 illustrate this property of the noncentral F-distribution. The simpler one-dimensional problems require the Student's t-distribution. See Exercise 5.10.

Definition 5.9 *Let $U \sim N(\mu, 1)$ and $V \sim \chi^2_k$. If U and V are independent, then $T = U/\sqrt{V/k}$ has the noncentral Student's t-distribution with k degrees of freedom and noncentrality μ, denoted $T \sim t_k(\mu)$. If $\mu = 0$, the distribution is generally known as Student's t, denoted by $T \sim t_k$, and whose density is given by*

$$p_T(t) = \textit{constant} \times (1 + t^2/k)^{-(k+1)/2},$$

where the complicated constant above is

$$\frac{\Gamma((k+1)/2)}{\Gamma(k/2)\sqrt{\pi k}}.$$

Note that if $T \sim t_k(\mu)$, then $T^2 \sim F_{1,k}(\frac{1}{2}\mu^2)$.

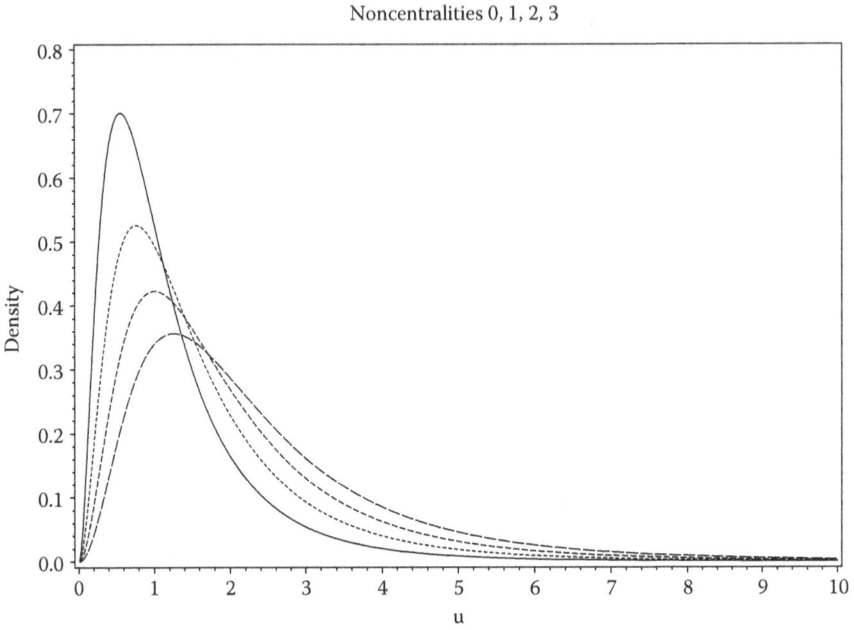

Figure 5.4: Noncentral F densities with 6 and 10 degrees of freedom. F densities have much heavier tails than chi-square densities.

5.4 Distribution of Quadratic Forms

The groundwork of the previous section prepares the way to find the distribution of sums of squares in the linear model with normally distributed errors. As we have written the linear model in terms of the errors as $\mathbf{y} = \mathbf{Xb} + \mathbf{e}$, the assumption we are placing on these errors is

$$\mathbf{e} \sim N_N(\mathbf{0}, \sigma^2 \mathbf{I}_N).$$

Notice that we could also write the model directly in terms of the response $\mathbf{y}$:

$$\mathbf{y} \sim N_N(\mathbf{Xb}, \sigma^2 \mathbf{I}_N),$$

so that the family of distributions for $\mathbf{y}$ is parameterized by $\mathbf{b}$ and σ^2.

Recall that the linear least squares problem led to the decomposition of the response vector into two pieces, one in the column space of the design matrix $\mathbf{X}$, and the second piece orthogonal to it:

$$\mathbf{y} = \mathbf{P_X y} + (\mathbf{I} - \mathbf{P_X})\mathbf{y}.$$

Using just moment assumptions in the Gauss–Markov model, the first piece $\mathbf{P_X y}$ was used in the estimation of coefficients $\mathbf{b}$, while the second piece was used to estimate the variance: $\hat{\sigma}^2 = \mathbf{y}^T(\mathbf{I} - \mathbf{P_X})\mathbf{y}/(N - r)$. Moreover, applying the Pythagorean Theorem gave a decomposition of the squared length of the response vector:

$$\mathbf{y}^T\mathbf{y} = \mathbf{y}^T\mathbf{P_X}\mathbf{y} + \mathbf{y}^T(\mathbf{I} - \mathbf{P_X})\mathbf{y}.$$

Now we have placed distributional assumptions on the errors, and our immediate goal is to construct the distribution of these sums of squares, which are quadratic forms in multivariate normal vectors with projection matrices.

Again, our approach is a number of small steps. While each one may look obvious at first glance, and, perhaps, making no apparent contribution to the goal, the reader is reminded of stopping to consider the consequences of each step.

Lemma 5.1 *A $p \times p$ symmetric matrix $\mathbf{A}$ is idempotent with rank s iff there exists a $p \times s$ matrix $\mathbf{G}$ with orthonormal columns, that is, $\mathbf{G}^T\mathbf{G} = \mathbf{I}_s$, such that $\mathbf{A} = \mathbf{G}\mathbf{G}^T$.*

Proof: (If) If $\mathbf{A} = \mathbf{G}\mathbf{G}^T$, then $\mathbf{A}^2 = \mathbf{G}(\mathbf{G}^T\mathbf{G})\mathbf{G}^T = \mathbf{G}\mathbf{G}^T = \mathbf{A}$. (Only if) If $\mathbf{A}$ is idempotent, then its eigenvalues are 0 or 1 and its rank is equal to the number of nonzero eigenvalues. Since $\mathbf{A}$ is symmetric, we can write its spectral decomposition as

$$\mathbf{A} = \mathbf{Q}\mathbf{\Lambda}\mathbf{Q}^T = [\mathbf{Q}_1 \quad \mathbf{Q}_2]\begin{bmatrix}\mathbf{I}_s & \mathbf{0}\\ \mathbf{0} & \mathbf{0}\end{bmatrix}\begin{bmatrix}\mathbf{Q}_1^T\\ \mathbf{Q}_2^T\end{bmatrix} = \mathbf{Q}_1\mathbf{Q}_1^T$$

where $\mathbf{Q}$ is the orthogonal matrix whose columns are eigenvectors of $\mathbf{A}$, partitioned by columns according to the rank of $\mathbf{A}$. Choose $\mathbf{G}$ as $\mathbf{Q}_1$. □

Example 5.1: Gram–Schmidt
From Exercise 2.17, if $\mathbf{X} = \mathbf{QR}$ where $\mathbf{Q}$ has orthonormal columns and $rank(\mathbf{X}) = rank(\mathbf{Q})$, then $\mathbf{P_X} = \mathbf{Q}\mathbf{Q}^T$. For many cases, the matrix $\mathbf{G}$ can be constructed via Gram–Schmidt. See also Exercises 2.22–2.24.

Example 5.2: Examples 2.6 and 2.10 Revisited
Recall the design matrix $\mathbf{X}$ from the simple linear regression model $y_i = \beta_0 + \beta_1 x_i + e_i$ with $x_i = i, i = 1, \ldots, 4 = N$.

$$\mathbf{X} = \begin{bmatrix}1 & 1\\ 1 & 2\\ 1 & 3\\ 1 & 4\end{bmatrix} \quad \text{and} \quad \mathbf{P_X} = \frac{1}{10}\begin{bmatrix}7 & 4 & 1 & -2\\ 4 & 3 & 2 & 1\\ 1 & 2 & 3 & 4\\ -2 & 1 & 4 & 7\end{bmatrix} = \mathbf{G}\mathbf{G}^T,$$

$$\text{where } \mathbf{G} = \begin{bmatrix}1/2 & -3/\sqrt{20}\\ 1/2 & -1/\sqrt{20}\\ 1/2 & 1/\sqrt{20}\\ 1/2 & 3/\sqrt{20}\end{bmatrix}.$$

Recall also the Gram–Schmidt orthonormalization on a similar matrix, which produced the first column of $\mathbf{G}$ as $\mathbf{Q}_{.1}$ and the second column as $\mathbf{Q}_{.2}$.

Result 5.14 *Let* $\mathbf{X} \sim N_p(\mu, \mathbf{I}_p)$ *and* $\mathbf{A}$ *be symmetric; then if* $\mathbf{A}$ *is idempotent with rank* s*, then* $\mathbf{X}^T\mathbf{A}\mathbf{X} \sim \chi^2_s(\phi = \frac{1}{2}\mu^T\mathbf{A}\mu)$.

Proof: Using Lemma 5.1, $\mathbf{A} = \mathbf{G}\mathbf{G}^T$. Then $\mathbf{G}^T\mathbf{X} \sim N_s(\mathbf{G}^T\mu, \mathbf{I}_s)$, and so

$$\mathbf{X}^T\mathbf{A}\mathbf{X} = \|\mathbf{G}^T\mathbf{X}\|^2 \sim \chi^2_s\left(\phi = \frac{1}{2}\mu^T\mathbf{A}\mu\right). \qquad \square$$

Result 5.15 *(General case) Let* $\mathbf{X} \sim N_p(\mu, \mathbf{V})$ *with* $\mathbf{V}$ *nonsingular, and let* $\mathbf{A}$ *be a symmetric matrix; then if* $\mathbf{AV}$ *is idempotent with rank* s*, then* $\mathbf{X}^T\mathbf{A}\mathbf{X} \sim \chi^2_s(\phi = \frac{1}{2}\mu^T\mathbf{A}\mu)$.

Proof: Factor positive definite $\mathbf{V} = \mathbf{\Gamma}\mathbf{\Gamma}^T$, and then construct $\mathbf{Y} = \mathbf{\Gamma}^{-1}\mathbf{X}$, so that $\mathbf{Y} = \mathbf{\Gamma}^{-1}\mathbf{X} \sim N_p(\eta, \mathbf{I}_p)$, where $\eta \equiv \mathbf{\Gamma}^{-1}\mu$, and $\mathbf{Y}^T\mathbf{B}\mathbf{Y} = \mathbf{X}^T\mathbf{A}\mathbf{X}$ for $\mathbf{B} = \mathbf{\Gamma}^T\mathbf{A}\mathbf{\Gamma}$. By Result 5.14, $\mathbf{Y}^T\mathbf{B}\mathbf{Y} \sim \chi^2_s(\phi)$, where $\phi = \frac{1}{2}\mu^T\mathbf{A}\mu = \frac{1}{2}\eta^T\mathbf{B}\eta$ if $\mathbf{B}$ is idempotent. Notice that since $\mathbf{AV}$ is idempotent, $\mathbf{AV} = \mathbf{AVAV}$ and $\mathbf{A} = \mathbf{AVA}$; pre- and post-multiplication give

$$\mathbf{B} = \mathbf{\Gamma}^T\mathbf{A}\mathbf{\Gamma} = \mathbf{\Gamma}^T\mathbf{AVA}\mathbf{\Gamma} = \mathbf{\Gamma}^T\mathbf{A}\mathbf{\Gamma}\mathbf{\Gamma}^T\mathbf{A}\mathbf{\Gamma} = \mathbf{B}^2,$$

so that $\mathbf{B}$ is idempotent. Note also that *rank* $(\mathbf{B})$ = *rank* $(\mathbf{AV})$. $\square$

It should be easy to see the application of these results to the case of the linear model with normally distributed errors, where $\mathbf{y} \sim N_N(\mathbf{Xb}, \sigma^2\mathbf{I}_N)$, using $\mathbf{V} = \sigma^2\mathbf{I}_N$. First, using $\mathbf{A} = (1/\sigma^2)(\mathbf{I} - \mathbf{P_X})$, we have

$$SSE/\sigma^2 = \mathbf{y}^T\mathbf{A}\mathbf{y} \sim \chi^2_{(N-r)},$$

since the noncentrality parameter is $\frac{1}{2}(\mathbf{Xb})^T(1/\sigma^2)(\mathbf{I} - \mathbf{P_X})(\mathbf{Xb}) = 0$. However, now using $\mathbf{A} = (1/\sigma^2)\mathbf{P_X}$, we have

$$SSR/\sigma^2 = \mathbf{y}^T\mathbf{A}\mathbf{y} \sim \chi^2_r\left(\frac{1}{2}(\mathbf{Xb})^T(\mathbf{Xb})/\sigma^2\right).$$

Notice that the noncentrality parameter above depends on both the size of the mean vector $\mathbf{Xb}$ and the variance σ^2. As for the joint distribution of these sums of squares, note that

$$\begin{bmatrix}\hat{\mathbf{y}}\\ \hat{\mathbf{e}}\end{bmatrix} = \begin{bmatrix}\mathbf{P_X}\\ \mathbf{I} - \mathbf{P_X}\end{bmatrix}\mathbf{y} \sim N_{2N}\left(\begin{bmatrix}\mathbf{Xb}\\ \mathbf{0}\end{bmatrix}, \begin{bmatrix}\sigma^2\mathbf{P_X} & \mathbf{0}\\ \mathbf{0} & \sigma^2(\mathbf{I} - \mathbf{P_X})\end{bmatrix}\right),$$

so that $SSE = \|\hat{\mathbf{e}}\|^2$ and $SSR = \|\hat{\mathbf{y}}\|^2$ are independent since $\hat{\mathbf{y}}$ and $\hat{\mathbf{e}}$ are independent. Completing these calculations is

$$F = \frac{\|\hat{\mathbf{y}}\|^2/r}{\|\hat{\mathbf{e}}\|^2/(N-r)} \sim F_{r,N-r}\left(\frac{1}{2}\|\mathbf{Xb}\|^2/\sigma^2\right). \qquad (5.5)$$

The independence of linear and quadratic forms in normal random vectors can be established with Result 5.16.

Result 5.16 *Let* $\mathbf{X} \sim N_p(\mu, \mathbf{V})$ *and* $\mathbf{A}$ *be symmetric with rank* s*; if* $\mathbf{BVA} = \mathbf{0}$*, then* $\mathbf{BX}$ *and* $\mathbf{X}^T\mathbf{AX}$ *are independent. Here* $\mathbf{B}$ *is* $q \times p$.

Proof: Construct the spectral decomposition of $\mathbf{A} = \mathbf{Q}_1\mathbf{\Lambda}_1\mathbf{Q}_1^T$ where $\mathbf{\Lambda}_1$ is $s \times s$ with $s = rank(\mathbf{A})$, so it's nonsingular. Construct then the following joint normal distribution:

$$\begin{bmatrix} \mathbf{BX} \\ \mathbf{Q}_1^T\mathbf{X} \end{bmatrix} = \begin{bmatrix} \mathbf{B} \\ \mathbf{Q}_1^T \end{bmatrix} \mathbf{X} \sim N_{q+s}\left(\begin{bmatrix} \mathbf{B}\mu \\ \mathbf{Q}_1^T\mu \end{bmatrix}, \begin{bmatrix} \mathbf{BVB}^T & \mathbf{BVQ}_1 \\ \mathbf{Q}_1^T\mathbf{VB}^T & \mathbf{Q}_1^T\mathbf{VQ}_1 \end{bmatrix} \right).$$

So if $\mathbf{BVQ}_1 = \mathbf{0}$, then $\mathbf{BX}$ and $\mathbf{Q}_1^T\mathbf{X}$ are independent, or $\mathbf{BX}$ and $\mathbf{X}^T\mathbf{AX}$ are independent. But if $\mathbf{BVA} = \mathbf{0}$, then $\mathbf{BVA} = \mathbf{0} = \mathbf{BVQ}_1\mathbf{\Lambda}_1\mathbf{Q}_1^T$. Now first postmultiply by $\mathbf{Q}_1$, then by $\mathbf{\Lambda}_1^{-1}$, so that if $\mathbf{BVA} = \mathbf{0}$, then $\mathbf{BVQ}_1 = \mathbf{0}$, and we have the independence of $\mathbf{BX}$ and $\mathbf{X}^T\mathbf{AX}$. See also Exercise 5.16. □

Corollary 5.4 *Let* $\mathbf{X} \sim N_p(\mu, \mathbf{V})$*,* $\mathbf{A}$ *be symmetric with rank* r*, and* $\mathbf{B}$ *be symmetric with rank* s*; if* $\mathbf{BVA} = \mathbf{0}$*, then* $\mathbf{X}^T\mathbf{AX}$ *and* $\mathbf{X}^T\mathbf{BX}$ *are independent.*

Proof: See Exercise 5.27. □

5.5 Cochran's Theorem

The fundamental step in the analysis of linear models is the decomposition of the sums of squares. Decomposing the response vector $\mathbf{y}$ into two orthogonal pieces, we obtain a version of the Pythagorean Theorem,

$$\mathbf{y}^T\mathbf{y} = \mathbf{y}^T\mathbf{P_X}\mathbf{y} + \mathbf{y}^T(\mathbf{I} - \mathbf{P_X})\mathbf{y}.$$

While some simpler steps paved the way for deriving the test for the general linear hypothesis, the more general decomposition is the goal of Cochran's Theorem. While this is often stated and proved in the greatest generality, the proof presented here provides the most commonly sought results; further related topics are covered in Exercises 5.18–5.20.

Theorem 5.1 *(Cochran's Theorem) Let* $\mathbf{y} \sim N_N(\mu, \sigma^2\mathbf{I}_N)$ *and let* $\mathbf{A}_i$*,* $i = 1, \ldots, k$ *be symmetric idempotent matrices with rank* s_i*. If* $\sum_{i=1}^k \mathbf{A}_i = \mathbf{I}_N$*, then* $(1/\sigma^2)\mathbf{y}^T\mathbf{A}_i\mathbf{y}$ *are independently distributed as* $\chi^2_{s_i}(\phi_i)$*, with* $\phi_i = \frac{1}{2\sigma^2}\mu^T\mathbf{A}_i\mu$ *and* $\sum_{i=1}^k s_i = N$.

Proof: Since $\mathbf{A}_i$ are idempotent, then by Lemma 5.1, we can write each one as $\mathbf{A}_i = \mathbf{Q}_i\mathbf{Q}_i^T$ where $\mathbf{Q}_i$ is a matrix of size $N \times s_i$ with rank s_i and orthonormal columns, that is, $\mathbf{Q}_i^T\mathbf{Q}_i = \mathbf{I}_{s_i}$. Now construct the matrix $\mathbf{Q}$ by stacking these $\mathbf{Q}_i$ side by side:

$$\mathbf{Q} = [\mathbf{Q}_1\ \mathbf{Q}_2\ \ldots\ \mathbf{Q}_k]$$

where each partition has N rows and s_i columns, so that $\mathbf{Q}$ has $S = \sum_{i=1}^k s_i$ columns. Note, however, that

$$\mathbf{Q}\mathbf{Q}^T = \mathbf{Q}_1\mathbf{Q}_1^T + \mathbf{Q}_2\mathbf{Q}_2^T + \cdots + \mathbf{Q}_k\mathbf{Q}_k^T = \mathbf{A}_1 + \mathbf{A}_2 + \cdots + \mathbf{A}_k = \mathbf{I}_N.$$

Hence also $trace\,(\mathbf{Q}\mathbf{Q}^T) = \sum_{i=1}^k trace\,(\mathbf{A}_i) = \sum_{i=1}^k s_i = S = trace\,(\mathbf{I}_N) = N$, so $S = N$, $\mathbf{Q}$ is square, $\mathbf{Q}^T$ is its inverse, and $\mathbf{Q}$ is an orthogonal matrix. Multiplying in the other order, we then have

$$\mathbf{Q}^T\mathbf{Q} = \begin{bmatrix} \mathbf{Q}_1^T \\ \mathbf{Q}_2^T \\ \ldots \\ \mathbf{Q}_k^T \end{bmatrix} [\mathbf{Q}_1\ \mathbf{Q}_2\ \ldots\ \mathbf{Q}_k] = \mathbf{I}_N = \begin{bmatrix} \mathbf{I}_{s_1} & \mathbf{0} & \mathbf{0} & \mathbf{0} \\ \mathbf{0} & \mathbf{I}_{s_2} & \mathbf{0} & \mathbf{0} \\ \ldots & \ldots & \ldots & \ldots \\ \mathbf{0} & \mathbf{0} & \mathbf{0} & \mathbf{I}_{s_k} \end{bmatrix}$$

so that $\mathbf{Q}_i^T\mathbf{Q}_j = \mathbf{0}$ for $i \neq j$. Since $\mathbf{Q}^T\mathbf{y} \sim N_N(\mathbf{Q}^T\mu, \sigma^2\mathbf{I}_N)$ with the block diagonal structure as above, $\mathbf{Q}_i^T\mathbf{y}$ are independent, with $\mathbf{Q}_i^T\mathbf{y} \sim N_{s_i}(\mathbf{Q}_i^T\mu, \sigma^2\mathbf{I}_{s_i})$. Similarly, $\mathbf{y}^T\mathbf{A}_i\mathbf{y} = \|\mathbf{Q}_i^T\mathbf{y}\|^2$ are independent, and dividing by σ, we have $(1/\sigma)\mathbf{Q}_i^T\mathbf{y} \sim N_{s_i}((1/\sigma)\mathbf{Q}_i^T\mu, \mathbf{I}_{s_i})$, and, finally, $\|(1/\sigma)\mathbf{Q}_i^T\mathbf{y}\|^2 = (1/\sigma^2)\mathbf{y}^T\mathbf{A}_i\mathbf{y} \sim \chi^2_{s_i}(\frac{1}{2\sigma^2}\mu^T\mathbf{A}_i\mu)$. □

Example 5.3: Cochran for ANOVA

Again, we'll return to ANOVA for the simplest application of Cochran's Theorem. Let $y_{ij} = \mu + \alpha_i + e_{ij}$ with e_{ij} *iid* $N(0, \sigma^2)$, or $\mathbf{y} \sim N_N(\mathbf{Xb}, \sigma^2\mathbf{I}_N)$ where $\mathbf{Xb}$ is

$$\begin{bmatrix} \mathbf{1}_{n_1} & \mathbf{1}_{n_1} & \mathbf{0} & \ldots & \mathbf{0} \\ \mathbf{1}_{n_2} & \mathbf{0} & \mathbf{1}_{n_2} & \ldots & \mathbf{0} \\ \ldots & \ldots & \ldots & & \ldots \\ \mathbf{1}_{n_a} & \mathbf{0} & \mathbf{0} & \ldots & \mathbf{1}_{n_a} \end{bmatrix} \begin{bmatrix} \mu \\ \alpha_1 \\ \ldots \\ \alpha_a \end{bmatrix}.$$

Partitioning $\mathbf{X} = [\mathbf{1}_N\ \mathbf{X}_1]$, we can construct the projection matrices $\mathbf{A}_1 = \mathbf{P_1}$, $\mathbf{A}_2 = \mathbf{P_X} - \mathbf{P}_1 = \mathbf{P_W}$, and $\mathbf{A}_3 = \mathbf{I} - \mathbf{P_X}$, where $\mathbf{W} = (\mathbf{I} - \mathbf{P}_1)\mathbf{X}_1$. Then we can see $\mathbf{A}_1 + \mathbf{A}_2 + \mathbf{A}_3 = \mathbf{I}$, and so we have the usual decomposition of sums of squares:

$$\mathbf{y}^T\mathbf{A}_1\mathbf{y} = SSM = N\overline{y}^2 \text{ and } SSM/\sigma^2 \sim \chi_1^2\left(\frac{1}{2}\eta^T\mathbf{P_1}\eta/\sigma^2\right)$$

$$\mathbf{y}^T\mathbf{A}_2\mathbf{y} = SSA_{cfm} = \sum_{i=1}^a n_i\overline{y}_{i.}^2 - N\overline{y}^2 \text{ and } SSA_{cfm}/\sigma^2 \sim \chi_{a-1}^2\left(\frac{1}{2}\eta^T\mathbf{P_W}\eta/\sigma^2\right)$$

$$\mathbf{y}^T\mathbf{A}_3\mathbf{y} = SSE = \sum_{i=1}^a\sum_{j=1}^{n_i}(y_{ij} - \overline{y}_{i.})^2 \text{ and } SSE/\sigma^2 \sim \chi_{N-a}^2\left(\frac{1}{2}\eta^T(\mathbf{I} - \mathbf{P_X})\eta/\sigma^2\right)$$

ANOVA Table

Source	df	Projection	SS	Noncentrality
Mean	1	$\mathbf{P}_1$	$SSM = N\overline{y}^2$	$\frac{1}{2}N(\mu+\overline{\alpha})^2/\sigma^2$
Group	$a-1$	$\mathbf{P}_\mathbf{W}$	$SSA_{cfm} = \sum_{i=1}^{a} n_i \overline{y}_{i.}^2 - N\overline{y}^2$	$\frac{1}{2}\sum_{i=1}^{a}(\alpha_i - \overline{\alpha})^2/\sigma^2$
Error	$N-a$	$\mathbf{I}-\mathbf{P}_\mathbf{X}$	$SSE = \sum_{i=1}^{a}\sum_{j=1}^{n_i}(y_{ij} - \overline{y}_{i.})^2$	0

and the arrangement in the ANOVA table can include two new columns, one for the projection matrix and one for the noncentrality parameter using a general form $E(\mathbf{y}) = \eta$. If the model is correct, that is, $\eta = \mathbf{Xb} \in \mathcal{C}(\mathbf{X})$, the noncentralities simplify, and SSE/σ^2 has a central χ^2 distribution.

5.6 Regression Models with Joint Normality

Up to this point, we have always considered the elements of the design matrix $\mathbf{X}$ to be fixed and known without error. This viewpoint is unquestioned for ANOVA and similar models, and appropriate for many other applications such as designed experiments where covariates are included. However, in observational studies, where individuals are sampled from a population, we may want to alter that viewpoint, seeing the response Y and covariate X as sampled from a joint population. We may still wish to understand the mean function of the response Y for given values of the explanatory variable X, that is, conditional on $X = x$.

To examine further the linear model with both variables sampled from a population, consider first the simplified version with a single covariate, that is, simple linear regression. We begin with the same model as we have been working with all along, that is,

$$Y = \beta_0 + \beta_1 x + e$$

but now we view this conditional on $X = x$. Using our normal linear model, we may write this as

$$(Y|X = x) \sim N(\beta_0 + \beta_1 x, \sigma^2) \tag{5.6}$$

as describing the conditional distribution of the response Y given that the explanatory variable X takes the value x. If now we assume that X follows its own normal distribution, say, $X \sim N(\mu_x, \Sigma_{xx})$, we can write the joint distribution as a bivariate normal,

$$\begin{bmatrix} Y \\ X \end{bmatrix} \sim N_2\left(\begin{bmatrix} \beta_0 + \beta_1\mu_x \\ \mu_x \end{bmatrix}, \begin{bmatrix} \sigma^2 + \beta_1^2\Sigma_{xx} & \beta_1\Sigma_{xx} \\ \beta_1\Sigma_{xx} & \Sigma_{xx} \end{bmatrix}\right). \tag{5.7}$$

Beginning from the other direction with some different notation, suppose (Y, X) were bivariate normal,

$$\begin{bmatrix} Y \\ X \end{bmatrix} \sim N_2\left(\begin{bmatrix} \mu_y \\ \mu_x \end{bmatrix}, \begin{bmatrix} \Sigma_{yy} & \Sigma_{yx} \\ \Sigma_{xy} & \Sigma_{xx} \end{bmatrix}\right), \tag{5.8}$$

and then using Exercise 5.23, we have the conditional distribution as

$$(Y|X = x) \sim N(\mu_y + \Sigma_{yx}\Sigma_{xx}^{-1}(x - \mu_x), \Sigma_{yy} - \Sigma_{yx}\Sigma_{xx}^{-1}\Sigma_{xy}). \tag{5.9}$$

We can see the correspondence between the parameterizations by matching up Equations (5.6) and (5.9); most importantly, we see that the slope parameter $\beta_1 = \Sigma_{yx}\Sigma_{xx}^{-1}$. Now when we estimate the slope, we use similar sample quantities:

$$\hat{\beta}_1 = \frac{S_{xy}}{S_{xx}} = \frac{\sum_i (x_i - \overline{x})(y_i - \overline{y})}{\sum_i (x_i - \overline{x})^2},$$

so that if two variables have a joint normal distribution, and we choose one as Y and the other as X, then the slope we are estimating is the slope of the linear conditional mean function $E(Y|X = x)$, and the variance we are estimating is the conditional variance $Var(Y|X = x)$. A peculiar property of the multivariate normal distribution is that this conditional variance does not depend on x.

The *correlation* between these two variables is often used to measure the strength of the relationship

$$\rho = \frac{Cov(X, Y)}{\sqrt{Var(X)Var(Y)}} = \frac{\Sigma_{xy}}{\sqrt{\Sigma_{xx}\Sigma_{yy}}}.$$

In the simple linear regression model, we find

$$\rho = \frac{\beta_1 \Sigma_{xx}}{\sqrt{\Sigma_{xx}(\sigma^2 + \beta_1^2 \Sigma_{xx})}} = \frac{\beta_1}{\sqrt{\beta_1^2 + \sigma^2/\Sigma_{xx}}},$$

showing that the correlation is zero when the slope is zero, and reaches its extremes ± 1 when either the slope β_1 grows large in absolute value, or the noise in the linear relationship σ^2 gets small relative to the variance of the explanatory variable Σ_{xx}.

From these expressions, we can see that the marginal or unconditional variance of Y is much larger, $\Sigma_{yy} = \sigma^2 + \beta_1^2 \Sigma_{xx}$, than the conditional variance, $Var(Y|X) = \sigma^2$. The usefulness of the explanatory variable X is to explain part of the variance in the response Y due to variation in X, reducing that variance to σ^2. If we look at this reduction in variance on a relative basis, we find a familiar expression:

$$\frac{\Sigma_{yx}\Sigma_{xx}^{-1}\Sigma_{xy}}{\Sigma_{yy}} = \frac{\Sigma_{yx}^2}{\Sigma_{xx}\Sigma_{yy}} = \rho^2 = \frac{\beta_1^2}{\beta_1^2 + \sigma^2/\Sigma_{xx}}.$$

Note that the size of that reduction depends on both the magnitude of the slope and the variation in X. Also note that if the slope were zero, then this reduction is also zero, owing from Y and X being independent, and $Var(Y|X) = Var(Y) = \sigma^2$.

We can generalize these concepts easily to multiple regression. Equation (5.8) only needs some parts to be redone in bold, but Equation (5.7) requires a little care. Partitioning off the intercept parameter by $\mathbf{b}^T = (\beta_0, \mathbf{b}_*^T)$, we can write (5.7) as

$$\begin{bmatrix} Y \\ \mathbf{X} \end{bmatrix} \sim N_p \left(\begin{bmatrix} \beta_0 + \mathbf{b}_*^T \mu_{\mathbf{x}} \\ \mu_x \end{bmatrix}, \begin{bmatrix} \sigma^2 + \mathbf{b}_*^T \Sigma_{xx} \mathbf{b}_* & \mathbf{b}_*^T \Sigma_{xx} \\ \Sigma_{xx} \mathbf{b}_* & \Sigma_{xx} \end{bmatrix} \right). \tag{5.10}$$

And as in the simple linear regression, our coefficient estimates are trying to estimate the coefficients of the affine $E(Y|\mathbf{X} = \mathbf{x})$, and the variance estimate aiming at $Var(Y|\mathbf{X} = \mathbf{x})$. The concept of correlation does not generalize that cleanly, except in the sense of variance reduction:

$$Var(Y|\mathbf{X} = \mathbf{x})/Var(Y).$$

The sample correlation coefficient r can be employed in the simple linear regression case, with

$$r = \frac{S_{xy}}{\sqrt{S_{xx} S_{yy}}} = \frac{\sum_i (x_i - \overline{x})(y_i - \overline{y})}{\sqrt{\sum_i (x_i - \overline{x})^2 \sum_i (y_i - \overline{y})^2}}.$$

We can show the relationship with the reduction in the variance estimate—or in the error sum of squares—with

$$r^2 = \frac{\mathbf{y}^T(\mathbf{P_X} - \mathbf{P_1})\mathbf{y}}{\mathbf{y}^T(\mathbf{I} - \mathbf{P_X})\mathbf{y}}. \tag{5.11}$$

In the multiple regression case, the right-hand side above is more often denoted as R^2 with the cryptic name *coefficient of determination*, and the expression is the sample correlation between the response and fitted values:

$$R^2 = \frac{\sum_i (\hat{y}_i - \overline{y})(y_i - \overline{y})}{\sqrt{\sum_i (\hat{y}_i - \overline{y})^2 \sum_i (y_i - \overline{y})^2}}. \tag{5.12}$$

See also Exercise 5.25.

Example 5.4: Errors in the Variables/Measurement Error
Consider now the simple linear regression problem, where the explanatory variable is measured with error. The simple linear regression model is retained, as in Equation (5.6), but we do not observe X, but $W = X + Z$ where $Z \sim N(0, \sigma_z^2)$ is independent of both X and Y. Then the joint distribution of the observed variables is

$$\begin{bmatrix} Y \\ W \end{bmatrix} \sim N_2 \left(\begin{bmatrix} \beta_0 + \beta_1 \mu_x \\ \mu_x \end{bmatrix}, \begin{bmatrix} \sigma^2 + \beta_1^2 \Sigma_{xx} & \beta_1 \Sigma_{xx} \\ \beta_1 \Sigma_{xx} & \Sigma_{xx} + \sigma_z^2 \end{bmatrix} \right), \tag{5.13}$$

so that if we construct a slope estimate using the explanatory variable measured with error, we find

$$E\left\{\frac{S_{wy}}{S_{ww}}\right\} = E\left\{\frac{\sum_i (w_i - \overline{w})(y_i - \overline{y})}{\sum_i (w_i - \overline{w})^2}\right\}$$

$$= \frac{\Sigma_{yw}}{\Sigma_{ww}} = \frac{\beta_1 \Sigma_{xx}}{\Sigma_{xx} + \sigma_z^2}$$

$$= \beta_1 \frac{1}{1 + \sigma_z^2 / \Sigma_{xx}}$$

and the usual slope estimate is biased downward.

5.7 Summary

- The multivariate normal distribution is defined in terms of linear combinations of independent standard normal random variables. Using moment generating functions and this simple definition, many of the commonly known properties of this distribution can be found.
- The chi-square distribution is defined in terms of sums of squares of independent normal random variables, and related distributions (noncentral chi-square, t, F) are also derived.
- The distributions of quadratic forms in vectors having the multivariate normal distribution are derived, with Cochran's Theorem providing a powerful and general summary.
- Regression models arising from observational studies can be expressed in terms of the joint multivariate normal distribution of the response and covariates.

5.8 Notes

- Results 5.1 and 5.2 deserve further explanation, as no proofs were given or cited. In reference works on probability theory, e.g., Billingsley [3] or Chung [7], these results are given in terms of characteristic functions, not moment generating functions. We have avoided the mathematics of complex variables, and hence characteristic functions in this book. However, these results carry over to moment generating functions based on the following argument. If the moment generating function $m_X(t)$ exists in an interval around the origin, then its characteristic function is $m_X(it)$. Viewing $E(exp\{zX\}) = f(z)$ as a function of a complex argument, then for real z we get the moment generating function and imaginary z the characteristic function. According to Lukacs [28], Section 7.1, this function $f(z)$ is analytic. If two moment generating functions $m_1(t) = m_2(t)$ coincide on an interval on the real line, then by analytic continuation, the corresponding functions must also coincide in a region of the complex plane $f_1(z) = f_2(z)$ that includes the imaginary axis, and so the characteristic functions coincide. As a result, these results stated for characteristic functions (uniqueness, independence) also hold for moment generating functions.

- We have defined the multivariate normal distribution in terms of moment generating functions to avoid the use of densities. This approach permits easy use of the singular normal distribution whose density doesn't exist.

- After considerable deliberation, we have chosen the definition of the noncentrality parameter in the noncentral chi-square distribution in Definition 5.6. The reader is again advised that different authors use different definitions, and they relate to the Poisson rate parameter. We chose ϕ as the noncentrality parameter to be the same as the Poisson rate parameter, and so the mean and variance of a noncentral chi-square variate are $p + 2\phi$ and $2p + 8\phi$, respectively. In Rao [37], for example, the noncentrality parameter is λ, the Poisson rate parameter is $\lambda/2$, and the mean and variance are $p + \lambda$ and $2p + 4\lambda$.

5.9 Exercises

5.1. * (Singular normal) Let $\mathbf{Y} = \mathbf{a} + \mathbf{BX}$ with $\mathbf{B}$ $(q \times p)$ and $q > p$, and $\mathbf{X} \sim N_p(\mathbf{0}, \mathbf{I}_p)$, and consider first the general case.

a. What is the rank of $\mathbf{V} = Cov(\mathbf{Y})$?
b. When $\mathbf{a} = \mathbf{0}$, in what space does the distribution of $\mathbf{Y}$ lie?
c. What is the dimension of the space?
d. Write $\mathcal{N}(\mathbf{V})$ in terms of $\mathbf{B}$.

For a concrete example, let $p = 2, q = 3, \mathbf{a} = \begin{bmatrix} 0 \\ -1 \\ 1 \end{bmatrix}$ and $\mathbf{B} = \begin{bmatrix} 1 & 2 \\ -1 & 0 \\ 0 & 1 \end{bmatrix}$.

e. Find a basis for $\mathcal{N}(\mathbf{V})$.
f. Find $Pr(2Y_1 - Y_2 > -2)$.
g. Find $Pr(Y_1 + Y_2 - 2Y_3 > 1)$.
h. Find $Pr(Y_1 + Y_2 - 2Y_3 = 0)$.
i. Repeat steps (f)–(h) with $\mathbf{a} = \mathbf{1}$.
j. Repeat with a new $\mathbf{B}$ as given below, replacing (g) and (h) with similar questions regarding the linear combination $Y_1 - 3Y_2 + Y_3$.

$$\mathbf{B} = \begin{bmatrix} 2 & 1 \\ 0 & 1 \\ -2 & 2 \end{bmatrix}$$

5.2. * Let $\mathbf{Z} \sim N_p(\mathbf{0}, \mathbf{I}_p)$, and let $\mathbf{A}$ be a $p \times p$ matrix such that $\mathbf{A}\mathbf{A}^T = \mathbf{V}$.

a. Show that if $\mathbf{Q}$ is an orthogonal matrix, then $\mathbf{QZ} \sim N_p(\mathbf{0}, \mathbf{I}_p)$.
b. Show that $(\mathbf{AQ})(\mathbf{AQ})^T = \mathbf{V}$.
c. Show that $\mathbf{X} = \mu + \mathbf{AQZ} \sim N_p(\mu, \mathbf{V})$.

5.3. * A peculiar property of the normal distribution permits a level of sloppiness in the exchange of conditional probability statements for convolutions.

a. Let $X \sim N(\mu, \sigma^2)$ and $(Y|X = x) \sim N(a + bx, \gamma^2)$. Show that the distribution of Y is $Y \sim N(a + b\mu, \gamma^2 + b^2\sigma^2)$.
b. Again let $X \sim N(\mu, \sigma^2)$, but $Y = a + bX + U$, where $U \sim N(0, \gamma^2)$ independent of X. Show that again the distribution of Y is $Y \sim N(a+b\mu, \gamma^2 + b^2\sigma^2)$.
This is not possible with other distributions.
c. Let X have the exponential distribution with scale parameter λ, so that the density is $\lambda e^{-x\lambda}$ for $x > 0$. Let $(Y|X = x)$ have the exponential distribution with scale parameter x and density xe^{-xy} for $y > 0$. Find the marginal density of Y.
d. Again let X have the exponential distribution with scale parameter λ, but $Y = X + U$ where U has an exponential distribution. Can you find a scale parameter (even a function of x) so that the marginal density of Y is the same as in part (c)?
e. Try this with other distributions and other conditioning mechanisms.

5.4. Prove Corollary 5.2(c) using basic probability calculus tools.

5.5. For what values of t does the mgf of the chi-square distribution exist?

5.6. (Another derivation of the central chi-square distribution)

a. Let $Z \sim N(0, 1)$; find the density of $U_1 = Z^2$.
b. Let U_1, U_2 be independent, each with the χ_1^2 distribution. Find the density of $U_1 + U_2$ directly using transformation rules.
c. Let $U_i, i = 1, \ldots, p$ be iid χ_1^2. Find the density of $U_1 + \cdots + U_p$.

5.7. * Find an expression for $Pr(U > c|U \sim \chi_{k+2}^2) - Pr(U > c|U \sim \chi_k^2)$ using integration by parts.

5.8. (Variation on Result 5.11) Show that if $U \sim \chi_p^2(\phi)$, then $Pr(U > c)$ is increasing in p for fixed $\phi \geq 0, c > 0$.

5.9. Find the density of the F-distribution.

5.10. Find the density of the noncentral Student's t-distribution. Also find its mean and variance.

5.11. Let $U \sim \chi_k^2(\phi)$; find its mean and variance. Confirm with Lemmas 4.1 and Result 4.6 (use $\sigma^2 = 1, \gamma_3 = 0$, and $\gamma_4 = 3$).

5.12. Let $\mathbf{X} \sim N_p(\mu, \mathbf{V})$ with $\mathbf{V}$ nonsingular, and let $U = \mathbf{X}^T\mathbf{A}\mathbf{X}$ for $\mathbf{A}$ symmetric.
   a. Show that the mgf for U is $m_U(t) = |\mathbf{I} - 2t\mathbf{A}\mathbf{V}|^{-1/2} exp\{-\frac{1}{2}[\mu^T\mathbf{V}^{-1}\mu - \mu^T(\mathbf{V} - 2t\mathbf{V}\mathbf{A}\mathbf{V})^{-1}\mu]\}$.
   b. Show that if $\mathbf{A}\mu = \mathbf{0}$, then $m_U(t) = |\mathbf{I} - 2t\mathbf{A}\mathbf{V}|^{-1/2}$.

5.13. Using the result of Exercise 5.12 above, prove Result 5.15.

5.14. Using the result of Exercise 5.12, show that
   a. $Var(\mathbf{X}^T\mathbf{A}\mathbf{X}) = 2tr(\mathbf{A}\mathbf{V})^2 + 4\mu^T\mathbf{A}\mathbf{V}\mathbf{A}\mu$
   b. (Easier) If $\mathbf{X} \sim N_p(\mathbf{0}, \mathbf{V})$, then $Var(\mathbf{X}^T\mathbf{A}\mathbf{X}) = 2tr(\mathbf{A}\mathbf{V})^2$.

5.15. Prove the converse to Result 5.14: Let $\mathbf{X} \sim N_p(\mu, \mathbf{V})$ and $\mathbf{A}$ is symmetric; then if $\mathbf{X}^T\mathbf{A}\mathbf{X} \sim \chi^2_s(\phi)$ for some ϕ, then $\mathbf{A}\mathbf{V}$ is idempotent with rank s.

5.16. Prove a converse to Result 5.16, assuming $\mathbf{V}$ to be nonsingular.

5.17. The density of the noncentral $\mathbf{F}$-distribution and Result 5.13 require some care, since the noncentral $\mathbf{F}$-distribution is not simply a Poisson mixture of central $\mathbf{F}$-distributions. Show that the density of the noncentral $\mathbf{F}$-distribution is

$$\sum_{j=0}^{\infty} \frac{\phi^j e^{-\phi}}{j!} \times \frac{\Gamma\left(\frac{p_1+p_2}{2} + j\right)(p_2/p_1)^{p_1/2}}{\Gamma\left(\frac{1}{2}p_1 + j\right)\Gamma\left(\frac{1}{2}p_2\right)} f^{(p_1/2+j)-1}\left(1 + \frac{p_1}{p_2}f\right)^{-(p_1+p_2)/2-j}$$

for $f > 0$ and zero otherwise

5.18. Prove the following lemmas, which are related to the proof of Cochran's Theorem:

   Lemma A: Let $\mathbf{A}$ be a $p \times p$ symmetric matrix. Then $\mathbf{A}$ is idempotent iff $rank(\mathbf{A}) + rank(\mathbf{I} - \mathbf{A}) = p$.
   Lemma B: If $\mathbf{A}$, $\mathbf{B}$, and $\mathbf{A} + \mathbf{B}$ are all idempotent, then $\mathbf{A}\mathbf{B} = \mathbf{0}$.
   Lemma C: $rank(\sum \mathbf{A}_i) \le \sum rank(\mathbf{A}_i)$.

5.19. Prove the following theorem related to Cochran's Theorem: Let $\mathbf{A}_i$, $i = 1, \ldots, k$ be $p \times p$ symmetric matrices such that $\sum \mathbf{A}_i = \mathbf{I}_p$. Then the following three conditions are equivalent:
   a. $\mathbf{A}_i\mathbf{A}_j = \mathbf{0}$ for $i \ne j$.
   b. $\mathbf{A}_i$ are all idempotent.
   c. $\sum rank(\mathbf{A}_i) = p$.

5.20. Prove the following converse to Cochran's Theorem: Let $\mathbf{y} \sim N_p(\mu, \sigma^2\mathbf{I}_p)$ and let $\mathbf{A}_i$, $i = 1, \ldots, k$ be symmetric matrices with rank n_i. If $(1/\sigma^2)\mathbf{y}^T\mathbf{A}_i\mathbf{y}$ are independently distributed as $\chi^2_{s_i}(\phi_i)$, then $n_i = s_i$, $i = 1, \ldots, k$ and $\sum n_i = p$.

5.21. Let the random variables X and Y, conditional on $Z = z$ be independent, and let the distribution of $(Y|Z = z)$ not depend on z. Prove that, unconditionally, X and Y are independent. (Assume for simplicity that the densities or mgf's exist.)

5.22. The noncentrality parameter λ of the noncentral t-distribution is not a location parameter. Show this by finding the density of $Y - \lambda$ when $Y \sim t_k(\lambda)$.

5.23. Let $\mathbf{X} \sim N_p(\mu, \mathbf{V})$ with $\mathbf{V}$ nonsingular, and partition as below:

$$\mathbf{X} = \begin{bmatrix} \mathbf{X}_1 \\ \mathbf{X}_2 \end{bmatrix} \begin{matrix} p_1 \\ p_2 \end{matrix}, \quad \mu = \begin{bmatrix} \mu_1 \\ \mu_2 \end{bmatrix} \begin{matrix} p_1 \\ p_2 \end{matrix}, \quad \mathbf{V} = \begin{bmatrix} \mathbf{V}_{11} & \mathbf{V}_{12} \\ \mathbf{V}_{21} & \mathbf{V}_{22} \end{bmatrix} \begin{matrix} p_1 \\ p_2 \end{matrix}$$

Show that the conditional distribution of $\mathbf{X}_1$ given $\mathbf{X}_2 = \mathbf{x}_2$ is multivariate normal with mean vector $\mu_1 + \mathbf{V}_{12}\mathbf{V}_{22}^{-1}(\mathbf{x}_2 - \mu_2)$ and covariance matrix $\mathbf{V}_{11} - \mathbf{V}_{12}\mathbf{V}_{22}^{-1}\mathbf{V}_{21}$. Hint: Use the partitioned inverse result from Exercise A.74.

5.24. In the normal linear model, that is, $\mathbf{y} \sim N_N(\mathbf{Xb}, \sigma^2\mathbf{I}_N)$, find the conditional distribution of $\mathbf{a}^T\mathbf{y}$ given $\mathbf{X}^T\mathbf{y}$. For simplicity, assume that $\mathbf{X}$ has full-column rank.

5.25. Show that R^2 given by (5.12) is also equal to the right-hand side of (5.11).

5.26. Recall the instrumental variables estimator of the slope in Exercise 4.6. We could follow this approach further. Consider again the simple linear regression problem

$$y_i = \beta_0 + \beta_1 x_i + e_i, \quad i = 1, \ldots, N,$$

where e_i are iid $N(0, \sigma^2)$. Consider the use of the matrix $\mathbf{Z}$ $(N \times 2)$ below:

$$\mathbf{Z} = \begin{bmatrix} 1 & z_1 \\ 1 & z_2 \\ \cdots & \cdots \\ 1 & z_N \end{bmatrix} \quad \text{and} \quad \mathbf{y} = \begin{bmatrix} y_1 \\ y_2 \\ \cdots \\ y_N \end{bmatrix}, \quad \mathbf{Xb} = \begin{bmatrix} 1 & x_1 \\ 1 & x_2 \\ \cdots & \cdots \\ 1 & x_N \end{bmatrix} \begin{bmatrix} \beta_0 \\ \beta_1 \end{bmatrix}$$

where, as before, $z_1, z_2, \ldots, z_N$ are known constants. Assume that the (2×2) matrix $\mathbf{Z}^T\mathbf{X}$ is nonsingular.

a. Show that $(\mathbf{Z}^T\mathbf{X})^{-1}\mathbf{Z}^T$ is a generalized inverse of $\mathbf{X}$.
b. The instrumental variables estimator $\tilde{\mathbf{b}}$ solves the equations $\mathbf{Z}^T\mathbf{Xb} = \mathbf{Z}^T\mathbf{y}$. Derive (show the steps) the distribution of $\tilde{\mathbf{b}}$.
c. Show that $\frac{1}{\sigma^2}\mathbf{y}^T(\mathbf{I}-\mathbf{P_Z})\mathbf{y}$ has a chi-square distribution and give its degrees of freedom and noncentrality parameter (and don't forget σ^2).
d. Show that if $\beta_1 = 0$, the noncentrality parameter above in (c) is zero.
e. Show that $\tilde{\mathbf{b}}$ and $(\mathbf{I} - \mathbf{P_Z})\mathbf{y}$ are independent. (Hence $\tilde{\mathbf{b}}$ and $\mathbf{y}^T(\mathbf{I} - \mathbf{P_Z})\mathbf{y}$ are independent.)
f. Let $\mathbf{k}^T = [0\ 1]$, so that $\mathbf{k}^T\tilde{\mathbf{b}} = \tilde{\beta}_1$ (this is the same slope estimate from before), and denote $Var(\tilde{\beta}_1) = \sigma^2 v$. If $\beta_1 = 0$, what is the distribution of the ratio F below?

$$F = \frac{(\tilde{\beta}_1)^2/v}{\mathbf{y}^T(\mathbf{I} - \mathbf{P_Z})\mathbf{y}/(N-2)}$$

g. The instrumental variables estimator $\tilde{\mathbf{b}}$ is given by $\tilde{\mathbf{b}} = (\mathbf{Z}^T\mathbf{X})^{-1}\mathbf{Z}^T\mathbf{y}$. Using this approach for estimation, the residuals (not the usual ones!) denoted by $\tilde{\mathbf{e}}$ are given by $\tilde{\mathbf{e}} = \mathbf{y} - \mathbf{X}\tilde{\mathbf{b}}$. Find the $(N \times N)$ matrix $\mathbf{P}$ so that we can write $\tilde{\mathbf{e}} = \mathbf{P}\mathbf{y}$.
h. Is the matrix $\mathbf{P}$ you found above in (g) symmetric? Is it idempotent? Show.
i. Give the joint distribution of $\tilde{\mathbf{b}}$ and $\tilde{\mathbf{e}}$. Are $\tilde{\mathbf{b}}$ and $\tilde{\mathbf{e}}$ independent? Show.
j. Can you show that $\frac{1}{\sigma^2}\|\tilde{\mathbf{e}}\|^2$ has a chi-square distribution (central or otherwise)? Either prove that it does (and get its degrees of freedom) or show where the conditions fail.

5.27. Prove Corollary 5.4.

5.28. Let $X_N, n = 1, 2, \ldots$ be a sequence of random variables where $X_N \sim \chi_p^2(\phi_N)$ where p is fixed and $\phi_N \to \infty$. Show that X_N (suitably centered and scaled) converges to a normal.

5.29. Numerical linear algebra views the linear model in statistics as the linear least squares problem, and one approach is to construct an $N \times N$ orthogonal matrix $\mathbf{U}$ (using Householder or Givens transformations—also see Exercises 2.22–2.24) such that

$$\mathbf{U}\mathbf{X} = \begin{bmatrix} \mathbf{U}_1 \\ \mathbf{U}_2 \end{bmatrix} \mathbf{X} = \begin{bmatrix} \mathbf{R} \\ \mathbf{0} \end{bmatrix}.$$

Let us consider our usual linear model with normal errors, that is, $\mathbf{y} \sim N_N(\mathbf{X}\mathbf{b}, \sigma^2\mathbf{I}_N)$, and to make life easier, assume that $\mathbf{X}$ has full-column rank, so that $\mathbf{R}$ is $p \times p$ and nonsingular. Note that since $\mathbf{U}$ is an orthogonal matrix, $\mathbf{U}^T\mathbf{U} = \mathbf{U}_1^T\mathbf{U}_1 + \mathbf{U}_2^T\mathbf{U}_2 = \mathbf{I}_N$ as well as

$$\begin{bmatrix} \mathbf{U}_1 \\ \mathbf{U}_2 \end{bmatrix} \begin{bmatrix} \mathbf{U}_1^T & \mathbf{U}_2^T \end{bmatrix} = \begin{bmatrix} \mathbf{I}_p & \mathbf{0} \\ \mathbf{0} & \mathbf{I}_{N-p} \end{bmatrix}.$$

a. Let $\mathbf{U}\mathbf{y} = \begin{bmatrix} \mathbf{z}_1 \\ \mathbf{z}_2 \end{bmatrix}$; find the joint distribution of $\mathbf{z}_1$ and $\mathbf{z}_2$.
b. It is easy to show that $\hat{\mathbf{b}} = \mathbf{R}^{-1}\mathbf{z}_1$. Find the joint distribution of $\hat{\mathbf{b}}$ and $\mathbf{z}_2$.
c. Are $\hat{\mathbf{b}}$ and $\mathbf{z}_2$ independent? Briefly state why or why not.
d. Find the distribution of $\|\mathbf{z}_2\|^2$. (Hint: don't forget to divide by σ^2.)
e. Relaxing the assumption that $\mathbf{X}$ has full-column rank, say $rank(\mathbf{X}) = s < p$, means here that $\mathbf{R}$ is $s \times p$ with full-row rank and a triangular structure as if rows of a triangular matrix were removed. Can you generalize the results above?

Chapter 6

Statistical Inference

6.1 Introduction

Consider the normal Gauss–Markov model given by $\mathbf{y} \sim N_N(\mathbf{Xb}, \sigma^2\mathbf{I}_N)$. Under the assumption of normality, we know the distribution of the best linear unbiased estimator $\mathbf{\Lambda}^T\hat{\mathbf{b}}$ of an estimable $\mathbf{\Lambda}^T\mathbf{b}$ is $N(\mathbf{\Lambda}^T\mathbf{b}, \sigma^2\mathbf{\Lambda}^T(\mathbf{X}^T\mathbf{X})^g\mathbf{\Lambda})$ and it is independent of $SSE/\sigma^2 \sim \chi^2_{N-r}$. We construct the unbiased estimator for σ^2 as $\hat{\sigma}^2 = SSE/(N-r)$ from Section 4.3. Beyond this point, we seek to establish some basic inferential results. First, we wish to find further properties of the usual estimators relying now on the normality assumption. The construction of hypothesis tests requires a few steps of its own: a test based on first principles, then the likelihood ratio test, and unifying the two while examining the effect of constraints. Construction of confidence intervals leads to the usual results on multiple comparisons.

6.2 Results from Statistical Theory

Although we have constructed best estimators from the linear model with only assumptions on the mean and variance, we must now consider what the additional assumption of normality may bring. We begin with the joint density of the data from $\mathbf{y} \sim N(\mathbf{Xb}, \sigma^2\mathbf{I}_N)$ and Corollary 5.5:

$$f(\mathbf{y}|\mathbf{b}, \sigma^2) = (2\pi)^{-N/2}(\sigma^2)^{-N/2}\, exp\left\{-\frac{1}{2\sigma^2}(\mathbf{y}-\mathbf{Xb})^T(\mathbf{y}-\mathbf{Xb})\right\}. \tag{6.1}$$

With just a little algebra, we can rewrite this density:

$$\begin{aligned} f(\mathbf{y}|\mathbf{b}, \sigma^2) =&(2\pi)^{-N/2}(\sigma^2)^{-N/2} exp\left\{-\frac{1}{2\sigma^2}(\mathbf{y}^T\mathbf{y}) + \frac{2}{2\sigma^2}(\mathbf{y}^T\mathbf{Xb})\right. \\ &\left. -\frac{1}{2\sigma^2}(\mathbf{b}^T\mathbf{X}^T\mathbf{Xb})\right\}. \end{aligned} \tag{6.2}$$

From the factorization theorem (Casella and Berger [6], Theorem 6.2.6), we can see that $T_1(\mathbf{y}) = \mathbf{y}^T\mathbf{y}$ and $T_2(y) = \mathbf{X}^T\mathbf{y}$ are sufficient statistics.

Result 6.1 *In the normal linear model* $\mathbf{y} \sim N(\mathbf{Xb}, \sigma^2\mathbf{I}_N)$ *with unknown parameters* $(\mathbf{b}, \sigma^2)$, $\mathbf{T}(\mathbf{y}) = (\mathbf{y}^T\mathbf{y}, \mathbf{X}^T\mathbf{y})$ *is a minimal sufficient statistic.*

Proof: From Equation (6.2) and the factorization theorem, $\mathbf{T}(\mathbf{y}) = (\mathbf{y}^T\mathbf{y}, \mathbf{X}^T\mathbf{y})$ are sufficient statistics. For minimal sufficiency ([6], Theorem 6.2.13), we must establish that $f(\mathbf{y}_1|\mathbf{b}, \sigma^2)/f(\mathbf{y}_2|\mathbf{b}, \sigma^2)$ is constant for all $(\mathbf{b}, \sigma^2)$ iff $\mathbf{T}(\mathbf{y}_1) = \mathbf{T}(\mathbf{y}_2)$. Examination of the ratio reveals

$$f(\mathbf{y}_1|\mathbf{b}, \sigma^2)/f(\mathbf{y}_2|\mathbf{b}, \sigma^2) = exp\left\{\frac{1}{2\sigma^2}(\mathbf{y}_2^T\mathbf{y}_2 - \mathbf{y}_1^T\mathbf{y}_1) + \frac{1}{\sigma^2}\mathbf{b}^T(\mathbf{X}^T\mathbf{y}_1 - \mathbf{X}^T\mathbf{y}_2)\right\}.$$

Clearly, if $\mathbf{T}(\mathbf{y}_1) = \mathbf{T}(\mathbf{y}_2)$, then the ratio is one and free of the parameters. For the converse, apply Result A.8. □

Corollary 6.1 *In the normal linear model* $\mathbf{y} \sim N(\mathbf{Xb}, \sigma^2\mathbf{I}_N)$ *with unknown parameters* $(\mathbf{b}, \sigma^2)$, $\mathbf{T}(\mathbf{y}) = (SSE, \mathbf{X}^T\mathbf{y})$ *is also a minimal sufficient statistic.*

Proof: See Exercise 6.1. □

Using the Gauss–Markov Theorem in Chapter 4, we established that the least squares estimator $\mathbf{\Lambda}^T\hat{\mathbf{b}}$ of an estimable $\mathbf{\Lambda}^T\mathbf{b}$ has the smallest variance among all linear, unbiased estimators. To show that it is the best among all unbiased estimators, we need to construct a complete sufficient statistic, and the best route is to show that the linear model admits an exponential family parameterization.

Result 6.2 *In the normal linear model* $\mathbf{y} \sim N_N(\mathbf{Xb}, \sigma^2\mathbf{I}_N)$ *with unknown parameters* $(\mathbf{b}, \sigma^2)$, $\mathbf{T}(\mathbf{y}) = (\mathbf{y}^T\mathbf{y}, \mathbf{X}^T\mathbf{y})$ *is a complete, sufficient statistic.*

Proof: First show that we can write the normal linear model in the exponential family. Let $w_1(\mathbf{b}, \sigma^2) = -\frac{1}{2\sigma^2}$ and $w_2(\mathbf{b}, \sigma^2) = \frac{1}{2\sigma^2}\mathbf{b}$; then we can write the density below:

$$f(\mathbf{y}|\mathbf{b}, \sigma^2) = (2\pi)^{-N/2}(\sigma^2)^{-N/2} exp\left\{-\frac{1}{2\sigma^2}(\mathbf{y}^T\mathbf{y}) + \frac{2}{2\sigma^2}(\mathbf{y}^T\mathbf{Xb}) - \frac{1}{2\sigma^2}(\mathbf{b}^T\mathbf{X}^T\mathbf{Xb})\right\} \tag{6.3}$$

in the exponential family with $T_1(\mathbf{y}) = \mathbf{y}^T\mathbf{y}$, and $T_2(\mathbf{y}) = \mathbf{X}^T\mathbf{y}$:

$$f(\mathbf{y}|\mathbf{b}, \sigma^2) = (2\pi)^{-N/2}(\sigma^2)^{-N/2} exp\left\{-\frac{1}{2\sigma^2}(\mathbf{b}^T\mathbf{X}^T\mathbf{Xb})\right\} \times exp\left\{w_1(\mathbf{b}, \sigma^2)T_1(\mathbf{y}) + w_2(\mathbf{b}, \sigma^2)T_2(\mathbf{y})\right\}. \tag{6.4}$$

Hence ([6], Theorem 6.2.25), $\mathbf{T}(\mathbf{y}) = (\mathbf{y}^T\mathbf{y}, \mathbf{X}^T\mathbf{y})$ is a complete, sufficient statistic. □

Corollary 6.2 *Under the normal Gauss–Markov model, the least squares estimator $\Lambda^T\hat{\mathbf{b}}$ of an estimable function $\Lambda^T\mathbf{b}$ has the smallest variance among all unbiased estimators.*

Proof: From Equation (6.2), $\mathbf{T}(\mathbf{y}) = (\mathbf{y}^T\mathbf{y}, \mathbf{X}^T\mathbf{y})$ is a complete, sufficient statistic. Since $\Lambda^T\hat{\mathbf{b}} = \Lambda^T(\mathbf{X}^T\mathbf{X})^g\mathbf{X}^T\mathbf{y}$ is a function of a complete sufficient statistic, it has the smallest variance of any estimator for its expectation, which is $\Lambda^T\mathbf{b}$. □

Maximum likelihood works well as a general principle for constructing estimators. In most cases maximum likelihood estimators have desirable properties, and the linear model is no exception.

Result 6.3 *In the normal linear model $\mathbf{y} \sim N_N(\mathbf{Xb}, \sigma^2\mathbf{I}_N)$ with unknown parameters $(\mathbf{b}, \sigma^2)$, $(\hat{\mathbf{b}}, SSE/N)$ is a maximum likelihood estimator of $(\mathbf{b}, \sigma^2)$ where $\hat{\mathbf{b}}$ solves the normal equations and $SSE = \mathbf{y}^T(\mathbf{I} - \mathbf{P_X})\mathbf{y}$.*

Proof: The likelihood function can be viewed in terms of the sum of squares function $Q(\mathbf{b}) = (\mathbf{y} - \mathbf{Xb})^T(\mathbf{y} - \mathbf{Xb})$:

$$L(b, \sigma^2) = (2\pi)^{-N/2}(\sigma^2)^{-N/2} exp\left\{-\frac{1}{2\sigma^2}Q(\mathbf{b})\right\}. \tag{6.5}$$

For each value of σ^2, a solution to the normal equations $\hat{\mathbf{b}}$ also minimizes $Q(\mathbf{b})$. For maximizing with respect to σ^2, taking the derivative of the concentrated or *profile log-likelihood function*,

$$logL(\hat{\mathbf{b}}, \sigma^2) = -(N/2)log(2\pi) - (N/2)log(\sigma^2) - \frac{1}{2\sigma^2}Q(\hat{\mathbf{b}}),$$

we have

$$-(N/2)\frac{1}{\sigma^2} - \frac{1}{2(\sigma^2)^2}Q(\hat{\mathbf{b}}).$$

Setting this to zero, we have the maximum (since the second derivative is negative) at

$$\hat{\sigma^2} = Q(\hat{\mathbf{b}})/N = SSE/N. \tag{6.6}$$

□

Corollary 6.3 *Under the normal linear model, the maximum likelihood estimator of an estimable function $\Lambda^T\mathbf{b}$ is $\Lambda^T\hat{\mathbf{b}}$, where $\hat{\mathbf{b}}$ solves the normal equations.*

Proof: This follows from the invariance property of maximum likelihood estimators ([6], Theorem 7.2.10). See also Exercise 6.3. □

It is important to note that the ML estimator of the variance differs in the denominator. The usual $\hat{\sigma}^2 = SSE/(N - r)$ is unbiased; the MLE for σ^2 is SSE/N, which is not unbiased. See also Exercise 6.4. Also note that Corollary 6.3 holds regardless of whether the solution to the normal equations is unique or not.

6.3 Testing the General Linear Hypothesis

Given the Gauss–Markov model with normal errors, that is, $\mathbf{y} \sim N_N(\mathbf{Xb}, \sigma^2 \mathbf{I}_N)$, we are now equipped to test the general linear hypothesis. Hypothesis testing in the normal linear model involves a number of issues arising from different viewpoints and principles. We will begin this discussion with a first principles approach for testing the most general problem, ending with the special case of testing a single parameter. In the section that follows, we will discuss the use of the likelihood ratio principle in testing this same general hypothesis. Then these two will be tied together in the context of estimation in a constrained parameter space.

The variety of hypotheses that we might want to test is quite extensive, for example:

- H_1: $b_j = 0$
- H_2: $b_1 = b_2 = b_3 = 0$
- H_3: $b_1 + b_3 = 1, b_2 = 3$
- H_4: $b_2 = b_3 = b_4$
- H_5: $\mathbf{b} \in \mathcal{C}(\mathbf{B})$

All of these examples can be written as a system of linear equations:

$$H : \mathbf{K}^T\mathbf{b} = \mathbf{m} \text{ versus } A : \mathbf{K}^T\mathbf{b} \neq \mathbf{m},$$

where $\mathbf{K}$ is $p \times s$ with full-column rank. The insistence on full-column rank for $\mathbf{K}$, or full-row rank for $\mathbf{K}^T$, follows from the avoidance of any redundancies in writing these hypotheses. Also, since usually $s > 1$, only two-sided alternatives will be considered, although some variations can be considered. In order to test the hypothesis H, we must have each component of $\mathbf{K}^T\mathbf{b}$ estimable. This means that each column of $\mathbf{K} \in \mathcal{C}(\mathbf{X}^T)$, or that we can write $\mathbf{K} = \mathbf{X}^T\mathbf{A}$ for some $\mathbf{A}$ $(N \times s)$ for $(p \times s)$ $\mathbf{K}$. Occasionally, some subtleties affect the correct determination of s.

Definition 6.1 *The general linear hypothesis H:* $\mathbf{K}^T\mathbf{b} = \mathbf{m}$ *is* testable *iff* $\mathbf{K}$ *has full-column rank and each component of* $\mathbf{K}^T\mathbf{b}$ *is estimable. If any of the components of* $\mathbf{K}^T\mathbf{b}$ *are not estimable, then the hypothesis is considered* nontestable.

Example 6.1: Column Space Hypothesis
Consider testing H_5: $\mathbf{b} \in \mathcal{C}(\mathbf{B})$. To write this as $\mathbf{K}^T\mathbf{b} = \mathbf{m}$, construct basis vectors for the orthogonal complement of $\mathcal{C}(\mathbf{B})$, so that $span\{\mathbf{c}^{(1)}, \ldots, \mathbf{c}^{(s)}\} = \mathcal{N}(\mathbf{B}^T)$. Then the vectors $\mathbf{c}^{(j)}$, $j = 1, \ldots, s$, form the columns of $\mathbf{K}$ and $\mathbf{m} = \mathbf{0}$.

Example 6.2: Interaction
Consider the two-way crossed model with interaction

$$y_{ijk} = \mu + \alpha_i + \beta_j + \gamma_{ij} + e_{ijk},$$

for $i = 1, \ldots, a$; $j = 1, \ldots, n$, and $k = 1, \ldots, n_{ij}$. We want to test for no interaction. At first glance, we could write $a \times n$ equations of the form $\gamma_{ij} = 0$, but these are not estimable functions. So what do we really mean by testing for interaction? From one point of view, the two-way crossed model with interaction has a design matrix with rank $a \times n$, and so there are only $a \times n$ linear independent estimable functions. Without interaction, the design matrix has rank $a + n - 1$. The difference is that we can only construct $a \times n - a - n + 1 = (a-1)(n-1)$ linearly independent functions that do not involve the main effect parameters: μ, α_i, β_j. These can be viewed as equations that characterize the additivity of the model without interaction:

$$E(y_{i_1 j_1 k} - y_{i_1 j_2 k} - y_{i_2 j_1 k} + y_{i_2 j_2 k}) = \gamma_{i_1 j_1} - \gamma_{i_1 j_2} - \gamma_{i_2 j_1} + \gamma_{i_2 j_2}.$$

In the model without interaction, these expectations are all zero. We can choose $(a-1)$ pairs of i_1 and i_2 and (independently) $(n-1)$ pairs of j_1 and j_2 and still retain linear independence of these estimable functions. As a result, in testing interaction in the two-way crossed model, $s = (a-1) \times (n-1)$. The remaining $a + n - 1$ components are confounded with the main effect parameters (see Exercises 6.9 and 6.10).

Example 6.3: General Subset
The general case of Example 6.2 is the most common testing situation, where we want to test H: $\mathbf{b}_1 = \mathbf{0}$,

$$\mathbf{Xb} = \begin{bmatrix} \mathbf{X}_0 & \mathbf{X}_1 \end{bmatrix} \begin{bmatrix} \mathbf{b}_0 \\ \mathbf{b}_1 \end{bmatrix} \begin{matrix} p_0 \\ p_1 \end{matrix}$$

where $rank(\mathbf{X}) = r$, $rank(\mathbf{X}_0) = r_0$, $rank((\mathbf{I} - \mathbf{P}_{\mathbf{X}_0})\mathbf{X}_1) = s$, with $s = r - r_0$. All of the components of b_1 are estimable only if $s = p_1$; this will hold if $\mathbf{X}$ has full-column rank. Similar to Example 6.2, we can only test s linearly independent functions of $\mathbf{b}_1$ alone, which are $(\mathbf{I} - \mathbf{P}_{\mathbf{X}_0})\mathbf{X}_1\mathbf{b}_1$. Notice that $E((\mathbf{I} - \mathbf{P}_{\mathbf{X}_0})\mathbf{y}) = (\mathbf{I} - \mathbf{P}_{\mathbf{X}_0})\mathbf{X}_1\mathbf{b}_1$ so that each component is estimable. See continuations of this example.

Example 6.4: Linear Trend in ANOVA
Consider the usual one-way ANOVA model $y_{ij} = \mu + \alpha_i + e_{ij}$ and test the hypothesis of a linear effect with the group covariate $x_i = i$. Taking the simple case with $a = 3$, this hypothesis can be written as $\alpha_i = \beta i$. The more common way to write this hypothesis is H:$\alpha_2 - \alpha_1 = \alpha_3 - \alpha_2$, or in terms of the quadratic contrast H:$\alpha_1 - 2\alpha_2 + \alpha_3 = 0$.

Based solely on first principles, we can construct a test for the general linear hypothesis in the following manner. If $\mathbf{K}^T\mathbf{b}$ is estimable, with s linearly independent estimable functions, then its BLUE is given by $\mathbf{K}^T\hat{\mathbf{b}} = \mathbf{K}^T(\mathbf{X}^T\mathbf{X})^g\mathbf{X}^T\mathbf{y}$, and its distribution is

$$\mathbf{K}^T\hat{\mathbf{b}} \sim N_s(\mathbf{K}^T(\mathbf{X}^T\mathbf{X})^g\mathbf{X}^T\mathbf{Xb}, \sigma^2\mathbf{K}^T(\mathbf{X}^T\mathbf{X})^g\mathbf{X}^T\mathbf{X}(\mathbf{X}^T\mathbf{X})^g\mathbf{K}) = N_s(\mathbf{K}^T\mathbf{b}, \sigma^2\mathbf{H}), \tag{6.7}$$

where $\mathbf{H} = \mathbf{K}^T(\mathbf{X}^T\mathbf{X})^g(\mathbf{X}^T\mathbf{X})(\mathbf{X}^T\mathbf{X})^g\mathbf{K} = \mathbf{K}^T(\mathbf{X}^T\mathbf{X})^g\mathbf{K}$. If $\mathbf{H}$ is nonsingular, then our work will be easy.

Result 6.4 *If $\mathbf{K}^T\mathbf{b}$ is estimable, then the $s \times s$ matrix $\mathbf{H} = \mathbf{K}^T(\mathbf{X}^T\mathbf{X})^g\mathbf{K}$ is nonsingular.*

Proof: To find the rank of $\mathbf{H}$ consider first the matrix $\mathbf{W} = \mathbf{X}(\mathbf{X}^T\mathbf{X})^g\mathbf{K}$, and note that $\mathbf{X}^T\mathbf{W} = \mathbf{X}^T\mathbf{X}(\mathbf{X}^T\mathbf{X})^g\mathbf{K} = \mathbf{K}$ and $\mathbf{H} = \mathbf{W}^T\mathbf{W}$. Following from Result A.1, we have the following bounds:

$$rank\,(\mathbf{W}) \geq rank\,(\mathbf{X}^T\mathbf{W}) = rank\,(\mathbf{K}) = s \geq rank\,(\mathbf{X}(\mathbf{X}^T\mathbf{X})^g\mathbf{K}) = rank\,(\mathbf{W}). \tag{6.8}$$

Hence $rank\,(\mathbf{W}) = s$, and so we also have $rank\,(\mathbf{W}) = rank\,(\mathbf{W}^T\mathbf{W}) = rank\,(\mathbf{H}) = s$. □

Since the covariance matrix $\mathbf{H}$ is nonsingular, the distribution of $\mathbf{K}^T\hat{\mathbf{b}}$ in (6.7) is nonsingular. The next two steps then are straightforward:

$$\mathbf{K}^T\hat{\mathbf{b}} - \mathbf{m} \sim N_s(\mathbf{K}^T\mathbf{b} - \mathbf{m}, \sigma^2\mathbf{H}),$$

and so from Result 5.10, we have

$$(\mathbf{K}^T\hat{\mathbf{b}} - \mathbf{m})^T(\sigma^2\mathbf{H})^{-1}(\mathbf{K}^T\hat{\mathbf{b}} - \mathbf{m}) \sim \chi_s^2(\phi)$$

where the noncentrality parameter is $\phi = \frac{1}{2}(\mathbf{K}^T\mathbf{b} - \mathbf{m})^T(\sigma^2\mathbf{H})^{-1}(\mathbf{K}^T\mathbf{b} - \mathbf{m})$. In the previous chapter, we showed that $\hat{\mathbf{y}} = \mathbf{P}_\mathbf{X}\mathbf{y}$ and $\hat{\mathbf{e}} = (\mathbf{I} - \mathbf{P}_\mathbf{X})\mathbf{y}$ are independent, so since $\mathbf{K}^T\hat{\mathbf{b}}$ depends only on $\hat{\mathbf{y}}$, then $\mathbf{K}^T\hat{\mathbf{b}}$ and *SSE* are independent. We can also prove independence more directly using Result 5.16 since $(\mathbf{K}^T(\mathbf{X}^T\mathbf{X})^g\mathbf{X}^T)(\sigma^2\mathbf{I}_N)(\mathbf{I}_N - \mathbf{P}_\mathbf{X}) = \mathbf{0}$. Adding the definition of the F-distribution, we have

$$F = \frac{(\mathbf{K}^T\hat{\mathbf{b}} - \mathbf{m})^T\mathbf{H}^{-1}(\mathbf{K}^T\hat{\mathbf{b}} - \mathbf{m})/s}{SSE/(N-r)} \sim F_{s,N-r}(\phi) \tag{6.9}$$

(note the cancellation of σ^2 above). Under H: $\mathbf{K}^T\mathbf{b} = \mathbf{m}$, the noncentrality parameter ϕ is zero and the F statistic in (6.9) has the central F-distribution; under A: $\mathbf{K}^T\mathbf{b} \neq \mathbf{m}$, the noncentrality parameter ϕ is positive and we have a noncentral F. From Result 5.13, a reasonable criterion for testing H is to reject H when the F statistic in (6.9) is too large. Since the distribution of F under H is known, we can find the critical value of $F_{s,N-r,\alpha}$, leading to the following procedure:

$$\text{Reject H if } F = \frac{(\mathbf{K}^T\hat{\mathbf{b}} - \mathbf{m})^T\mathbf{H}^{-1}(\mathbf{K}^T\hat{\mathbf{b}} - \mathbf{m})/s}{SSE/(N-r)} > F_{s,N-r,\alpha}, \tag{6.10}$$

which will be a test with level α, and increasing power when the alternative is true. In other words, following first principles, or "this seems like a good idea," we have found an unbiased test for the general linear hypothesis.

Example 6.5: Testing Equality in One-Way ANOVA
Consider the usual one-way ANOVA model $y_{ij} = \mu + \alpha_i + e_{ij}$. The most common hypothesis to be tested is that the effects α_i are all equal. One way of writing this test is

$$\mathrm{H}_1 : \alpha_1 - \alpha_2 = 0, \alpha_1 - \alpha_3 = 0, \ldots, \alpha_1 - \alpha_a = 0$$

while another is

$$\mathrm{H}_2 : \alpha_1 - \alpha_2 = 0, \alpha_2 - \alpha_3 = 0, \ldots, \alpha_{a-1} - \alpha_a = 0.$$

We'll choose the first way of writing the hypothesis and work through the results, first with $a = 3$, then the general case:

$$\mathbf{X} = \begin{bmatrix} \mathbf{1}_{n_1} & \mathbf{1}_{n_1} & \mathbf{0} & \ldots & \mathbf{0} \\ \mathbf{1}_{n_2} & \mathbf{0} & \mathbf{1}_{n_2} & \ldots & \mathbf{0} \\ \mathbf{1}_{n_3} & \mathbf{0} & \mathbf{0} & \ldots & \mathbf{0} \\ \ldots & \ldots & \ldots & \ldots & \ldots \\ \mathbf{1}_{n_a} & \mathbf{0} & \ldots & \mathbf{0} & \mathbf{1}_{n_a} \end{bmatrix} \quad \text{and} \quad \mathbf{b} = \begin{bmatrix} \mu \\ \alpha_1 \\ \alpha_2 \\ \ldots \\ \alpha_a \end{bmatrix}.$$

Taking the case $a = 3$, we have

$$(\mathbf{X}^T\mathbf{X})^g = \begin{bmatrix} 0 & 0 & 0 & 0 \\ 0 & 1/n_1 & 0 & 0 \\ 0 & 0 & 1/n_2 & 0 \\ 0 & 0 & 0 & 1/n_3 \end{bmatrix}$$

so

$$\hat{\mathbf{b}} = \begin{bmatrix} 0 \\ \overline{y}_{1.} \\ \overline{y}_{2.} \\ \overline{y}_{3.} \end{bmatrix}$$

$$\mathbf{K}^T\hat{\mathbf{b}} - \mathbf{m} = \begin{bmatrix} 0 & 1 & -1 & 0 \\ 0 & 1 & 0 & -1 \end{bmatrix} \begin{bmatrix} 0 \\ \overline{y}_{1.} \\ \overline{y}_{2.} \\ \overline{y}_{3.} \end{bmatrix} - \begin{bmatrix} 0 \\ 0 \end{bmatrix} = \begin{bmatrix} \overline{y}_{1.} - \overline{y}_{2.} \\ \overline{y}_{1.} - \overline{y}_{3.} \end{bmatrix}$$

Now

$$\mathbf{K}^T(\mathbf{X}^T\mathbf{X})^g\mathbf{K} = \begin{bmatrix} \frac{1}{n_1} + \frac{1}{n_2} & \frac{1}{n_1} \\ \frac{1}{n_1} & \frac{1}{n_1} + \frac{1}{n_3} \end{bmatrix},$$

its inverse is

$$\frac{1}{N} \begin{bmatrix} n_2(n_1 + n_3) & -n_2 n_3 \\ -n_2 n_3 & n_3(n_1 + n_2) \end{bmatrix}$$

and the quadratic form

$$(\mathbf{K}^T\hat{\mathbf{b}} - \mathbf{m})^T \left[\mathbf{K}^T(\mathbf{X}^T\mathbf{X})^g\mathbf{K}\right]^{-1} (\mathbf{K}^T\hat{\mathbf{b}} - \mathbf{m})$$
$$= \begin{bmatrix}\overline{y}_{1.} - \overline{y}_{2.} & \overline{y}_{1.} - \overline{y}_{3.}\end{bmatrix} \frac{1}{N} \begin{bmatrix} n_2(n_1+n_3) & -n_2n_3 \\ -n_2n_3 & n_3(n_1+n_2)\end{bmatrix} \begin{bmatrix}\overline{y}_{1.} - \overline{y}_{2.} \\ \overline{y}_{1.} - \overline{y}_{3.}\end{bmatrix}$$
$$= \sum_{i=1}^{3} n_i\left(\overline{y}_{i.} - \overline{y}_{..}\right)^2,$$

but only after about six more lines of algebra. The diligent reader should supply the missing steps.

Taking the general case,

$$(\mathbf{X}^T\mathbf{X})^g = \begin{bmatrix} 0 & 0 & 0 & \dots & 0 \\ 0 & 1/n_1 & 0 & \dots & 0 \\ 0 & 0 & 1/n_2 & \dots & 0 \\ \dots & \dots & \dots & \dots & \dots \\ 0 & 0 & 0 & \dots & 1/n_a \end{bmatrix}$$

so

$$\hat{\mathbf{b}} = \begin{bmatrix} 0 \\ \overline{y}_{1.} \\ \overline{y}_{2.} \\ \dots \\ \overline{y}_{a.} \end{bmatrix}$$

and

$$\mathbf{K}^T\hat{\mathbf{b}} - \mathbf{m} = \begin{bmatrix} 0 & 1 & -1 & 0 & \dots & 0 \\ 0 & 1 & 0 & -1 & \dots & 0 \\ & \dots & & & \dots & \\ 0 & 1 & 0 & 0 & 0 & -1 \end{bmatrix} \begin{bmatrix} 0 \\ \overline{y}_{1.} \\ \overline{y}_{2.} \\ \dots \\ \overline{y}_{a.} \end{bmatrix} - \begin{bmatrix} 0 \\ 0 \\ \dots \\ 0 \end{bmatrix}$$
$$= \begin{bmatrix}\mathbf{0} & \mathbf{1} & -\mathbf{I}_{a-1}\end{bmatrix} \begin{bmatrix} 0 \\ \overline{y}_{1.} \\ \overline{y}_{2.} \\ \dots \\ \overline{y}_{a.} \end{bmatrix} = \begin{bmatrix} \overline{y}_{1.} - \overline{y}_{2.} \\ \overline{y}_{1.} - \overline{y}_{3.} \\ \dots \\ \overline{y}_{1.} - \overline{y}_{a.} \end{bmatrix}.$$

Now partition

$$(\mathbf{X}^T\mathbf{X})^g = \begin{bmatrix} 0 & 0 & 0 \\ 0 & 1/n_1 & 0 \\ 0 & 0 & \mathbf{D}_* \end{bmatrix}$$

where the partitioning is $(1, 1, a-1)$ in both rows (as for $\mathbf{K}^T$) and columns, and $\mathbf{D}_* = diag(1/n_2, \ldots, 1/n_a)$. Note that $\mathbf{D}_*$ is an $(a-1)$ diagonal matrix. Using the partitioned form, then $\mathbf{K}^T(\mathbf{X}^T\mathbf{X})^g\mathbf{K} = \mathbf{D}_* + \frac{1}{n_1}\mathbf{1}\mathbf{1}^T$ (it's $(a-1)\times(a-1)$); its inverse is $\mathbf{D}_*^{-1} - \frac{1}{N}\mathbf{D}_*^{-1}\mathbf{1}\mathbf{1}^T\mathbf{D}_*^{-1}$ (from Exercise A.72) and the quadratic form is

$$\begin{aligned}
&(\mathbf{K}^T\hat{\mathbf{b}} - \mathbf{m})^T[\mathbf{K}^T(\mathbf{X}^T\mathbf{X})^g\mathbf{K}]^{-1}(\mathbf{K}^T\hat{\mathbf{b}} - \mathbf{m}) \\
&\quad = (\mathbf{K}^T\hat{\mathbf{b}} - \mathbf{m})^T\mathbf{D}_*^{-1}(\mathbf{K}^T\hat{\mathbf{b}} - \mathbf{m}) - \frac{1}{N}(\mathbf{K}^T\hat{\mathbf{b}} - \mathbf{m})^T\mathbf{D}_*^{-1}\mathbf{1}\mathbf{1}^T\mathbf{D}_*^{-1}(\mathbf{K}^T\hat{\mathbf{b}} - \mathbf{m}) \\
&\quad = \sum_{i=2}^{a}(\bar{y}_{1.} - \bar{y}_{i.})^2 - \frac{1}{N}\left[\sum_{i=2}^{a} n_i(\bar{y}_{1.} - \bar{y}_{i.})\right]^2.
\end{aligned}$$

It still takes some algebra to find a familiar expression:

$$\begin{aligned}
&\sum_{i=1}^{a} n_i(\bar{y}_{i.} - \bar{y}_{..})^2 \\
&\quad = \sum_{i=1}^{a} n_i(\bar{y}_{i.} - \bar{y}_{1.} + \bar{y}_{1.} - \bar{y}_{..})^2 \\
&\quad = \sum_{i=1}^{a} n_i(\bar{y}_{i.} - \bar{y}_{1.})^2 + 2\sum_{i=1}^{a} n_i(\bar{y}_{i.} - \bar{y}_{1.})(\bar{y}_{1.} - \bar{y}_{..}) + \sum_{i=1}^{a} n_i(\bar{y}_{1.} - \bar{y}_{..})^2 \\
&\quad = \sum_{i=1}^{a} n_i(\bar{y}_{i.} - \bar{y}_{1.})^2 + 2(\bar{y}_{1.} - \bar{y}_{..})\sum_{i=1}^{a} n_i(\bar{y}_{i.} - \bar{y}_{1.}) + N(\bar{y}_{1.} - \bar{y}_{..})^2 \\
&\quad = \sum_{i=2}^{a} n_i(\bar{y}_{i.} - \bar{y}_{1.})^2 + 2(\bar{y}_{1.} - \bar{y}_{..})(N\bar{y}_{..} - N\bar{y}_{1.}) + N(\bar{y}_{1.} - \bar{y}_{..})^2 \\
&\quad = \sum_{i=2}^{a} n_i(\bar{y}_{i.} - \bar{y}_{1.})^2 - N(\bar{y}_{1.} - \bar{y}_{..})^2.
\end{aligned}$$

It should be obvious that $\mathbf{y}^T(\mathbf{I} - \mathbf{P_X})\mathbf{y} = SSE = \sum_{i=1}^{a}\sum_{j=1}^{n_i}(y_{ij} - \bar{y}_{i.})^2$, and so the general linear hypothesis form of the usual ANOVA test for equal treatment effects becomes

$$\begin{aligned}
F &= \frac{(\mathbf{K}^T\hat{\mathbf{b}} - \mathbf{m})^T\left[\mathbf{K}^T(\mathbf{X}^T\mathbf{X})^g\mathbf{K}\right]^{-1}(\mathbf{K}^T\hat{\mathbf{b}} - \mathbf{m})/(a-1)}{\mathbf{y}^T(\mathbf{I} - \mathbf{P_X})\mathbf{y}/(N-a)} \\
&= \frac{\sum_{i=1}^{a} n_i(\bar{y}_{i.} - \bar{y}_{..})^2/(a-1)}{\sum_{i=1}^{a}\sum_{j=1}^{n_i}(y_{ij} - \bar{y}_{i.})^2/(N-a)}
\end{aligned}$$

which is our familiar ANOVA test statistic to be compared to $F_{(a-1),(N-a),\alpha}$.

Now that we've constructed this F-test in an unfamiliar form, we're interested in establishing that it has some useful properties that would make it preferable to any other test. In particular, we would like the test to be invariant to certain changes in $\mathbf{K}$ and $\mathbf{m}$ that would have no logical or mathematical consequences. Clearly, the hypothesis $2\mathbf{K}^T\mathbf{b} = 2\mathbf{m}$ is not logically different, and a test that is affected by such a simple change would not be desirable.

Example 6.5: continued

Consider again the usual one-way ANOVA model $y_{ij} = \mu + \alpha_i + e_{ij}$, but note that the most common hypothesis to be tested is that the effects α_i are all equal. One way of writing this test in the form $\mathbf{K}^T\mathbf{b} = \mathbf{m}$ is

$$\mathrm{H}_1 : \alpha_1 - \alpha_2 = 0, \alpha_1 - \alpha_3 = 0, \ldots, \alpha_1 - \alpha_a = 0$$

while another is

$$\mathrm{H}_2 : \alpha_1 - \alpha_2 = 0, \alpha_2 - \alpha_3 = 0, \ldots, \alpha_{a-1} - \alpha_a = 0.$$

The insistence on full-column rank for $\mathbf{K}$ precludes, say, $\alpha_a - \alpha_1 = 0$, in the latter case.

While there may be two logically equivalent ways of expressing the test, will they lead to the same test procedure even though their expression of $\mathbf{K}$ is different? More precisely, suppose we have two expressions of these tests, $\mathbf{K}^T\mathbf{b} = \mathbf{m}$ and $\mathbf{K}_*^T\mathbf{b} = \mathbf{m}_*$, such that

$$\mathcal{S} = \{\mathbf{b} : \mathbf{K}^T\mathbf{b} = \mathbf{m}\} = \mathcal{S}_* = \left\{\mathbf{b} : \mathbf{K}_*^T\mathbf{b} = \mathbf{m}_*\right\}.$$

Can we show that they would both give the same numerator sum of squares in the F-test? Looking at the geometry, since $\mathcal{S}$ is parallel to $\mathcal{N}(\mathbf{K}^T)$, and $\mathcal{S}_*$ is parallel to $\mathcal{N}(\mathbf{K}_*^T)$, then the two nullspaces must be parallel to each other, and we should be showing $\mathcal{C}(\mathbf{K}) = \mathcal{C}(\mathbf{K}_*)$. The algebra (Result A.13) says that the points in $\mathcal{S}$ can be expressed as

$$\mathbf{K}(\mathbf{K}^T\mathbf{K})^{-1}\mathbf{m} + (\mathbf{I} - \mathbf{K}(\mathbf{K}^T\mathbf{K})^{-1}\mathbf{K}^T)\mathbf{z} \tag{6.11}$$

since $\mathbf{K}(\mathbf{K}^T\mathbf{K})^{-1}$ is a generalized inverse of $\mathbf{K}^T$, remembering that $\mathbf{K}$ has full-column rank. Now all points in $\mathcal{S}$ given by (6.11) should satisfy $\mathbf{K}_*^T\mathbf{b} = \mathbf{m}_*$, so that

$$\mathbf{K}_*^T\{\mathbf{K}(\mathbf{K}^T\mathbf{K})^{-1}\mathbf{m} + (\mathbf{I} - \mathbf{K}(\mathbf{K}^T\mathbf{K})^{-1}\mathbf{K}^T)\mathbf{z}\} = \mathbf{m}_* \tag{6.12}$$

for all $\mathbf{z}$, so (Result A.8) that $\mathbf{K}_*^T = \mathbf{K}_*^T\mathbf{P_K}$, or $\mathbf{P_K}\mathbf{K}_* = \mathbf{K}_*$, or $\mathcal{C}(\mathbf{K}_*) \subseteq \mathcal{C}(\mathbf{K})$. Reversing the roles of $\mathbf{K}$ and $\mathbf{K}_*$, $\mathbf{m}$ and $\mathbf{m}_*$, will then provide the subsetting in the opposite direction, leading to $\mathcal{C}(\mathbf{K}) = \mathcal{C}(\mathbf{K}_*)$, or that there exists a nonsingular matrix $\mathbf{Q}$ such that $\mathbf{K}\mathbf{Q}^T = \mathbf{K}_*$. Also following from Result A.8 and (6.12), we have $\mathbf{m}_* = \mathbf{Q}\mathbf{m}$, and $\mathbf{Q} = \mathbf{K}_*^T\mathbf{K}(\mathbf{K}^T\mathbf{K})^{-1}$. Doing the algebra on the numerator sum of squares of the F-statistic, we see that testing with $\mathbf{K}_*^T\mathbf{b} = \mathbf{m}_*$, we have

$$\begin{aligned}
&(\mathbf{K}_*^T\hat{\mathbf{b}} - \mathbf{m}_*)^T \left[\mathbf{K}_*^T(\mathbf{X}^T\mathbf{X})^g\mathbf{K}_*\right]^{-1} (\mathbf{K}_*^T\hat{\mathbf{b}} - \mathbf{m}_*)\\
&= (\mathbf{Q}\mathbf{K}^T\hat{\mathbf{b}} - \mathbf{Q}\mathbf{m})^T \left[\mathbf{Q}\mathbf{K}^T(\mathbf{X}^T\mathbf{X})^g\mathbf{K}\mathbf{Q}^T\right]^{-1} (\mathbf{Q}\mathbf{K}^T\hat{\mathbf{b}} - \mathbf{Q}\mathbf{m})\\
&= (\mathbf{K}^T\hat{\mathbf{b}} - \mathbf{m})^T \left[\mathbf{K}^T(\mathbf{X}^T\mathbf{X})^g\mathbf{K}\right]^{-1} (\mathbf{K}^T\hat{\mathbf{b}} - \mathbf{m})
\end{aligned}$$

which is the same numerator sum of squares when testing $\mathbf{K}^T\mathbf{b} = \mathbf{m}$. Therefore, the F-statistic is invariant to equivalent specifications of the hypothesis $\mathbf{K}^T\mathbf{b} = \mathbf{m}$. Moreover, "equivalent" means nonsingular linear transformations.

Example 6.5: continued

Consider the usual one-way ANOVA model $y_{ij} = \mu + \alpha_i + e_{ij}$ with $a = 3$. The first hypothesis can be written as

$$\begin{bmatrix} 0 & 1 & -1 & 0 \\ 0 & 1 & 0 & -1 \end{bmatrix} \begin{bmatrix} \mu \\ \alpha_1 \\ \alpha_2 \\ \alpha_3 \end{bmatrix} = \begin{bmatrix} 0 \\ 0 \end{bmatrix}$$

while the second uses $\mathbf{K}_* = \mathbf{K}\mathbf{Q}^T$ or $\mathbf{K}_*^T = \mathbf{Q}\mathbf{K}^T$:

$$\begin{bmatrix} 0 & 1 & -1 & 0 \\ 0 & 0 & 1 & -1 \end{bmatrix} \begin{bmatrix} \mu \\ \alpha_1 \\ \alpha_2 \\ \alpha_3 \end{bmatrix} = \begin{bmatrix} 0 \\ 0 \end{bmatrix}$$

where

$$\mathbf{Q} = \begin{bmatrix} 1 & 0 \\ -1 & 1 \end{bmatrix}.$$

Before leaving this section on testing, the simpler hypothesis, where $s = 1$, leads to the familiar t-test. Suppose we want to test the hypothesis H: $\mathbf{k}^T\mathbf{b} = m$ where $\mathbf{k}$ is a vector and m is now a scalar. We can entertain both a one-sided alternative and the usual two-sided alternative:

$$\mathrm{A}_1 : \mathbf{k}^T\mathbf{b} > m, \text{ or } \mathrm{A}_2 : \mathbf{k}^T\mathbf{b} \neq m.$$

As before, we must insist on $\mathbf{k}^T\mathbf{b}$ being estimable. The route for constructing the test statistic should be obvious: find the best estimator for $\mathbf{k}^T\mathbf{b}$ and see how far the hypothesized value m is from the estimate.

The BLUE for estimable $\mathbf{k}^T\mathbf{b}$ is $\mathbf{k}^T\hat{\mathbf{b}}$ where $\hat{\mathbf{b}}$ solves the usual normal equations. Its distribution is given by

$$\mathbf{k}^T\hat{\mathbf{b}} = \mathbf{k}^T(\mathbf{X}^T\mathbf{X})^g\mathbf{X}^T\mathbf{y} \sim N(\mathbf{k}^T\mathbf{b}, \sigma^2\mathbf{k}^T(\mathbf{X}^T\mathbf{X})^g\mathbf{k}),$$

so that

$$(\mathbf{k}^T\hat{\mathbf{b}} - m)/\sqrt{\sigma^2\mathbf{k}^T(\mathbf{X}^T\mathbf{X})^g\mathbf{k}} \sim N((\mathbf{k}^T\mathbf{b} - m)/\sqrt{\sigma^2\mathbf{k}^T(\mathbf{X}^T\mathbf{X})^g\mathbf{k}}, 1).$$

Filling in some algebraic details, we have

$$t = \frac{\mathbf{k}^T\hat{\mathbf{b}} - m}{\sqrt{\sigma^2\mathbf{k}^T(\mathbf{X}^T\mathbf{X})^g\mathbf{k}}} \Big/ \sqrt{\frac{\hat{\sigma}^2}{\sigma^2}} = \frac{\mathbf{k}^T\hat{\mathbf{b}} - m}{\sqrt{\hat{\sigma}^2\mathbf{k}^T(\mathbf{X}^T\mathbf{X})^g\mathbf{k}}} \sim t_{N-r}(\mu) \tag{6.13}$$

where the noncentrality parameter is

$$\mu = (\mathbf{k}^T\mathbf{b} - m)/\sqrt{\sigma^2\mathbf{k}^T(\mathbf{X}^T\mathbf{X})^g\mathbf{k}}. \tag{6.14}$$

For the one-sided alternative, A_1: $\mathbf{k}^T\mathbf{b} > m$, the test: Reject if $t > c_\alpha$ provides a level α test for $\alpha = Pr(T > c_\alpha | T \sim t_{N-r})$. For the two-sided case, A_2: $\mathbf{k}^T\mathbf{b} \neq m$, the test: Reject if $|t| > c_{\alpha/2}$ then provides a level α test.

6.4 The Likelihood Ratio Test and Change in *SSE*

In Section 6.3, we derived the test of the general linear hypothesis based solely on first principles by constructing a test statistic that measures a departure from the hypothesis. Since we know its distribution when the hypothesis is true, we can construct a test with the correct level. With Result 5.13, we know that a test that rejects when F is large will have power greater than the level, and increasing power with further departures from the hypothesis. However, we did not set out to construct a test that was the best or most powerful under certain conditions, although we have shown that the test was invariant to nonsingular linear transformations.

The likelihood ratio test, nonetheless, cannot claim to be the best test, but is motivated by the simple criterion of rejecting when the likelihood under the hypothesis is small. More precisely, we assume the usual normal linear model,

$$\mathbf{y} \sim N_N(\mathbf{X}\mathbf{b}, \sigma^2\mathbf{I}_N),$$

and we want to test the hypothesis H: $\mathbf{K}^T\mathbf{b} = \mathbf{m}$ with $\mathbf{K}^T\mathbf{b}$ estimable (columns of $\mathbf{K}$ are in $\mathcal{C}(\mathbf{X}^T)$). The parameter space under the hypothesis can be written as

$$\Omega_0 = \{(\mathbf{b}, \sigma^2) : \mathbf{K}^T\mathbf{b} = \mathbf{m}, \sigma^2 > 0\},$$

and the union of the hypothesis and alternative ($\mathbf{K}^T\mathbf{b} \neq \mathbf{m}$) is given by

$$\Omega = \{(\mathbf{b}, \sigma^2) : \mathbf{b} \in R^p, \sigma^2 > 0\}.$$

The likelihood function for the normal linear model is

$$L(\mathbf{b}, \sigma^2) = (2\pi\sigma^2)^{-N/2} \, exp\left\{-\frac{1}{2}Q(\mathbf{b})/\sigma^2\right\}$$

where $Q(\mathbf{b})$ is the error sum of squares function $Q(\mathbf{b}) = (\mathbf{y} - \mathbf{X}\mathbf{b})^T(\mathbf{y} - \mathbf{X}\mathbf{b})$. The test criterion ϕ, known as the *likelihood ratio*, is the ratio of the maxima under the two sets:

$$\phi(\mathbf{y}) = \max{}_{\Omega_0} L(\mathbf{b}, \sigma^2) / \max{}_{\Omega} L(\mathbf{b}, \sigma^2),$$

and the test rejects the hypothesis when $\phi(\mathbf{y})$ is too small, say, reject when $\phi(\mathbf{y}) < c$. The problem with the LRT, of course, is finding the critical value c so that $Pr(\phi(\mathbf{y}) < c | \Omega_0) \geq \alpha$.

Finding $\phi(\mathbf{y})$ involves maximizing the likelihood over each of these sets, and the easy route is to maximize with respect to σ^2 for given $\mathbf{b}$. Taking logs and then differentiating with respect to σ^2 we have

$$\frac{\partial logL}{\partial \sigma^2} = -\frac{1}{2}N/\sigma^2 + \frac{1}{2}Q(\mathbf{b})/\sigma^4$$

which is zero when $\sigma^2 = Q(\mathbf{b})/N$. The value of the likelihood for this value of σ^2 is then

$$L(\mathbf{b}, Q(\mathbf{b})/N) = (2\pi Q(\mathbf{b})/N)^{-N/2} exp\{-N/2\},$$

and to complete the task all that is needed is to minimize $Q(\mathbf{b})$. For Ω, this is simply minimized at the usual $\hat{\mathbf{b}}$; for Ω_0, let $\hat{\mathbf{b}}_H$ denote the value of $\mathbf{b}$ that minimized $Q(\mathbf{b})$ over Ω_0 (and postponing how to find it). With lots of constants canceling, we find $\phi(\mathbf{y}) = [Q(\hat{\mathbf{b}}_H)/Q(\hat{\mathbf{b}})]^{-N/2}$. The likelihood ratio test is then to reject H when $\phi(\mathbf{y})$ is too small, that is, $\phi(\mathbf{y}) < c$, and we can do some algebra to find c in terms of something more familiar:

$$\begin{aligned}
\phi(\mathbf{y}) = \left[Q(\hat{\mathbf{b}}_H)/Q(\hat{\mathbf{b}})\right]^{-N/2} &< c, \text{ or} \\
Q(\hat{\mathbf{b}}_H)/Q(\hat{\mathbf{b}}) &> c^{-2/N}, \text{ or} \\
\left[Q(\hat{\mathbf{b}}_H) - Q(\hat{\mathbf{b}})\right]/Q(\hat{\mathbf{b}}) &> (c^{-2/N} - 1), \text{ or} \\
\frac{\left[Q(\hat{\mathbf{b}}_H) - Q(\hat{\mathbf{b}})\right]/s}{Q(\hat{\mathbf{b}})/(N-r)} &> \frac{N-r}{s}(c^{-2/N} - 1). \qquad (6.15)
\end{aligned}$$

So the likelihood ratio test for the general linear hypothesis H: $\mathbf{K}^T\mathbf{b} = \mathbf{m}$ leads to the familiar expression in terms of the error sum of squares for the full model Ω and the error sum of squares for the model restricted by the hypothesis Ω_0. While this expression on the left-hand side of Equation (6.15) looks like the familiar F-test, we hesitate to name that quantity F. At this point, we can show for some particular cases, as the two examples below, that this expression does have the F-distribution. The task of showing that this holds in general will be postponed until the next section where we will show that the expression above is the same F-statistic obtained in Section 6.3. This will then show that the test for the general linear hypothesis based on first principles and derived in Section 6.3 is a likelihood ratio test.

Example 6.5: continued

To find the likelihood ratio test for the one-way ANOVA model, we begin by finding the unrestricted *SSE*, which we have obtained earlier as

$$Q(\hat{\mathbf{b}}) = \sum_i \sum_j (y_{ij} - \bar{y}_{i.})^2 = \mathbf{y}^T(\mathbf{I} - \mathbf{P_X})\mathbf{y}.$$

To find $Q(\hat{\mathbf{b}}_H)$, we have two approaches. From a geometric view, we see that for $\mathbf{b} \in \boldsymbol{\Omega}_0 = \{(\mu, \alpha's, \sigma^2) : \alpha_1 = \cdots = \alpha_a\}$, the mean vector $\mathbf{Xb} \in \mathcal{C}(\mathbf{1}_N)$, and so

$$Q(\hat{\mathbf{b}}_H) = \mathbf{y}^T(\mathbf{I} - \mathbf{P}_1)\mathbf{y} = \sum_i \sum_j (y_{ij} - \bar{y}_{..})^2.$$

From a more algebraic viewpoint, we have

$$\begin{aligned} min_{\Omega_0} \sum_i \sum_j (y_{ij} - (\mu + \alpha_i))^2 &= min_{\mu, \alpha_i = \alpha} \sum_i \sum_j (y_{ij} - (\mu + \alpha_i)))^2 \\ &= min_{\mu + \alpha} \sum_i \sum_j (y_{ij} - (\mu + \alpha))^2 \\ &= \sum_i \sum_j (y_{ij} - \overline{y}_{..})^2. \end{aligned}$$

Some familiar algebra gives us

$$Q(\hat{\mathbf{b}}_H) = Q(\hat{\mathbf{b}}) + \mathbf{y}^T(\mathbf{P_X} - \mathbf{P}_1)\mathbf{y} = Q(\hat{\mathbf{b}}) + \sum_i n_i(y_{i.} - \overline{y}_{..})^2$$

and the familiar F-statistic:

$$F = \frac{\left[Q(\hat{\mathbf{b}}_H) - Q(\hat{\mathbf{b}})\right]/(a-1)}{Q(\hat{\mathbf{b}})/(N-a)} = \frac{\sum_i n_i(y_{i.} - \overline{y}_{..})^2/(a-1)}{\sum_i \sum_j (y_{ij} - \overline{y}_{i.})^2/(N-a)}$$

whose distribution was established in Chapter 5.

Example 6.3: continued
Recall this most common testing situation, where we want to test H: $\mathbf{b}_1 = \mathbf{0}$,

$$\mathbf{Xb} = \begin{bmatrix} \mathbf{X}_0 & \mathbf{X}_1 \end{bmatrix} \begin{bmatrix} \mathbf{b}_0 \\ \mathbf{b}_1 \end{bmatrix},$$

where $rank\,(\mathbf{X}) = r$, $rank\,(\mathbf{X}_0) = r_0$, $rank\,((\mathbf{I} - \mathbf{P}_{\mathbf{X}_0})\mathbf{X}_1) = s$, with $s = r - r_0$. In this situation, it's easy to compute $Q(\hat{\mathbf{b}}_H)$ and $Q(\hat{\mathbf{b}})$:

$$Q(\hat{\mathbf{b}}_H) = \mathbf{y}^T(\mathbf{I} - \mathbf{P}_{\mathbf{X}_0})\mathbf{y} = SSE(reduced)$$

$$Q(\hat{\mathbf{b}}) = \mathbf{y}^T(\mathbf{I} - \mathbf{P_X})\mathbf{y} = SSE(full).$$

And so we can also compute

$$\begin{aligned} Q(\hat{\mathbf{b}}_H) - Q(\hat{\mathbf{b}}) &= \mathbf{y}^T(\mathbf{I} - \mathbf{P}_{\mathbf{X}_0})\mathbf{y} - \mathbf{y}^T(\mathbf{I} - \mathbf{P_X})\mathbf{y} \\ &= \mathbf{y}^T(\mathbf{P_X} - \mathbf{P}_{\mathbf{X}_0})\mathbf{y} = SSE(reduced) - SSE(full) \end{aligned}$$

Now we can apply Cochran's Theorem to the following sums of squares:

$$\mathbf{y}^T\mathbf{P}_{\mathbf{X}_0}\mathbf{y}$$

$$\mathbf{y}^T(\mathbf{P_X} - \mathbf{P}_{\mathbf{X}_0})\mathbf{y}$$

$$\mathbf{y}^T(\mathbf{I} - \mathbf{P_X})\mathbf{y}$$

since they are quadratic forms in symmetric idempotent matrices that sum to the identity matrix. So these three pieces are independent with noncentral chi-square distributions, with noncentralities $(\mathbf{Xb})^T \mathbf{P}_{\mathbf{X}_0}(\mathbf{Xb})/(2\sigma^2)$, $(\mathbf{Xb})^T(\mathbf{P}_\mathbf{X} - \mathbf{P}_{\mathbf{X}_0})(\mathbf{Xb})/(2\sigma^2)$, and 0. As a result our usual test statistic,

$$F = \frac{\left[Q(\hat{\mathbf{b}}_H) - Q(\hat{\mathbf{b}})\right]/s}{Q(\hat{\mathbf{b}})/(N-r)} = \frac{\mathbf{y}^T(\mathbf{P}_\mathbf{X} - \mathbf{P}_{\mathbf{X}_0})\mathbf{y}/s}{\mathbf{y}^T(\mathbf{I} - \mathbf{P}_\mathbf{X})\mathbf{y}/(N-r)}$$

has an F-distribution with s and $N - r$ degrees of freedom with noncentrality $(\mathbf{Xb})^T(\mathbf{P}_\mathbf{X} - \mathbf{P}_{\mathbf{X}_0})(\mathbf{Xb})/(2\sigma^2)$. If the hypothesis H is true, then $\mathbf{Xb} \in \mathcal{C}(\mathbf{X}_0)$ (since $\mathbf{b}_1 = \mathbf{0}$), and so the F-statistic has the central F-distribution when the hypothesis is true. Note that we can construct the likelihood ratio test statistic, and establish its distribution, without resolving the estimable components of the hypothesis.

6.5 First Principles Test and LRT

In Section 6.2 we presented a first principles approach to testing the general linear hypothesis H: $\mathbf{K}^T\mathbf{b} = \mathbf{m}$. The only apparent property that this test can advertise is that it is unbiased. The likelihood ratio test carries the appeal of maximum likelihood, but requires maximization of the likelihood in a restricted parameter space. For simple problems, such as testing H: $\beta_1 = 1$, $\beta_2 = \beta_3$, the first principles approach appears to be the easiest to implement. For subset problems, as we see in Example 6.5 for one-way ANOVA, this same approach is cumbersome, and LRT is so much easier.

We left the discussion of the LRT with just the supposition that a minimum of the error sum of squares function $\mathbf{Q}(\mathbf{b})$ can be found subject to the constraint of the hypothesis. However, we can apply the results of Section 3.9 where we looked at the least squares problem subject to linear constraints. There we found that the solution to the restricted normal equations (RNEs) (3.10) gives the minimizing value of $\mathbf{b}$ subject to the constraint, and that RNEs are consistent.

We resolve the dilemma of two potential approaches for constructing a test statistic by showing that the F-test for the general linear hypothesis we obtained earlier from first principles is the same as the full versus reduced F-test, which arises from the likelihood ratio principle. The importance of this result demands its designation as a theorem.

Theorem 6.1 *If $\mathbf{K}^T\mathbf{b}$ is a set of linearly independent estimable functions, and $\hat{\mathbf{b}}_H$ is part of a solution to the RNEs with constraint $\mathbf{K}^T\mathbf{b} = \mathbf{m}$, then*

$$\begin{aligned} Q(\hat{\mathbf{b}}_H) - Q(\hat{\mathbf{b}}) &= (\hat{\mathbf{b}}_H - \hat{\mathbf{b}})^T\mathbf{X}^T\mathbf{X}(\hat{\mathbf{b}}_H - \hat{\mathbf{b}}) \\ &= (\mathbf{K}^T\hat{\mathbf{b}} - \mathbf{m})^T\left[\mathbf{K}^T(\mathbf{X}^T\mathbf{X})^g\mathbf{K}\right]^{-1}(\mathbf{K}^T\hat{\mathbf{b}} - \mathbf{m}). \end{aligned}$$

Proof: Recall that when we were testing H: $\mathbf{K}^T \mathbf{b} = \mathbf{m}$, we showed that $[\mathbf{K}^T (\mathbf{X}^T \mathbf{X})^g \mathbf{K}]$ was nonsingular when $\mathbf{K}^T \mathbf{b}$ is a set of linearly independent estimable functions. Also note

$$Q(\hat{\mathbf{b}}_H) - Q(\hat{\mathbf{b}}) = (\hat{\mathbf{b}} - \hat{\mathbf{b}}_H)^T \mathbf{X}^T \mathbf{X} (\hat{\mathbf{b}} - \hat{\mathbf{b}}_H) \tag{6.16}$$

since the cross-product $(\hat{\mathbf{b}} - \hat{\mathbf{b}}_H)^T \mathbf{X}^T (\mathbf{y} - \mathbf{X}\hat{\mathbf{b}}) = \mathbf{0}$. Now from the RNEs with $\mathbf{K}$ in place of $\mathbf{P}$, $\mathbf{m}$ in place of δ, consider

$$\mathbf{X}^T \mathbf{X} (\hat{\mathbf{b}} - \hat{\mathbf{b}}_H) = \mathbf{X}^T \mathbf{y} - (\mathbf{X}^T \mathbf{y} - \mathbf{K}\hat{\theta}_H) = \mathbf{K}\hat{\theta}_H, \tag{6.17}$$

where $\hat{\theta}_H$ is the Lagrange multiplier. Premultiplying (6.17) by $\mathbf{K}^T (\mathbf{X}^T \mathbf{X})^g$ gives

$$\mathbf{K}^T (\mathbf{X}^T \mathbf{X})^g \mathbf{X}^T \mathbf{X} (\hat{\mathbf{b}} - \hat{\mathbf{b}}_H) = \mathbf{K}^T (\hat{\mathbf{b}} - \hat{\mathbf{b}}_H) = \mathbf{K}^T (\mathbf{X}^T \mathbf{X})^g \mathbf{K}\hat{\theta}_H$$

since $\mathbf{K}^T \mathbf{b}$ is estimable. Since $[\mathbf{K}^T (\mathbf{X}^T \mathbf{X})^g \mathbf{K}]$ is nonsingular from Result 6.4, we have

$$\hat{\theta}_H = [\mathbf{K}^T (\mathbf{X}^T \mathbf{X})^g \mathbf{K}]^{-1} (\mathbf{K}^T \hat{\mathbf{b}} - \mathbf{m}), \tag{6.18}$$

since $\mathbf{K}^T \hat{\mathbf{b}}_H = \mathbf{m}$, and finally, premultiplying (6.17) by $(\hat{\mathbf{b}} - \hat{\mathbf{b}}_H)^T$ we then have

$$Q(\hat{\mathbf{b}}_H) - Q(\hat{\mathbf{b}}) = (\hat{\mathbf{b}} - \hat{\mathbf{b}}_H)^T \mathbf{K}\hat{\theta}_H = (\hat{\mathbf{b}} - \hat{\mathbf{b}}_H)^T \mathbf{K}[\mathbf{K}^T (\mathbf{X}^T \mathbf{X})^g \mathbf{K}]^{-1} (\mathbf{K}^T \hat{\mathbf{b}} - \mathbf{m}),$$

which gives our result, noting again that $\mathbf{K}^T \hat{\mathbf{b}}_H = \mathbf{m}$.   □

Corollary 6.4   *If $\mathbf{K}^T \mathbf{b}$ is a set of linearly independent estimable functions, $\hat{\mathbf{b}}$ a solution to the usual normal equations, then the first part $\hat{\mathbf{b}}_H$ of a solution to the RNE with constraint $\mathbf{K}^T \mathbf{b} = \mathbf{m}$ can be found by solving for $\mathbf{b}$ in the equations*

$$\mathbf{X}^T \mathbf{X} \mathbf{b} = \mathbf{X}^T \mathbf{y} - \mathbf{K}[\mathbf{K}^T (\mathbf{X}^T \mathbf{X})^g \mathbf{K}]^{-1} (\mathbf{K}^T \hat{\mathbf{b}} - \mathbf{m}). \tag{6.19}$$

Proof: The expression from the RNE is

$$\mathbf{X}^T \mathbf{X} \hat{\mathbf{b}}_H + \mathbf{K}\hat{\theta}_H = \mathbf{X}^T \mathbf{y},$$

and (6.18) gives the Lagrange multiplier, leading to solving for $\mathbf{b}$ in (6.19).   □

The reader should note that the first departure from the generality of the form of the restricted parameter case comes above in Theorem 6.1; until then we made no direct connection between testing and the constrained parameter space.

Recall in Section 3.7, we consider imposing constraints on the solution to the normal equations. In the context of this section, the result was that if we imposed enough completely nonestimable constraints, we could get a unique solution to the normal equations. Here, we're considering a slight variation in imposing constraints of the form $\mathbf{P}^T \mathbf{b} = \delta$, where we do not insist that the right-hand-side vector δ is zero. Recall again what we mean by jointly nonestimable—each component of $\mathbf{P}^T\mathbf{b}$ is nonestimable, and no linear combination is estimable—so $\mathcal{C}(\mathbf{P}) \cap \mathcal{C}(\mathbf{X}^T) = \{\mathbf{0}\}$.

Result 6.5 *If $\mathbf{P}^T\mathbf{b}$ is a set of linearly independent, jointly nonestimable functions, and $\hat{\mathbf{b}}_H$ is part of a solution to the RNE with constraint $\mathbf{P}^T\mathbf{b} = \delta$, then $Q(\hat{\mathbf{b}}_H) = Q(\hat{\mathbf{b}})$ and $\hat{\theta} = \mathbf{0}$.*

Proof: The first part of the RNE is

$$\mathbf{X}^T\mathbf{X}\hat{\mathbf{b}}_H + \mathbf{P}\hat{\theta}_H = \mathbf{X}^T\mathbf{y}, \text{ or } \mathbf{P}\hat{\theta}_H = \mathbf{X}^T\mathbf{y} - \mathbf{X}^T\mathbf{X}\hat{\mathbf{b}}_H$$

and notice that $\mathbf{X}^T\mathbf{y} - \mathbf{X}^T\mathbf{X}\hat{\mathbf{b}}_H$ is in $\mathcal{C}(\mathbf{X}^T)$. Since $\mathcal{C}(\mathbf{P}) \cap \mathcal{C}(\mathbf{X}^T) = \{\mathbf{0}\}$, then $\mathbf{P}\hat{\theta}_H$ must be the zero vector; with full-column-rank $\mathbf{P}$, then $\hat{\theta}_H$ must also be zero. Hence $\hat{\mathbf{b}}_H$ is another solution to the normal equations, and from Result 2.3, $Q(\hat{\mathbf{b}}_H) = Q(\hat{\mathbf{b}})$. ∎

6.6 Confidence Intervals and Multiple Comparisons

Consider our usual linear model with distributional assumptions, that is, $\mathbf{y} \sim N_N(\mathbf{Xb}, \sigma^2\mathbf{I}_N)$. If $\lambda^T\mathbf{b}$ is estimable, then we have the distribution of our least squares estimator

$$\lambda^T\hat{\mathbf{b}} \sim N(\lambda^T\mathbf{b}, \sigma^2\lambda^T(\mathbf{X}^T\mathbf{X})^g\lambda).$$

Estimating σ^2 with the usual $\hat{\sigma}^2 = SSE/(N-r)$, then

$$t = \frac{\lambda^T\hat{\mathbf{b}} - \lambda^T\mathbf{b}}{\hat{\sigma}\sqrt{\lambda^T(\mathbf{X}^T\mathbf{X})^g\lambda}}$$

has the central Student's t-distribution with $(N-r)$ df. We can then construct a confidence interval for estimable $\lambda^T\mathbf{b}$ based on this pivotal, employing

$$Pr\left(\left|\frac{\lambda^T\hat{\mathbf{b}} - \lambda^T\mathbf{b}}{\hat{\sigma}\sqrt{\lambda^T(\mathbf{X}^T\mathbf{X})^g\lambda}}\right| \le t_{N-r,\alpha/2}\right) = 1-\alpha$$

to obtain the interval

$$\lambda^T\hat{\mathbf{b}} - t_{N-r,\alpha/2}\sqrt{\hat{\sigma}^2\lambda^T(\mathbf{X}^T\mathbf{X})^g\lambda} \le \lambda^T\mathbf{b} \le \lambda^T\hat{\mathbf{b}} + t_{N-r,\alpha/2}\sqrt{\hat{\sigma}^2\lambda^T(\mathbf{X}^T\mathbf{X})^g\lambda}.$$

To extend this to several estimable functions $\lambda_j^T\mathbf{b}$, linearly independent and estimable, consider the vector $\tau = \mathbf{\Lambda}^T\mathbf{b}$ found by stacking the vectors λ_j as columns of the matrix $\mathbf{\Lambda} = (\lambda_1|\lambda_2|\ldots|\lambda_s)$, so that $\tau_j = \lambda_j^T\mathbf{b}$. We can then construct *one-at-a-time* intervals for each component τ_j using the same familiar distributional results:

$$\hat{\tau} = \mathbf{\Lambda}^T\hat{\mathbf{b}} \sim N(\mathbf{\Lambda}^T\mathbf{b} = \tau, \sigma^2\mathbf{\Lambda}^T(\mathbf{X}^T\mathbf{X})^g\mathbf{\Lambda} = \sigma^2\mathbf{H}).$$

The confidence interval for $\tau_j = \lambda_j^T \mathbf{b}$ then takes the form

$$l_j \leq \tau_j \leq u_j \tag{6.20}$$

where

$$l_j = \hat{\tau}_j - t_{N-r,\alpha/2}\sqrt{\hat{\sigma}^2 \mathbf{H}_{jj}} \tag{6.21}$$

and

$$u_j = \hat{\tau}_j + t_{N-r,\alpha/2}\sqrt{\hat{\sigma}^2 \mathbf{H}_{jj}}. \tag{6.22}$$

These intervals satisfy

$$Pr(l_j \leq \tau_j \leq u_j) = 1 - \alpha$$

for each j, but if we define the rectangular solid, or box $B = \{t : l_j \leq \tau_j \leq u_j \text{ for } j = 1, \ldots, s\}$, then we find

$$Pr(\tau \in B) < 1 - \alpha.$$

The result is the origin of the *multiple comparisons* or *simultaneous confidence intervals* problem. If we want these confidence statements about each component to hold simultaneously, then these one-at-a-time intervals are too small. The solutions to the confidence interval problem replace $t_{N-r,\alpha/2}$ with some larger number so that new, bigger box B_* will have the right coverage, that is, $Pr(\tau \in B_*) \geq 1 - \alpha$.

The multiple comparisons form arises from testing a joint hypothesis H: $\mathbf{K}^T\mathbf{b} = \mathbf{m}$ where τ_j is a component of $\tau = \mathbf{K}^T\mathbf{b}$. In situations where the rejection of the joint hypothesis may be presumed, our interest may lie in which of the components of τ contributed to rejection. To test the hypothesis, we may form

$$t_j = \left[\hat{\tau}_j - m_j\right] \Big/ \sqrt{\hat{\sigma}^2 (\mathbf{K}^T(\mathbf{X}^T\mathbf{X})^g \mathbf{K})_{jj}}$$

and reject the hypothesis H if any $|t_j|$ is too large. Using $t_{N-r,\alpha/2}$ leads to the same problem, as

$$Pr(|t_j| < t_{N-r,\alpha/2} \text{ for all } j) = Pr(\tau \in B) < 1 - \alpha$$

or

$$Pr(|t_j| \geq t_{N-r,\alpha/2} \text{ for at least one } j) = Pr(\tau \notin B) > \alpha,$$

and the solution is the same: replace $t_{N-r,\alpha/2}$ by some larger number. The goal in the multiple comparisons problem is permit testing of the hypothesis H in a componentwise fashion, while ensuring that the overall probability of rejection does not exceed α. While this approach to hypothesis testing may be more convenient than the F-test described in this chapter, the gain of convenience comes with a loss of power and invariance (see Exercise 6.19).

The two simplest routes for choosing replacements for $t_{N-r,\alpha/2}$ are known by the names *Bonferroni* and *Scheffé*. As you may guess, the third to be discussed, the *studentized range* method due to Tukey, is quite effective in the balanced one-way ANOVA problem. We will begin with an elementary probability inequality attributed to Bonferroni.

Result 6.6 *(Bonferroni inequalities) Let E_j be a collection of events in the same sample space, then*

$$(i) \quad Pr(\cup E_j) = Pr(\text{ at least one }) \leq \textstyle\sum_j Pr(E_j)$$

$$(ii) \quad Pr(\cap E_j) = Pr(\text{ all }) \geq 1 - \textstyle\sum_j Pr(E_j^c)$$

Proof: Construct the disjoint events B_j and use $Pr(\cup B_j) = \sum_j Pr(B_j)$, where

$$B_1 = E_1$$

$$B_2 = E_1^c \cap E_2$$

$$B_3 = E_1^c \cap E_2^c \cap E_3$$

$$\ldots$$

$$B_j = E_1^c \cap E_2^c \cap \ldots \cap E_{j-1}^c \cap E_j$$

and note $\cup B_j = \cup E_j$ and $B_j \subseteq E_j$. Then $Pr(\cup E_j) = Pr(\cup B_j) = \sum_j Pr(B_j) \leq \sum_j Pr(E_j)$, which is (i). For (ii), $Pr(\cap E_j) = 1 - Pr((\cap E_j)^c) = 1 - Pr(\cup E_j^c)$, and then use (i) with events E_j^c. □

For the simultaneous confidence interval problem, the Bonferroni solution is to replace $t_{N-r,\alpha/2}$ with $t_{N-r,\alpha/(2s)}$—merely dividing the level by the number of intervals constructed. Here the event E_j is

$$\left\{\tau_j \in \left[\hat{\tau}_j - t_{N-r,\alpha/(2s)}\sqrt{\hat{\sigma}^2 \mathbf{H}_{jj}},\ \hat{\tau}_j + t_{N-r,\alpha/(2s)}\sqrt{\hat{\sigma}^2 \mathbf{H}_{jj}}\right]\right\}, \tag{6.23}$$

and so

$$Pr(\tau \in B_*) = Pr(\cap E_j) \geq 1 - \sum_j Pr(E_j^c) = 1 - s \times (\alpha/s) = 1 - \alpha,$$

so the Bonferroni guarantees coverage of at least $1 - \alpha$. This will satisfy the requirement; sometimes that probability can be substantially higher. One of the drawbacks of the Bonferroni approach is that the number of intervals needs to be specified in advance. The strengths of the Bonferroni method are that it is quite easy to use and can give smaller intervals than some of its competitors.

Example 6.6: Pairwise Differences in One-Way ANOVA
Return to the usual one-way ANOVA model $y_{ij} = \mu + \alpha_i + e_{ij}$, where we are interested in constructing simultaneous confidence intervals for all pairwise differences $\alpha_i - \alpha_k$.

For $a = 4$, there are six pairs $(i, k) : (1, 2), (1, 3), (1, 4), (2, 3), (2, 4), (3, 4)$, so we would choose $s = 6$ above.

The second route is the Scheffé method, which has its own advantages and disadvantages. Instead of considering confidence statements for just the components $\tau_j, j = 1, \ldots, s$ of τ, consider constructing a confidence interval $C(\mathbf{u}, c)$ for any linear combination $\mathbf{u}^T\tau$ of the form

$$C(\mathbf{u}, c) = \left[\mathbf{u}^T\hat{\tau} - c\hat{\sigma}\sqrt{\mathbf{u}^T\mathbf{H}\mathbf{u}}, \mathbf{u}^T\hat{\tau} + c\hat{\sigma}\sqrt{\mathbf{u}^T\mathbf{H}\mathbf{u}}\right], \tag{6.24}$$

where $\mathbf{H} = \mathbf{\Lambda}^T(\mathbf{X}^T\mathbf{X})^g\mathbf{\Lambda}$ and $\hat{\sigma}^2 = SSE/(N - r)$, as usual. As we've noted earlier, the issue is how to choose c so that the intervals have the right coverage. Since in this case, we wish to consider any linear combination $\mathbf{u}$, we will need to choose c such that

$$Pr(\mathbf{u}^T\tau \in C(\mathbf{u}, c) \text{ for all } \mathbf{u}) = 1 - \alpha.$$

Let's start some analysis with the simple step

$$Pr(\mathbf{u}^T\tau \in C(\mathbf{u}, c)) = Pr\left(\left|\frac{\mathbf{u}^T\hat{\tau} - \mathbf{u}^T\tau}{\hat{\sigma}\sqrt{\mathbf{u}^T\mathbf{H}\mathbf{u}}}\right| \le c\right).$$

Then expand it to the problem of interest:

$$\begin{aligned} Pr(\mathbf{u}^T\tau \in C(\mathbf{u}, c) \text{ for all } \mathbf{u}) &= Pr\left(\max_{\mathbf{u}}\left|\frac{\mathbf{u}^T\hat{\tau} - \mathbf{u}^T\tau}{\hat{\sigma}\sqrt{\mathbf{u}^T\mathbf{H}\mathbf{u}}}\right| \le c\right) \\ &= Pr\left(\max_{\mathbf{u}}\frac{\left[\mathbf{u}^T(\hat{\tau} - \tau)\right]^2}{\hat{\sigma}^2\mathbf{u}^T\mathbf{H}\mathbf{u}} \le c^2\right) \\ &= Pr((\hat{\tau} - \tau)^T\mathbf{H}^{-1}(\hat{\tau} - \tau)/\hat{\sigma}^2 \le c^2) \\ &= Pr((\hat{\tau} - \tau)^T\mathbf{H}^{-1}(\hat{\tau} - \tau)/(s\hat{\sigma}^2) \le c^2/s) \end{aligned}$$

owing to the extended or general form of the Cauchy–Schwarz inequality:

$$\max_{\mathbf{u}\neq\mathbf{0}}\frac{(\mathbf{u}^T\mathbf{w})^2}{\mathbf{u}^T\mathbf{A}\mathbf{u}} = \mathbf{w}^T\mathbf{A}^{-1}\mathbf{w}. \tag{6.25}$$

Since $(\hat{\tau} - \tau)^T\mathbf{H}^{-1}(\hat{\tau} - \tau)/(s\hat{\sigma}^2)$ has an F-distribution with s and $(N - r)$ degrees of freedom (Exercise 6.15), choose c such that

$$c^2/s = F_{s,N-r,\alpha}, \text{ or } c = \sqrt{sF_{s,N-r,\alpha}}. \tag{6.26}$$

The advantage of the Scheffé approach is that the number of intervals does not need to be specified in advance, but can be determined *after* the data have been examined. This has been called technically *post-hoc analysis* or more familiarly a *fishing license*, since one can "fish" through the results, looking for significant effects.

The disadvantage is that, as may be expected, the Scheffé intervals are often larger than its competitors.

An early form of the Scheffé approach is known as the Working–Hotelling [23] method for simple linear regression for constructing confidence intervals for $\beta_0+x\beta_1$, with $s = 2$ and $\tau = (\beta_0, \beta_1)$, using

$$c = \sqrt{2F_{2,N-r,\alpha}}. \tag{6.27}$$

Example 6.6: continued
For constructing simultaneous confidence intervals for all pairwise differences $\alpha_i - \alpha_k$ in the usual one-way ANOVA model $y_{ij} = \mu + \alpha_i + e_{ij}$, with $a = 4$, we could choose $\tau = (\alpha_1 - \alpha_2, \alpha_1 - \alpha_3, \alpha_1 - \alpha_4)$ using just the necessary $s = 3$ components instead of the full complement of $r = 4$ linearly estimable functions.

The strong motivation for these simultaneous confidence intervals—which are really confidence regions—is that they are much more convenient than the ellipsoid that would arise from inverting the usual hypothesis test. Conveying the information in the ellipsoid is much more awkward than just a few intervals. This extreme convenience carries over to the multiple comparison side, where we are interested, essentially, in testing hypotheses of equality. Making the Bonferroni or Scheffé corrections permits construction of tests that are very simple to implement. Consider the construction of the Bonferroni "box" B_* with limits given by (6.23). This box can be employed to test the general linear hypothesis H: $\mathbf{\Lambda}^T\mathbf{b} = \tau^*$ by rejecting H if τ^* is not in the box B_*. This test is easy to implement:

$$\text{reject H if } |\tau_j^* - \hat{\tau}_j| > t_{N-r,\alpha/(2s)}\sqrt{\hat{\sigma}^2\mathbf{H}_{jj}} \text{ for any } j.$$

While extremely convenient, this test has level

$$Pr(\tau^* \notin B_*) \leq \alpha,$$

and it will be conservative. This approach for testing is *not* invariant to linear transformations, so that the hypothesis H: $\mathbf{C\Lambda}^T\mathbf{b} = \mathbf{C}\tau^*$ with nonsingular matrix $\mathbf{C}$ may *not* lead to the same test. Following the same route but using the Scheffé limits reproduces the usual F-test when all possible linear combinations are entertained. Essentially, the Scheffé approach arises from inverting the F-test to construct the simultaneous confidence interval and will be invariant to linear transformations.

Example 6.7: Noninvariance
Consider again the usual one-way ANOVA model $y_{ij} = \mu + \alpha_i + e_{ij}$, but now with $a = 3$ and $n_i = n$ (balanced), and choose $\tau = (\alpha_1 - \alpha_2, \alpha_1 - \alpha_3)$. Then the Bonferroni confidence intervals using (6.23) give the interval endpoints $\overline{y}_{1.} - \overline{y}_{2.} \pm t_{N-3,\alpha/4}\hat{\sigma}\sqrt{2/n}$ and $\overline{y}_{1.} - \overline{y}_{3.} \pm t_{N-3,\alpha/4}\hat{\sigma}\sqrt{2/n}$. To simplify matters, let $t_{N-3,\alpha/4}\hat{\sigma}\sqrt{2/n} = d$, say. Then the Bonferroni test of H: $\alpha_1 - \alpha_2 = 0$ and $\alpha_1 - \alpha_3 = 0$ would be to reject H

if $|\overline{y}_{1.} - \overline{y}_{2.}| > d$ or $|\overline{y}_{1.} - \overline{y}_{3.}| > d$. Alternatively, testing the logically equivalent hypothesis H: $\alpha_1 - \alpha_2 = 0$ and $\alpha_2 - \alpha_3 = 0$ would reject H if $|\overline{y}_{1.} - \overline{y}_{2.}| > d$ or $|\overline{y}_{2.} - \overline{y}_{3.}| > d$. Taking something simple like $d = 3$, then if $(\overline{y}_{1.}, \overline{y}_{2.}, \overline{y}_{3.}) = (1, 3, 5)$, we would reject using the first test, but not the second. See Exercise 6.19.

Tukey [47] designed the third approach, applicable only to balanced one-way ANOVA problems, for constructing confidence intervals for either (i) all pairwise differences of treatment effects or (ii) constrasts. The key is the compilation of "Studentized range" tables, for the statistic W as follows. Let Z_j, $j = 1, \ldots, k$ be iid $N(0, 1)$ random variables, and let $U \sim \chi^2_\nu$ and independent of Z_j's. Then define the random variable W as

$$W = \frac{max_j Z_j - min_j Z_j}{\sqrt{U/\nu}}.$$

Again, the key is the availability of tables of the distribution of W, that is, for various values of k, ν, and α, we have $q^*_{k,\nu}(\alpha)$ such that $Pr(W > q^*_{k,\nu}(\alpha)) = \alpha$.

The usefulness of the distribution of W lies in the relationship

$$\max_j Z_j - \min_j Z_j = \max_{i,j}(Z_i - Z_j).$$

To apply this to the simultaneous confidence interval problem, consider the following situation:

$$(i) \quad \hat{\tau}_j \text{ independent } N(\tau_j, c^2\sigma^2),\ j = 1, \ldots, s$$

$$(ii) \quad \text{independently, } (N - r)\hat{\sigma}^2/\sigma^2 \sim \chi^2_{N-r}$$

then

$$\begin{aligned}
Pr&\left(\frac{\max_j(\hat{\tau}_j - \tau_j)/(\sigma c) - \min_j(\hat{\tau}_j - \tau_j)/(\sigma c)}{\sqrt{\hat{\sigma}^2/\sigma^2}} \le q^*_{s,N-r}(\alpha)\right) = 1 - \alpha \\
&= Pr(\max_j(\hat{\tau}_j - \tau_j) - \min_j(\hat{\tau}_j - \tau_j) \le c\hat{\sigma}q^*) \\
&= Pr(|(\hat{\tau}_i - \tau_i) - (\hat{\tau}_j - \tau_j)| \le c\hat{\sigma}q^* \text{ for all } i, j) \\
&= Pr((\hat{\tau}_i - \hat{\tau}_j) - c\hat{\sigma}q^* \le \tau_i - \tau_j \le (\hat{\tau}_i - \hat{\tau}_j) + c\hat{\sigma}q^* \text{ for all } i, j), \qquad (6.28)
\end{aligned}$$

which provides simultaneous confidence intervals for all $s(s-1)/2$ pairwise differences $(\tau_i - \tau_j)$. For the *balanced-only* one-way ANOVA problem, we have $\overline{y}_{i.} \sim N(\mu + \alpha_i, \sigma^2/n)$ for $i = 1, \ldots, a$, and independently $n(a-1)\hat{\sigma}^2/\sigma^2 \sim \chi^2_{n(a-1)}$, so that we have the following simultaneous confidence intervals for all pairwise treatment differences:

$$(\overline{y}_{i.} - \overline{y}_{j.}) - \frac{\hat{\sigma}}{\sqrt{n}} q^*_{a,n(a-1)} \le \alpha_i - \alpha_j \le (\overline{y}_{i.} - \overline{y}_{j.}) + \frac{\hat{\sigma}}{\sqrt{n}} q^*_{a,n(a-1)} \qquad (6.29)$$

at level $100 \times (1-\alpha)\%$. This result can be extended to cover all contrasts by employing the following lemma:

Lemma 6.1 *If* $|\tau_i - \tau_j| \le h$ *for all* i, j, *and* $\sum_i u_i = 0$, *then* $|\sum_i u_i \tau_i| \le h \times \frac{1}{2} \sum_i |u_i|$.

Proof: See Exercise 6.17. □

The confidence intervals above in (6.28) can be extended to include the following intervals in (6.30), with the same overall coverage:

$$\sum_i u_i \overline{y}_{i.} - \frac{\hat{\sigma}}{\sqrt{n}} q^*_{a,n(a-1)} \times \frac{1}{2} \sum_i |u_i| \le \sum_i u_i \tau_i$$
$$\le \sum_i u_i \overline{y}_{i.} + \frac{\hat{\sigma}}{\sqrt{n}} q^*_{a,n(a-1)} \times \frac{1}{2} \sum_i |u_i| \tag{6.30}$$

as long as $\sum u_i = 0$, that is, a contrast.

A method known as *Tukey–Kramer* extends this approach to the unbalanced problem, using confidence intervals of the form

$$(\overline{y}_{i.} - \overline{y}_{j.}) - \hat{\sigma} \sqrt{\frac{n_i^{-1} + n_j^{-1}}{2}} q^*_{a,N-a} \le \alpha_i - \alpha_j$$
$$\le (\overline{y}_{i.} - \overline{y}_{j.}) + \hat{\sigma} \sqrt{\frac{n_i^{-1} + n_j^{-1}}{2}} q^*_{a,N-a}, \tag{6.31}$$

using the correct degrees of freedom from the pooled variance estimate. According to a result of Hayter [20], the coverage is smallest when the n_i's are equal, so that these confidence intervals have the correct coverage.

The field of simultaneous inference for the one-way ANOVA problem is a battleground of ease of use, availability of tables, and coverage. While the case of all pairwise comparisons, for which the Tukey approach applies, is the most common, comparisons against a control or against the best permit myriad potential improvements. See Hsu [24] for a survey, although further interest in recent years has been spurred by problems arising in molecular genetics where a can be extraordinarily large.

6.7 Identifiability

The concept of estimability is tied to the existence of moments and, historically, to the assumption of normal errors. The concepts of identifiability and observational equivalence are more general and do not require the distribution of $\mathbf{y}$ to even have a first moment.

Definition 6.2 *Two parameter vectors* $\theta^{(1)}$ *and* $\theta^{(2)}$ *are* observationally equivalent, *denoted* $\theta^{(1)} \sim \theta^{(2)}$, *iff the distribution of the response is the same for both parameter* *vectors. More technically, the two vectors are not observationally equivalent if there exists a set A such that $Pr(\mathbf{y} \in A|\theta^{(1)}) \neq Pr(\mathbf{y} \in A|\theta^{(2)})$; they are observationally equivalent if no such set A exists.*

Definition 6.3 *A function $g(\theta)$ is an* identifying function *iff $g(\theta^{(1)}) = g(\theta^{(2)})$ iff $\theta^{(1)} \sim \theta^{(2)}$.*

Definition 6.4 *A function $g(\theta)$ is* identified *iff $\theta^{(1)} \sim \theta^{(2)}$ implies $g(\theta^{(1)}) = g(\theta^{(2)})$. The contraposition makes this clearer: if $g(\theta^{(1)}) \neq g(\theta^{(2)})$, then the distributions are different.*

To fit the linear model into this framework requires some reconciliation. While the definitions for identification examine the family of distributions of the response vector $\mathbf{y}$, the usual linear model more naturally considers the assumptions on the distribution of the error $\mathbf{e}$. Falling back on the assumption of normal errors gains little. For a middle ground, consider the family of distributions known as a *location family*.

Definition 6.5 *A family of distributions $F(\mathbf{y}|\theta)$ is called a* location family *with location parameter θ if $F(\mathbf{y}|\theta) = F_0(\mathbf{y} - \theta)$ for some distribution F_0.*

The linear model describes a location family, where the distribution of $\mathbf{y}$ depends on the parameter vector $\mathbf{b}$ only through its location vector $\mathbf{Xb}$, so that two parameter vectors are observationally equivalent iff they lead to the same location vector for $\mathbf{y}$. Clearly now, $\mathbf{b}^{(1)} \sim \mathbf{b}^{(2)}$ iff $\mathbf{Xb}^{(1)} = \mathbf{Xb}^{(2)}$, so that the identifying function is $\mathbf{Xb}$. Moreover, under the assumption that the first moment exists, estimability and identifiability coincide.

Whether a linear function $\lambda^T\mathbf{b}$ is identified now depends on whether $\mathbf{Xb}^{(1)} = \mathbf{Xb}^{(2)}$ implies $\lambda^T\mathbf{b}^{(1)} = \lambda^T\mathbf{b}^{(2)}$. This is obviously true if $\lambda \in \mathcal{C}(\mathbf{X}^T)$. Taking a geometric viewpoint, this requirement is equivalent to $\mathbf{X}(\mathbf{b}^{(1)} - \mathbf{b}^{(2)}) = \mathbf{0}$ implies $\lambda^T(\mathbf{b}^{(1)} - \mathbf{b}^{(2)}) = \mathbf{0}$, or that $\lambda \perp \mathcal{N}(\mathbf{X})$. In the case of a full-column-rank design matrix $\mathbf{X}$ where $\mathcal{N}(\mathbf{X}) = \{\mathbf{0}\}$, all nontrivial functions of $\mathbf{b}$ are identified.

Example 6.8: Identifiability

Let Y_i be iid from a binomial distribution with unknown parameters m and p, so that $\theta = (m, p)$. The viewpoint of estimability only uses means, so here we have $E(Y_i) = mp$, and only the product $mp = \lambda$ is estimable. If we examine points in the parameter space with the same mean λ, we get the manifold $\theta = (\lambda/p, p)$. Merely by taking $A = \{\mathbf{0}\}$, we have

$$Pr(Y = \mathbf{0}|(\lambda/p, p)) = Pr(Y_i = 0, i = 1, \ldots, N|(\lambda/p, p)) = (1 - p)^{\lambda N/p},$$

which is different for each value of p, so not all points in the parameter space giving the same mean have the same distribution. In contrast to estimability that relies on means only, identifiability uses not only the variance $mp(1 - p)$, which would clearly distinguish among these points with the same mean, but the whole distribution.

Example 6.9: Two-Sample Problem with Cauchy Distribution
Let $X_i, i = 1, \ldots, m$, be iid from a Cauchy distribution with unknown location parameter $\mu + \alpha$ and scale σ; $Y_i, i = 1, \ldots, n$, are iid from a Cauchy distribution with location parameter $\mu + \beta$ and same scale σ. The joint density is

$$\sigma^{m+n} \prod_{i=1}^{m} \left[\sigma^2 + (x_i - \mu - \alpha)^2\right]^{-1} \prod_{i=1}^{n} \left[\sigma^2 + (y_i - \mu - \beta)^2\right]^{-1},$$

and while clearly $\alpha - \beta$ is identifiable, it is not estimable, since means do not exist for either X_i or Y_i.

Example 6.10: Estimability via Variance
Consider the usual simple linear regression model

$$y_i = \beta_0 + \beta_1 x_i + e_i,$$

and while we will assume the errors e_i are iid, we will not assume a normal distribution, but instead the exponential with parameter λ, so that $Ee_i = \lambda$ and $Var(e_i) = \lambda$. This leads to

$$E(y_i) = \beta_0 + \beta_1 x_i + \lambda,$$

so we see that neither β_0 nor λ is (linearly) estimable, while their sum $\beta_0 + \lambda$ will be (linearly) estimable. However, since $Var(y_i) = \lambda$, then

$$E\frac{SSE}{df} = E\frac{\mathbf{y}^T(\mathbf{I} - \mathbf{P_X})\mathbf{y}}{N - 2} = \lambda$$

so that while λ is not linearly estimable, we can find a quadratic function of $\mathbf{y}$ that is unbiased for λ. So λ and also β_0 are estimable, but not *linearly* estimable.

Example 6.11: Latent Variables
Consider a multiple regression model with a latent variable w_i. Here w_i will not be observed, but a parametric model will be posed for it that may have observable covariates. So the main model with the response y_i is

$$y_i = \beta_0 + \beta_1 x_i + \beta_2 w_i + e_i$$

where we will first assume a simple covariance structure for the error e_i: $Ee_i = 0$, $Var(e_i) = \sigma^2$, $Cov(e_i, e_j) = 0$ for $i \neq j$. The model for the latent variable w_i follows a simple linear regression model with covariate z_i:

$$w_i = \gamma_0 + \gamma_1 z_i + f_i;$$

for simplicity, assume the same simple (Gauss–Markov) error structure for f_i as $Ef_i = 0$, $Var(f_i) = \eta^2$, $Cov(f_i, f_j) = 0$ for $i \neq j$. Then we have the unconditional mean model for the observed response y_i as

$$E(y_i) = \beta_0 + \beta_1 x_i + \beta_2(\gamma_0 + \gamma_1 z_i)$$

and variances

$$Var(y_i) = \sigma^2 + \beta_2^2 \eta^2.$$

Are all seven parameters $(\beta_0, \beta_1, \beta_2, \gamma_0, \gamma_1, \sigma^2, \eta^2)$ identifiable? We should be able to estimate three regression coefficients, $\beta_0 + \beta_2\gamma_0$, β_1, and $\beta_2\gamma_1$, and the variance $\sigma^2 + \beta_2^2\eta^2$, so clearly some parameters are not identifiable. In latent variable models such as these, a common fix is to set β_2 equal to some value, say 1. The motivation for such an apparently arbitrary decision is that often unobservable variables are hypothetical constructs and have no units of measurement. As a result, an arbitrary rescaling of their value has no effect. Setting β_2 would have the effect of setting a unit of measurement according to its effect on the response.

6.8 Summary

- The best linear unbiased estimators $\lambda^T \hat{\mathbf{b}}$ constructed in Chapter 4 under the Gauss–Markov assumptions involving only moments have the smallest variance among all unbiased estimators.

- A test of the general linear hypothesis H: $\mathbf{K}^T\mathbf{b} = \mathbf{m}$ can be constructed following first principles using the result

$$F = \frac{(\mathbf{K}^T\hat{\mathbf{b}} - \mathbf{m})^T \mathbf{H}^{-1} (\mathbf{K}^T\hat{\mathbf{b}} - \mathbf{m})/s}{SSE/(N-r)} \sim F_{s,N-r}(\lambda).$$

  Rejecting for large values of the test statistic F produces an unbiased test: its power is greater under the alternative.

- A test based on the likelihood ratio principle leads to the familiar statistic

$$F = \frac{[Q(\hat{\mathbf{b}}_H) - Q(\hat{\mathbf{b}})]/s}{Q(\hat{\mathbf{b}})/(N-r)}$$

  which can be shown to have the $F_{s,N-r}(\lambda)$ distribution in some simple circumstances.

- The first principles F-statistic and the likelihood ratio test statistic are the same—in other words, the first principles test is a likelihood ratio test.

- We construct confidence intervals for several linearly estimable functions that simultaneously have the desired coverage.

- We define identifiability as an alternative to estimability to handle cases where the error distribution does not have moments.

6.9 Notes

1. In testing the general linear hypothesis, Theorem 6.1 gave the results in the case where $\mathbf{K}^T\mathbf{b}$ was estimable. Result 6.5 told what happens if it is completely nonestimable, as in Section 3.7. In the case of a mixture of estimable and non-estimable parts, no clean general result is available. However, see Section 7.6.

2. The concept of identifiability arose in econometrics (see, e.g., Kmenta [25]) in the examination of *simultaneous equation models*. For example, quantity sold in a period may depend on price and other covariates, and vice versa, following different relationships for supply and demand:

$$p_i = \beta_0 + \beta_1 q_i + \beta_2 x_i + u_i$$
$$q_i = \gamma_0 + \gamma_1 p_i + \gamma_2 z_i + v_i.$$

   With some algebra to put the responses on the left-hand side, we get a multivariate problem

$$\begin{bmatrix} 1 & -\beta_1 \\ -\gamma_1 & 1 \end{bmatrix} \begin{bmatrix} p_i \\ q_i \end{bmatrix} = \begin{bmatrix} \beta_0 & \beta_2 & 0 \\ \gamma_0 & 0 & \gamma_1 \end{bmatrix} \begin{bmatrix} 1 \\ x_i \\ z_i \end{bmatrix} + \begin{bmatrix} u_i \\ v_i \end{bmatrix}$$

   in the form $\mathbf{\Gamma y}_i - \mathbf{B x}_i = \mathbf{e}_i$ with parameter matrices $\mathbf{\Gamma}$, $\mathbf{B}$, and $\mathbf{\Sigma} = Cov(\mathbf{e}_i)$. Premultiplication by a matrix $\mathbf{G}$ gives new parameter matrices $\mathbf{G\Gamma}$, $\mathbf{GB}$, and $\mathbf{G\Sigma G}^T$ with no change in the distribution. Putting restrictions on these parameter matrices, say, forcing particular elements of $\mathbf{\Gamma}$ or $\mathbf{B}$ to be zero, restricts parameter matrices that give the same distributions by restricting $\mathbf{G}$. Identifiability is achieved when the only possible $\mathbf{G}$ is the identity matrix.

3. Testing with inequality constraints, or a similarly constrained parameter space, faces many difficulties. Instead of the linear restricted normal equations, the likelihood ratio test leads to an optimization problem known as *nonnegative least squares*, which has no closed form (see Lawson and Hanson [26]). In addition, when its solution lies on a boundary of the space, the distribution of the likelihood ratio statistic is peculiar.

6.10 Exercises

6.1. Prove Corollary 6.2.

6.2. * Show that the second derivative of the log-likelihood function under the normal Gauss–Markov model with respect to σ^2 is negative:

$$\log L(\mathbf{b}, \sigma^2) = -(N/2)\log(2\pi) - (N/2)\log(\sigma^2) - \frac{1}{2\sigma^2} Q(\mathbf{b}).$$

6.3. * In constructing maximum likelihood estimators in Result 6.3 and Corollary 6.3, does it matter whether $\mathbf{X}$ has full-column rank or not?

6.4. Recall that the usual, unbiased estimator for the variance is $\hat{\sigma}^2 = SSE/(N - r)$ where $SSE = \mathbf{y}^T(\mathbf{I} - \mathbf{P_X})\mathbf{y}$ while the MLE is SSE/N. Find the estimator in the class SSE/c (that is, find c) that minimizes mean square error, $E\left[SSE/c - \sigma^2\right]^2$.

6.5. Let $Y_i, i = 1, \ldots, N$ be iid exponential(λ), that is, each has density $f(y) = \lambda^{-1}e^{-y/\lambda}$ for $y > 0$.

a. Find $E(Y_i^k)$ for $k = 1, 2, 3, 4$.
b. Find the variance of the usual variance estimator, $Var\left(\sum_i (Y_i - \overline{Y})^2/(N-1)\right)$.
c. Compare this to the variance of the MLE for λ.

6.6. Least squares is intimately related to the normal distribution. For $y_i = \mathbf{b}^T\mathbf{x}_i + e_i$ where e_i are $iid\ N(0, \sigma^2)$, the MLE $\hat{\mathbf{b}}$ for the coefficient vector minimizes the sum of squares function $Q(\mathbf{b})$. For these other error distributions, find a similar criterion whose minimizing vector gives the MLE:

a. e_i iid Uniform($-\sigma, \sigma$).
b. e_i are iid from the logistic distribution with density $f(x) = e^{-x}/(1 + e^{-x})^2$.
c. e_i are iid from the double exponential distribution with density $f(x) = exp\{-|x|/\sigma\}/(2\sigma)$.

Hint: Two of these three cases lead to simple results.

6.7. Recall Exercise 4.1; find the MLE of p and compare it to the BLUE.

6.8. Consider applying the subset framework of Example 6.3 to testing treatment effects in the general one-way ANOVA model, $y_{ij} = \mu + \alpha_i + e_{ij}$, so that $\mathbf{X}_0 = \mathbf{1}_N$, $\mathbf{b}_0$ is the single component μ, $\mathbf{b}_1 = (\alpha_1, \ldots, \alpha_a)^T$, and $\mathbf{X}_1$ has the usual corresponding structure. Find the linearly independent functions of $(\mathbf{I} - \mathbf{P}_1)\mathbf{X}_1\mathbf{b}_1$ that are testable.

6.9. In Example 6.2, the two-way crossed model with interaction, $y_{ijk} = \mu + \alpha_i + \beta_j + \gamma_{ij} + e_{ijk}$, we considered which functions of the interactions were not estimable. Construct the remaining $a+n-1$ linearly independent nonestimable functions and show how they are confounded with the main effects.

6.10. Consider again the two-way crossed model with interaction, $y_{ijk} = \mu + \alpha_i + \beta_j + \gamma_{ij} + e_{ijk}$. Following Example 6.3, construct the design matrices for the full and reduced models for testing whether the interaction is zero, give their ranks, and thus establish the usual test statistic as a likelihood ratio test with its proper degrees of freedom. For simplicity, choose $a = 3$, $n = 2$, and $n_{ij} = 2$.

6.11. Consider the simple linear regression problem with $x_i = i$, and $N = 5$. Find the power of the F-test for testing whether the slope is zero when testing at level $\alpha = 0.05$ and the slope takes values 0.1, 0.2, and 0.3.

6.12. Suppose we design our F-test in Equation (6.10) to reject small values instead of large. Following the same scenario as in Exercise 6.11, find the power under the same alternatives.

6.13. For Example 6.7, construct a similar Scheffé-style multiple comparison test.

6.14. a. Prove the usual Cauchy–Schwarz inequality

$$(\mathbf{u}^T\mathbf{w})^2 \leq \|\mathbf{u}\| \times \|\mathbf{w}\|$$

by finding the scalar α that minimizes $\|\mathbf{u} - \alpha\mathbf{w}\|^2$ and examining the case where the minimum is zero.

b. Obtain the generalized Cauchy–Schwarz inequality (6.25) by applying the same steps to

$$(\mathbf{u} - \alpha\mathbf{A}^{-1}\mathbf{w})^T\mathbf{A}(\mathbf{u} - \alpha\mathbf{A}^{-1}\mathbf{w})$$

6.15. Fill in all of the steps in the Scheffé approach to show that $(\hat{\tau} - \tau)^T\mathbf{H}^{-1}(\hat{\tau} - \tau)/(s\hat{\sigma}^2)$ has an F-distribution with s and $(N - r)$ degrees of freedom.

6.16. Revise the generalized Cauchy–Schwarz inequality to cover the case where $\mathbf{u}$ varies only in $\mathcal{C}(\mathbf{B})$ by finding an upper bound for $(\mathbf{u}^T\mathbf{w})^2/\mathbf{u}^T\mathbf{A}\mathbf{u}$ for $\mathbf{u} \neq \mathbf{0}$, $\mathbf{u} \in \mathcal{C}(\mathbf{B})$.

6.17. Prove Lemma 6.1 (not easy). Hint: Write $\sum_i u_i = \sum_{u_i>0} u_i - \sum_{u_i<0} u_i$ and let $\sum_{u_i>0} u_i = g$, say, then also $g = -\sum_{u_i<0} u_i$.

6.18. For the one-way ANOVA model with $n_i = n = 5$ and $a = 3$, compute the expected length (or squared length) of simultaneous confidence intervals computed using Bonferroni, Scheffe, and Tukey's methods for all pairwise differences $\alpha_i - \alpha_k$.

6.19. For the one-way ANOVA model with $n_i = n = 5$ and $a = 3$, compute the power when $\alpha_1 = \alpha_2 = 1$ and $\alpha_3 = 0$ when testing the hypothesis of equality in the following ways:

a. Use the usual F-test, writing the hypothesis as $\alpha_1 - \alpha_2 = 0$ and $\alpha_1 - \alpha_3 = 0$.
b. Use the usual F-test, writing the hypothesis as $\alpha_1 - \alpha_3 = 0$ and $\alpha_2 - \alpha_3 = 0$.
c. Let $\tau_1 = \alpha_1 - \alpha_2$ and $\tau_2 = \alpha_1 - \alpha_3$, and use a Bonferroni correction, rejecting if either component is significant. That is, form $t_j = \hat{\tau}_j/se(\hat{\tau}_j)$ and reject if $|t_j| > t(\alpha/4)$ for $j = 1, 2$.
d. Repeat (c) but with $\alpha_1 - \alpha_3$ and $\alpha_2 - \alpha_3$. (Hint: Yes, (a) and (b) should be the same.)

6.20. In Example 6.10, write the joint density of $y_1, y_2, \ldots, y_N$ and show that $(\beta_0, \beta_1, \lambda)$ are identified.

6.21. In Example 6.10, construct the quadratic function of $\mathbf{y}$ that is an unbiased estimator of β_0.

6.22. In Example 6.11, find the identifying functions.

6.23. In Example 6.11, show that if we add the restriction that $\beta_2 = 1$, then the remaining parameters are identified.

6.24. Consider the usual one-way *unbalanced* ANOVA model, $y_{ij} = \mu + \alpha_i + e_{ij}$, with $E(e_{ij}) = 0$, $Var\,(e_{ij}) = \sigma^2$, e_{ij} uncorrelated with $i = 1, \ldots, a;\ j = 1, \ldots, n_i$. Commonly, we impose the constraint $\sum_i n_i \alpha_i = 0$ because it leads to the appealing effect estimates $\hat{\alpha}_i = \overline{y}_{i.} - \overline{y}_{..}$

a. Write this constraint in the form $\mathbf{P}^T\mathbf{b} = \mathbf{0}$.
b. Show that $\mathbf{P}^T\mathbf{b}$ is not estimable.
c. Find all solutions to the normal equations in terms of the $\overline{y}_{i.}$'s and any necessary arbitrary constants.
d. Find a solution that satisfies the constraint.
e. Show that your solution in (d) solves the augmented system below for $\mathbf{b}$ and find θ.

$$\begin{bmatrix} \mathbf{X}^T\mathbf{X} & \mathbf{P} \\ \mathbf{P}^T & \mathbf{0} \end{bmatrix} \begin{bmatrix} \mathbf{b} \\ \theta \end{bmatrix} = \begin{bmatrix} \mathbf{X}^T\mathbf{y} \\ \mathbf{0} \end{bmatrix}$$

f. Using row/column operations, show that the $(a + 2) \times (a + 2)$ matrix above is nonsingular.
g. A third route: Show that your solution in (d) solves the nonsingular system $(\mathbf{X}^T\mathbf{X} + \mathbf{P}\mathbf{P}^T)\mathbf{b} = \mathbf{X}^T\mathbf{y}$. (Hint: Don't write out the matrix!)
h. Show that the matrix in (g) is nonsingular by showing that this matrix has rank $a + 1$: $\begin{bmatrix} \mathbf{X} \\ \mathbf{P}^T \end{bmatrix}$

6.25. Four rods, labeled P, Q, R, and S, are nominally 500 mm in length and are to be measured for their actual length. Designate their deviations from 500 mm by p, q, r, s so that the length of rod P is $(500 + p)$ mm. The rods can only be measured two at a time. The results of the measurements and some analysis are:

Rods	Length(mm)
P and Q	999
P and R	1,003
P and S	1,002
Q and R	1,004
Q and S	1,003
R and S	1,001

$$\mathbf{y} = \begin{bmatrix} -1 \\ 3 \\ 2 \\ 4 \\ 3 \\ 1 \end{bmatrix}, \quad \mathbf{Xb} = \begin{bmatrix} 1 & 1 & 0 & 0 \\ 1 & 0 & 1 & 0 \\ 1 & 0 & 0 & 1 \\ 0 & 1 & 1 & 0 \\ 0 & 1 & 0 & 1 \\ 0 & 0 & 1 & 1 \end{bmatrix} \begin{bmatrix} p \\ q \\ r \\ s \end{bmatrix},$$

$$(\mathbf{X}^T\mathbf{X})^{-1} = \frac{1}{12} \begin{bmatrix} 5 & -1 & -1 & -1 \\ -1 & 5 & -1 & -1 \\ -1 & -1 & 5 & -1 \\ -1 & -1 & -1 & 5 \end{bmatrix},$$

$$\mathbf{X}^T\mathbf{y} = \begin{bmatrix} 4 \\ 6 \\ 8 \\ 6 \end{bmatrix}, \quad \hat{\mathbf{b}} = \begin{bmatrix} 0 \\ 1 \\ 2 \\ 1 \end{bmatrix}, \quad \mathbf{y}^T\mathbf{y} = 40, \overline{y} = 2.$$

Let's make the usual linear models distributional assumptions, $\mathbf{y} \sim N_6(\mathbf{Xb}, \sigma^2\mathbf{I}_6)$.

a. Give an unbiased estimate of σ^2 and its degrees of freedom.
b. Test the hypothesis that all four rods are equal in length versus the alternative that they are not equal. First give $\mathbf{K}$, $\mathbf{m}$, $\mathbf{K}^T\hat{\mathbf{b}} - \mathbf{m}$, and the degrees of freedom of the F-statistic, then compute the F-statistic (hint: three equal signs).
c. Let $\mathcal{T} = \{\mathbf{b} : \mathbf{K}^T\mathbf{b} = \mathbf{m}\}$. Find the vector $\mathbf{c}$ so that $\mathcal{T} = span\{\mathbf{c}\}$.
d. Using z as the parameter, write the restricted model as $\mathbf{y} = \mathbf{Xc}z + \mathbf{e}$, and find $\hat{z}$ that minimizes $\|\mathbf{y} - \mathbf{Xc}z\|^2$. (Hint: If this isn't easy, you're doing it wrong.)
e. Taking $\hat{\mathbf{b}}_H = \mathbf{c}\hat{z}$, find the complete solution to the restricted normal equations

$$\begin{bmatrix} \mathbf{X}^T\mathbf{X} & \mathbf{K} \\ \mathbf{K}^T & \mathbf{0} \end{bmatrix} \begin{bmatrix} \mathbf{b} \\ \theta \end{bmatrix} = \begin{bmatrix} \mathbf{X}^T\mathbf{y} \\ \mathbf{m} \end{bmatrix}.$$

f. Is $\theta = 0$ in the equations in (e)?

Chapter 7

Further Topics in Testing

7.1 Introduction

The theory presented in Chapter 6 regarding testing the general linear hypothesis conflicts somewhat with common practice. The first principles method for testing presented in Section 6.3 is not the most commonly used. In the discussion of the likelihood ratio test in Section 6.4, the proposed solution to the problem of minimizing the error sum of squares function $Q(\mathbf{b})$, solving the restricted normal equations (3.10), is not a very popular approach. The two most common methods of hypothesis testing are fitting full and reduced models using reparameterizations or by using type I or sequential sums of squares. These two approaches and two other common testing situations are explored in this chapter.

7.2 Reparameterization

The most common method of testing a complicated hypothesis is to do the likelihood ratio test by fitting both the full model, unrestricted by the hypothesis, and the reduced model, which incorporates the restrictions of the hypothesis. The difference in their error sum of squares then forms the numerator sum of squares for the F-test: $Q(\hat{\mathbf{b}}_H) - Q(\hat{\mathbf{b}})$. Sometimes the form of the general linear hypothesis makes it difficult to see how to fit the reduced model.

While one route to fitting the reduced model is to solve the RNEs, as mentioned in Section 6.5, that is not the most inviting path to take. Usually we reparameterize in terms of the remaining parameters.

Example 7.1: Reparameterization in Multiple Regression
Consider the multiple regression model with three covariates and a complicated hypothesis H: $\beta_1 + \beta_2 = 1$, $\beta_3 = 0$. The full, unrestricted model looks like

$$y_i = \beta_0 + \beta_1 x_{i1} + \beta_2 x_{i2} + \beta_3 x_{i3} + e_i, i = 1, \ldots, N.$$

To fit the reduced model, we usually impose the hypothesis by reparameterization, substituting $\beta_2 = 1 - \beta_1$ and $\beta_3 = 0$ into the model to obtain

$$\begin{aligned} y_i &= \beta_0 + \beta_1 x_{i1} + (1 - \beta_1)x_{i2} + (0)x_{i3} + e_i \\ &= \beta_0 + \beta_1 x_{i1} + x_{i2} - \beta_1 x_{i2} + e_i \end{aligned}$$

and finishing with the simple linear regression model

$$y_i - x_{i2} = \beta_0 + \beta_1(x_{i1} - x_{i2}) + e_i,$$

with a new response variable and explanatory variable. However, we could have done a different reparameterization, substituting instead $\beta_1 = 1 - \beta_2$, and obtaining a different response variable and changing the sign of the explanatory variable. See Exercise 7.1.

As Example 7.1 shows, this route, while correct, raises questions about when this approach is legal, and why different choices lead to the same results—the same estimators and the same error sum of squares. The reader is warned that the subsequent discussion is technical, tedious, and cumbersome, but necessary to support this common practice.

The reparameterization approach is really just the construction of all possible solutions to the linear hypothesis equations $\mathbf{K}^T\mathbf{b} = \mathbf{m}$, which is a reparameterization of the restricted parameter space $\mathcal{T} = \{\mathbf{b} : \mathbf{K}^T\mathbf{b} = \mathbf{m}\}$. Following Section A.4, given a generalized inverse of $\mathbf{K}^T$, we can construct all solutions to these equations, taking the form

$$\mathbf{b}(\mathbf{z}) = \mathbf{K}^{gT}\mathbf{m} + (\mathbf{I} - \mathbf{K}^{gT}\mathbf{K}^T)\mathbf{z}.$$

The problem of minimizing the sum of squares over the restricted parameter space $\mathcal{T} = \{\mathbf{b} : \mathbf{K}^T\mathbf{b} = \mathbf{m}\}$ then takes a new form,

$$\min_{\mathbf{b}\in\mathcal{T}} \|\mathbf{y} - \mathbf{X}\mathbf{b}(\mathbf{z})\|^2 = \min_{\mathbf{z}} \|\mathbf{y} - \mathbf{X}\mathbf{K}^{gT}\mathbf{m} - \mathbf{X}(\mathbf{I} - \mathbf{K}^{gT}\mathbf{K}^T)\mathbf{z}\|^2, \tag{7.1}$$

where $\mathbf{y} - \mathbf{X}\mathbf{K}^{gT}\mathbf{m}$ is the new response vector and $\mathbf{X}(\mathbf{I} - \mathbf{K}^{gT}\mathbf{K}^T)\mathbf{z}$ is the product of the new design matrix and new parameter vector. In practice, however, we exploit the simple form of the projection matrix $(\mathbf{I} - \mathbf{K}^{gT}\mathbf{K}^T)$. The usual way we construct a generalized inverse is to permute to get a nonsingular matrix in the upper-left corner, invert it, and permute back. In the case of $\mathbf{K}^T$, which has full-row rank, this means just permuting columns: $\mathbf{K}^T\mathbf{Q} = [\mathbf{C}\ \ \mathbf{D}]$, where $\mathbf{Q}$ is a permutation matrix. Permuting so that $\mathbf{D}$ is nonsingular (more convenient for the examples below) and using Result A.13 we construct

$$\mathbf{K}^{gT} = \mathbf{Q}\begin{bmatrix} \mathbf{0} \\ \mathbf{D}^{-1} \end{bmatrix},$$

and so

$$(\mathbf{I} - \mathbf{K}^{gT}\mathbf{K}^T) = \mathbf{Q}\begin{bmatrix} \mathbf{I}_{p-s} & \mathbf{0} \\ -\mathbf{D}^{-1}\mathbf{C} & \mathbf{0} \end{bmatrix}\mathbf{Q}^T.$$

Partitioning $\mathbf{XQ} = [(\mathbf{XQ})_1 \ (\mathbf{XQ})_2]$ into its first $p - s$ and last s columns, the new response vector can be rewritten,

$$\mathbf{y} - \mathbf{XK}^{gT}\mathbf{m} = \mathbf{y} - \mathbf{XQ}\begin{bmatrix}\mathbf{0}\\ \mathbf{D}^{-1}\end{bmatrix}\mathbf{m} = \mathbf{y} - (\mathbf{XQ})_2\mathbf{D}^{-1}\mathbf{m},$$

in light of this generalized inverse. Since only the first $p - s$ components of the new permuted parameter vector $\mathbf{Q}^T\mathbf{z}$ come into play, we can retain those and relabel them in terms of the original components of $\mathbf{b}$, creating a reduced parameter space with dimension $(p - s)$ of $\mathbf{b}_1^* = [\mathbf{I}_{p-s} \ \ \mathbf{0}]\mathbf{Q}^T\mathbf{z}$. With the new design matrix $(\mathbf{XQ})_1 - (\mathbf{XQ})_2\mathbf{D}^{-1}\mathbf{C}$ and parameter vector $\mathbf{b}_2^*$ we can write

$$\mathbf{X}(\mathbf{I} - \mathbf{K}^{gT}\mathbf{K}^T)\mathbf{z} = \mathbf{XQ}\begin{bmatrix}\mathbf{I}_{p-s} & \mathbf{0}\\ -\mathbf{D}^{-1}\mathbf{C} & \mathbf{0}\end{bmatrix}\mathbf{Q}^T\mathbf{z} = \left[(\mathbf{XQ})_1 - (\mathbf{XQ})_2\mathbf{D}^{-1}\mathbf{C}\right]\mathbf{b}_2^*.$$

Example 7.1: continued

Consider the multiple regression model with three covariates and a complicated hypothesis H: $\beta_1 + \beta_2 = 1, \beta_3 = 0$. The hypothesis can be written as

$$\mathbf{K}^T\mathbf{b} = \begin{bmatrix}0 & 1 & 1 & 0\\ 0 & 0 & 0 & 1\end{bmatrix}\begin{bmatrix}\beta_0\\ \beta_1\\ \beta_2\\ \beta_3\end{bmatrix} = \begin{bmatrix}1\\ 0\end{bmatrix} = \mathbf{m}.$$

To obtain the first reparameterization, no permutation is needed, so that

$$\mathbf{K}^T\mathbf{Q} = [\mathbf{C} \quad \mathbf{D}] = \begin{bmatrix}0 & 1 & 1 & 0\\ 0 & 0 & 0 & 1\end{bmatrix}$$

and $\mathbf{D} = \mathbf{I}_2 = \mathbf{D}^{-1}$. The rows of $\mathbf{XQ}$ are then $[1 \ x_{i1} \ x_{i2} \ x_{i3}]$, and the new parameter is $\mathbf{b}_2^* = [\mathbf{I}_{p-s} \ \ \mathbf{0}]\mathbf{Q}^T\mathbf{b}$ written (note the substitution of $\mathbf{b}$ for $\mathbf{z}$) as

$$\mathbf{b}_2^* = \begin{bmatrix}\beta_0\\ \beta_1\end{bmatrix} = [\mathbf{I}_{p-s} \quad \mathbf{0}]\mathbf{Q}^T\mathbf{b} = \begin{bmatrix}1 & 0 & 0 & 0\\ 0 & 1 & 0 & 0\end{bmatrix}\begin{bmatrix}1 & 0 & 0 & 0\\ 0 & 1 & 0 & 0\\ 0 & 0 & 1 & 0\\ 0 & 0 & 0 & 1\end{bmatrix}\begin{bmatrix}\beta_0\\ \beta_1\\ \beta_2\\ \beta_3\end{bmatrix}.$$

Hence we parameterize in terms of β_0 and β_1, giving $\mathbf{X}(\mathbf{I} - \mathbf{K}^{gT}\mathbf{K}^T)\mathbf{z} = [(\mathbf{XQ})_1 - (\mathbf{XQ})_2\mathbf{D}^{-1}\mathbf{C}]\mathbf{b}_2^*$ with its ith component of the form

$$\left([1 \quad x_{i1}] - [x_{i2} \quad x_{i3}]\begin{bmatrix}1 & 0\\ 0 & 1\end{bmatrix}\begin{bmatrix}0 & 1\\ 0 & 0\end{bmatrix}\right)\begin{bmatrix}\beta_0\\ \beta_1\end{bmatrix} = [1 \quad x_{i1} - x_{i2}]\begin{bmatrix}\beta_0\\ \beta_1\end{bmatrix}.$$

The ith component of the new response vector $(\mathbf{y} - (\mathbf{XQ})_2\mathbf{D}^{-1}\mathbf{m})_i$ is $y_i - x_{i2}$.

Example 7.2: Linear Trend in One-Way ANOVA

Recall the usual ANOVA model in Example 6.4, with $y_{ij} = \mu + \alpha_i + e_{ij}$ and taking the simple case of $a = 3$. There we wanted to test a linear effect with covariate $x_i = i$.

Now the easy route would be to immediately reparameterize to get the reduced model, $E(y_{ij}) = b_0 + b_1 x_i$, but a more roundabout approach is to begin with the hypothesis stated as H: $\alpha_1 - 2\alpha_2 + \alpha_3 = 0$. If we follow the reparameterization approach above, do we get the simple linear regression model with $x_i = i$? First write the hypothesis as

$$[0 \quad 1 \quad -2 \quad 1] \begin{bmatrix} \mu \\ \alpha_1 \\ \alpha_2 \\ \alpha_3 \end{bmatrix},$$

and again, permutation is unnecessary. Then we have $\mathbf{b}_2^* = [\mathbf{I}_{p-s} \quad \mathbf{0}]\mathbf{Q}^T\mathbf{b}$ as

$$\mathbf{b}_2^* = \begin{bmatrix} \mu \\ \alpha_1 \\ \alpha_2 \end{bmatrix} = \begin{bmatrix} 1 & 0 & 0 & 0 \\ 0 & 1 & 0 & 0 \\ 0 & 0 & 1 & 0 \end{bmatrix} \begin{bmatrix} 1 & 0 & 0 & 0 \\ 0 & 1 & 0 & 0 \\ 0 & 0 & 1 & 0 \\ 0 & 0 & 0 & 1 \end{bmatrix} \begin{bmatrix} \mu \\ \alpha_1 \\ \alpha_2 \\ \alpha_3 \end{bmatrix}.$$

Since $\mathbf{m} = \mathbf{0}$, the response vector is unchanged, and taking a toy balanced problem with $n_i = 2$, we see the form of $\mathbf{XQ}$ remains unchanged,

$$\begin{bmatrix} 1 & 1 & 0 & 0 \\ 1 & 1 & 0 & 0 \\ 1 & 0 & 1 & 0 \\ 1 & 0 & 1 & 0 \\ 1 & 0 & 0 & 1 \\ 1 & 0 & 0 & 1 \end{bmatrix} \begin{bmatrix} 1 & 0 & 0 & 0 \\ 0 & 1 & 0 & 0 \\ 0 & 0 & 1 & 0 \\ 0 & 0 & 0 & 1 \end{bmatrix} = \begin{bmatrix} 1 & 1 & 0 & 0 \\ 1 & 1 & 0 & 0 \\ 1 & 0 & 1 & 0 \\ 1 & 0 & 1 & 0 \\ 1 & 0 & 0 & 1 \\ 1 & 0 & 0 & 1 \end{bmatrix},$$

and the form of the new design matrix $[(\mathbf{XQ})_1 - (\mathbf{XQ})_2\mathbf{D}^{-1}\mathbf{C}]\mathbf{b}_2^*$,

$$\left[\begin{bmatrix} 1 & 1 & 0 \\ 1 & 1 & 0 \\ 1 & 0 & 1 \\ 1 & 0 & 1 \\ 1 & 0 & 0 \\ 1 & 0 & 0 \end{bmatrix} - \begin{bmatrix} 0 \\ 0 \\ 0 \\ 0 \\ 1 \\ 1 \end{bmatrix} [0 \quad 1 \quad -2] \right] \begin{bmatrix} \mu \\ \alpha_1 \\ \alpha_2 \end{bmatrix} = \begin{bmatrix} 1 & 1 & 0 \\ 1 & 1 & 0 \\ 1 & 0 & 1 \\ 1 & 0 & 1 \\ 1 & -1 & 2 \\ 1 & -1 & 2 \end{bmatrix} \begin{bmatrix} \mu \\ \alpha_1 \\ \alpha_2 \end{bmatrix}.$$

The first column gives the intercept, and the second and third columns, linearly dependent, give a shifted covariate.

7.3 Applying Cochran's Theorem for Sequential SS

The real power of Cochran's Theorem is revealed by sequentially partitioning the explanatory variables into subsets, or the design matrix $\mathbf{X}$ into submatrices $\mathbf{X}_j$, $j = 0, \ldots, k$, as

$$\mathbf{X} = [\mathbf{X}_0 \mid \mathbf{X}_1 \mid \mathbf{X}_2 \mid \ldots \mid \mathbf{X}_k]$$

where each subset matrix $\mathbf{X}_j$ is $N \times p_j$, and corresponding partitioning with $\mathbf{b}$, that is,

$$\mathbf{b} = \begin{bmatrix} \mathbf{b}_0 \\ \mathbf{b}_1 \\ \dots \\ \mathbf{b}_k \end{bmatrix} \begin{matrix} p_0 \\ p_1 \\ \dots \\ p_k \end{matrix},$$

so that $\mathbf{Xb} = \mathbf{X}_0\mathbf{b}_0 + \mathbf{X}_1\mathbf{b}_1 + \ldots + \mathbf{X}_k\mathbf{b}_k$. Now in Cochran's Theorem, each projection matrix was orthogonal to the others; here, they will be constructed sequentially to be orthogonal to all previous matrices in the following way. Denote the nested subset matrix as $\mathbf{X}_j^*$ where

$$\mathbf{X}_j^* = [\mathbf{X}_0 \mid \mathbf{X}_1 \mid \mathbf{X}_2 \mid \ \dots \ \mid \mathbf{X}_j]$$

for $j = 0, \ldots, k$, and note that $\mathbf{X}_0^* = \mathbf{X}_0$ and $\mathbf{X}_k^* = \mathbf{X}$. A useful notation is the regression sum of squares function:

$$R(\mathbf{b}_0, \ldots, \mathbf{b}_j) = \mathbf{y}^T \mathbf{P}_{\mathbf{X}_j^*} \mathbf{y},$$

that is, the argument for R is a list of parameters, and the value is a quadratic form in $\mathbf{y}$ for the projection matrix from the columns of the design matrix corresponding to that list of parameters. Since $\mathcal{C}(\mathbf{X}_{j-1}^*) \subseteq \mathcal{C}(\mathbf{X}_j^*)$, we know that $\mathbf{P}_{\mathbf{X}_j^*} - \mathbf{P}_{\mathbf{X}_{j-1}^*}$ is the projection matrix onto the part of $\mathcal{C}(\mathbf{X}_j)$ that is orthogonal to $\mathbf{X}_{j-1}^*$, or all of previous subsets of columns. The rank of this projection matrix will be the incremental rank gained by adding $\mathbf{X}_j$ to $\mathbf{X}_{j-1}^*$ to form $\mathbf{X}_j^*$, and denoted by r_j. Now we can apply Cochran's Theorem to the linear model $\mathbf{y} \sim N_N(\mathbf{Xb}, \sigma^2 \mathbf{I}_N)$ with the following sequence of symmetric projection matrices:

$$\begin{aligned}
\mathbf{A}_0 &= \mathbf{P}_{\mathbf{X}_0} \\
\mathbf{A}_1 &= \mathbf{P}_{\mathbf{X}_1^*} - \mathbf{P}_{\mathbf{X}_0} \\
\mathbf{A}_2 &= \mathbf{P}_{\mathbf{X}_2^*} - \mathbf{P}_{\mathbf{X}_1^*} \\
&\dots \\
\mathbf{A}_k &= \mathbf{P}_{\mathbf{X}} - \mathbf{P}_{\mathbf{X}_{k-1}^*} \\
\mathbf{A}_{k+1} &= \mathbf{I}_N - \mathbf{P}_{\mathbf{X}}.
\end{aligned}$$

Each matrix $\mathbf{A}_j$, $j = 0, \ldots, k+1$, is symmetric and idempotent, with rank r_j (defining $r_{k+1} = N - rank\,(\mathbf{X}) = N - r$), and

$$\sum_{j=0}^{k+1} \mathbf{A}_j = \mathbf{I}_N.$$

This leads to the following ANOVA table (Table 7.1).

TABLE 7.1 ANOVA Table for Sequential SS

Source	df	Projection	SS	noncentrality
$\mathbf{b}_0$	$r(\mathbf{X}_0)$	$\mathbf{P}_{\mathbf{X}_0}$	$R(\mathbf{b}_0)$	$(2\sigma^2)^{-1}(\mathbf{Xb})^T\mathbf{P}_{\mathbf{X}_0}(\mathbf{Xb})$
$\mathbf{b}_1$ after $\mathbf{b}_0$	$r(\mathbf{X}_1^*) - r(\mathbf{X}_0)$	$\mathbf{P}_{\mathbf{X}_1^*} - \mathbf{P}_{\mathbf{X}_0}$	$R(\mathbf{b}_0, \mathbf{b}_1) - R(\mathbf{b}_0)$	$(2\sigma^2)^{-1}(\mathbf{Xb})^T(\mathbf{P}_{\mathbf{X}_1^*} - \mathbf{P}_{\mathbf{X}_0})(\mathbf{Xb})$
...	...			
$\mathbf{b}_j$ after $\mathbf{b}_0, \ldots, \mathbf{b}_{j-1}$	$r(\mathbf{X}_j^*) - r(\mathbf{X}_{j-1}^*)$	$\mathbf{P}_{\mathbf{X}_j^*} - \mathbf{P}_{\mathbf{X}_{j-1}^*}$	$R(\mathbf{b}_0, \ldots, \mathbf{b}_j) - R(\mathbf{b}_0, \ldots, \mathbf{b}_{j-1})$	$(2\sigma^2)^{-1}(\mathbf{Xb})^T(\mathbf{P}_{\mathbf{X}_j^*} - \mathbf{P}_{\mathbf{X}_{j-1}^*})(\mathbf{Xb})$
...	...			
$\mathbf{b}_k$ after $\mathbf{b}_0, \ldots, \mathbf{b}_{k-1}$	$r(\mathbf{X}_k^*) - r(\mathbf{X}_{k-1}^*)$	$\mathbf{P}_{X_k^*} - \mathbf{P}_{\mathbf{X}_{k-1}^*}$	$R(\mathbf{b}_0, \ldots, \mathbf{b}_k) - R(\mathbf{b}_0, \ldots, \mathbf{b}_{k-1})$	$(2\sigma^2)^{-1}(\mathbf{Xb})^T(\mathbf{P}_{\mathbf{X}} - \mathbf{P}_{\mathbf{X}_{k-1}^*})(\mathbf{Xb})$
Error	$N - r(X)$	$\mathbf{I} - \mathbf{P}_{\mathbf{X}}$	$\mathbf{y}^T\mathbf{y} - R(\mathbf{b})$	0
Total	N	$\mathbf{I}$	$\mathbf{y}^T\mathbf{y}$	$(2\sigma^2)^{-1}(\mathbf{Xb})^T(\mathbf{Xb})$

The result of Cochran's Theorem is a complete decomposition of the sum of squares $\mathbf{y}^T\mathbf{y}$ where each piece $\mathbf{y}^T\mathbf{A}_j\mathbf{y}$ is independent and

$$(1/\sigma^2)\mathbf{y}^T\mathbf{A}_j\mathbf{y} \sim \chi^2_{r_j}\left(\frac{(\mathbf{Xb})^T\mathbf{A}_j(\mathbf{Xb})}{2\sigma^2}\right)$$

where $r_j = rank\,(\mathbf{A}_j) = rank\,(\mathbf{X}_j^*) - rank\,(\mathbf{X}_{j-1}^*)$, for $j = 0, \ldots, k$. The last piece $(1/\sigma^2)\mathbf{y}^T\mathbf{A}_j\mathbf{y} \sim \chi^2_{N-r}$ for the error sum of squares. The differences in sums of squares, $R(\mathbf{b}_0, \ldots, \mathbf{b}_j) - R(\mathbf{b}_0, \ldots, \mathbf{b}_{j-1}) = \mathbf{y}^T\mathbf{A}_j\mathbf{y}$, are called *sequential sums of squares* or *type I SS* in SAS. The sequential sums of squares account for the reduction in the error sum of squares with the inclusion of a set of explanatory variables, or the addition of columns to the design matrix. We will use these for testing the most common cases of the general linear hypothesis.

This sequential construction of idempotent matrices can then be applied to our proof of Cochran's Theorem. For each of the idempotent matrices A_j above, most of them of the form

$$\mathbf{A}_j = \mathbf{P}_{\mathbf{X}_j^*} - \mathbf{P}_{\mathbf{X}_{j-1}^*},$$

construct the $N \times r_j$ matrix $\mathbf{Q}_j$ with orthonormal columns such that $\mathbf{A}_j = \mathbf{Q}_j\mathbf{Q}_j^T$ and $\mathbf{Q}_j^T\mathbf{Q}_j = \mathbf{I}_{r_j}$. Gathering them together, we can do more with them. Construct the matrix $\mathbf{Q}$ as

$$\mathbf{Q} = [\mathbf{Q}_0 \mid \mathbf{Q}_1 \mid \mathbf{Q}_2 \mid \ \ldots \ \mid \mathbf{Q}_k \mid \mathbf{Q}_{k+1}]$$

where each partition has N rows and r_j columns, so $\mathbf{Q}$ has $N = \sum_{j=0}^{k+1} r_j$ columns, as we found with Cochran's Theorem. Now let's return to our original least squares problem of minimizing

$$\begin{aligned}\|\mathbf{y} - \mathbf{Xb}\|^2 &= (\mathbf{y} - \mathbf{Xb})^T(\mathbf{y} - \mathbf{Xb}) \\ &= (\mathbf{y} - \mathbf{Xb})^T\mathbf{Q}\mathbf{Q}^T(\mathbf{y} - \mathbf{Xb}) = \|\mathbf{Q}^T(\mathbf{y} - \mathbf{Xb})\|^2\end{aligned}$$

where, so far, not much appears to have happened. But now watch what happens with $\mathbf{Q}^T\mathbf{y}$:

$$\mathbf{Q}^T\mathbf{y} = \begin{bmatrix} \mathbf{Q}_0^T\mathbf{y} \\ \mathbf{Q}_1^T\mathbf{y} \\ \ldots \\ \mathbf{Q}_{k+1}^T\mathbf{y} \end{bmatrix} \sim N_N(\mathbf{Q}^T\mathbf{Xb}, \sigma^2\mathbf{I}_N),$$

so that each piece $\mathbf{Q}_j^T\mathbf{y}$ is independent, with $\mathbf{Q}_j^T\mathbf{y} \sim N_{r_j}(\mathbf{Q}_j^T\mathbf{Xb}, \sigma^2\mathbf{I}_{r_j})$. The structure of $\mathbf{Q}^T\mathbf{Xb}$ is revealing:

$$\mathbf{Q}^T\mathbf{Xb} = \begin{bmatrix} \mathbf{R}_{00} & \mathbf{R}_{01} & \ldots & \mathbf{R}_{0k} \\ \mathbf{0} & \mathbf{R}_{11} & \ldots & \mathbf{R}_{1k} \\ \ldots & \ldots & & \ldots \\ \mathbf{0} & \mathbf{0} & \mathbf{0} & \mathbf{R}_{kk} \\ \mathbf{0} & \mathbf{0} & \mathbf{0} & \mathbf{0} \end{bmatrix} \begin{bmatrix} \mathbf{b}_0 \\ \mathbf{b}_1 \\ \ldots \\ \mathbf{b}_k \end{bmatrix} = \begin{bmatrix} \mathbf{R} \\ \mathbf{0} \end{bmatrix} \mathbf{b} \qquad (7.2)$$

so that its structure is essentially block upper triangular, with an extra block row of zeros, and diagonal blocks $\mathbf{R}_{jj}$ with full-row rank r_j. Note that the partitioning of columns is $p_0, p_1, \ldots, p_k$ and the partitioning of rows is $r_0, r_1, \ldots, r_k, r_{k+1}$. The extra (last) row of block zeros ($N - r$ rows) corresponds to $\mathbf{Q}_{k+1}$, and

$$\mathbf{A}_{k+1} = \mathbf{I} - \mathbf{P}_{\mathbf{X}} = \mathbf{Q}_{k+1}\mathbf{Q}_{k+1}^T,$$

so that $\mathbf{Q}_{k+1}^T\mathbf{X} = \mathbf{0}$. In a similar fashion, $\mathbf{A}_j$ is orthogonal to the preceding columns of $\mathbf{X}$ so that

$$\mathbf{Q}_j^T\mathbf{X}_i = \mathbf{0} \quad \text{for} \quad i = 1, \ldots, j-1.$$

The diagonal blocks $\mathbf{R}_{jj}$ are $r_j \times p_j$ and correspond to

$$\mathbf{R}_{jj}^T\mathbf{R}_{jj} = \mathbf{X}_j^T(\mathbf{I} - \mathbf{P}_{\mathbf{X}_{j-1}^*})\mathbf{X}_j = \mathbf{X}_j^T(\mathbf{P}_{X_j^*} - \mathbf{P}_{X_{j-1}^*})\mathbf{X}_j,$$

hence their rank r_j.

Example 7.3: Sequential Partitioning in Multiple Regression
Consider the following multiple regression problem, $y_i = b_0 + b_1x_{i1} + b_2x_{i2} + e_i$, with $N = 4$, and the design matrix $\mathbf{X}$ below partitioned as $\mathbf{X} = [\mathbf{X}_0 \,|\, \mathbf{X}_1 \,|\, \mathbf{X}_2]$, so that

$$\mathbf{X} = \begin{bmatrix} 1 & 0 & 2 \\ 1 & 3 & 3 \\ 1 & 2 & 2 \\ 1 & 3 & 1 \end{bmatrix} \quad \text{and} \quad \mathbf{Q}_0 = \frac{1}{2}\begin{bmatrix} 1 \\ 1 \\ 1 \\ 1 \end{bmatrix}, \quad (\mathbf{I} - \mathbf{P}_1)\mathbf{X}_1 = \begin{bmatrix} -2 \\ 1 \\ 0 \\ 1 \end{bmatrix}$$

and so

$$\mathbf{Q}_1 = \frac{1}{6}\begin{bmatrix} -2 \\ 1 \\ 0 \\ 1 \end{bmatrix}.$$

For the last step,

$$(\mathbf{I} - \mathbf{P}_{\mathbf{X}_1^*})\mathbf{X}_2 = \begin{bmatrix} 0 \\ 1 \\ 0 \\ -1 \end{bmatrix},$$

so we have

$$\mathbf{Q}_2 = \frac{1}{2}\begin{bmatrix} 0 \\ 1 \\ 0 \\ -1 \end{bmatrix}.$$

For the matrix $\mathbf{R}$, $\mathbf{R}_{00} = \sqrt{4} = 2$, $R_{01} = \mathbf{Q}_0^T\mathbf{X}_1 = 4$, and so we have $R_{11}^2 = 6$ or $R_{11} = \sqrt{6}$. The last elements of $\mathbf{R}$ are $R_{02} = 4$, $R_{12} = 0$, and $R_{22} = \sqrt{2}$. Finally, note the factorization result:

$$\mathbf{X}^T\mathbf{X} = \begin{bmatrix} 4 & 8 & 8 \\ 8 & 22 & 16 \\ 8 & 16 & 18 \end{bmatrix} = \mathbf{R}^T\mathbf{R} = \begin{bmatrix} 2 & 0 & 0 \\ 4 & \sqrt{6} & 0 \\ 4 & 0 & \sqrt{2} \end{bmatrix} \begin{bmatrix} 2 & 4 & 4 \\ 0 & \sqrt{6} & 0 \\ 0 & 0 & \sqrt{2} \end{bmatrix}.$$

Example 7.4: Sequential Partitioning in Two-Way Crossed Model
Consider the two-way crossed model with no interaction, $y_{ijk} = \mu + \alpha_i + \beta_j + e_{ijk}$ with $a = 2, b = 3$, and $n = 2$ for simplicity. Partition the design matrix with the first column corresponding to X_0, the next two columns form $\mathbf{X}_1$ corresponding to the α's, and the last three columns form $\mathbf{X}_2$.

$$y = \begin{bmatrix} y_{111} \\ y_{112} \\ y_{121} \\ y_{122} \\ y_{131} \\ y_{132} \\ y_{211} \\ y_{212} \\ y_{221} \\ y_{222} \\ y_{231} \\ y_{232} \end{bmatrix}, \quad \mathbf{Xb} = \begin{bmatrix} 1 & 1 & 0 & 1 & 0 & 0 \\ 1 & 1 & 0 & 1 & 0 & 0 \\ 1 & 1 & 0 & 0 & 1 & 0 \\ 1 & 1 & 0 & 0 & 1 & 0 \\ 1 & 1 & 0 & 0 & 0 & 1 \\ 1 & 1 & 0 & 0 & 0 & 1 \\ 1 & 0 & 1 & 1 & 0 & 0 \\ 1 & 0 & 1 & 1 & 0 & 0 \\ 1 & 0 & 1 & 0 & 1 & 0 \\ 1 & 0 & 1 & 0 & 1 & 0 \\ 1 & 0 & 1 & 0 & 0 & 1 \\ 1 & 0 & 1 & 0 & 0 & 1 \end{bmatrix} \begin{bmatrix} \mu \\ \alpha_1 \\ \alpha_2 \\ \beta_1 \\ \beta_2 \\ \beta_3 \end{bmatrix}$$

Then

$$\mathbf{Q}_0 = \frac{1}{\sqrt{abn}}\mathbf{1}_{abn}, \ (\mathbf{I} - \mathbf{P}_1)\mathbf{X}_1 = \frac{1}{2}\begin{bmatrix} \mathbf{1}_{bn} & -\mathbf{1}_{bn} \\ -\mathbf{1}_{bn} & \mathbf{1}_{bn} \end{bmatrix},$$

and so

$$\mathbf{Q}_1 = \frac{1}{\sqrt{abn}}\begin{bmatrix} \mathbf{1}_{bn} \\ -\mathbf{1}_{bn} \end{bmatrix}.$$

For the matrix $\mathbf{R}$,

$$R_{00} = \sqrt{abn}, \mathbf{R}_{01} = \mathbf{Q}_0^T\mathbf{X}_1 = \sqrt{\frac{bn}{a}}a\mathbf{1}^T = \sqrt{3}[1 \quad 1],$$

and so we have

$$\mathbf{R}_{11}^T\mathbf{R}_{11} = bn\left(\mathbf{I}_a - \frac{1}{a}\mathbf{1}_a\mathbf{1}_a^T\right) = 3\begin{bmatrix} 1 & -1 \\ -1 & 1 \end{bmatrix},$$

so that its factor is

$$\mathbf{R}_{11} = \sqrt{3}[1 \quad -1]$$

showing the dependence in $\mathbf{X}_1$. For the last step,

$$\mathbf{R}_{02} = \mathbf{Q}_0^T\mathbf{X}_2 = \sqrt{\frac{an}{b}}\mathbf{1}_b^T = \frac{2\sqrt{3}}{3}[1 \quad 1 \quad 1], \quad \mathbf{R}_{12} = \begin{bmatrix} 0 & 0 & 0 \\ 0 & 0 & 0 \end{bmatrix},$$

and the big step is

$$(\mathbf{I} - \mathbf{P}_{\mathbf{X}_1^*})\mathbf{X}_2 = \frac{1}{3}\begin{bmatrix} 2\mathbf{1}_n & -\mathbf{1}_n & -\mathbf{1}_n \\ -\mathbf{1}_n & 2\mathbf{1}_n & -\mathbf{1}_n \\ -\mathbf{1}_n & -\mathbf{1}_n & 2\mathbf{1}_n \\ 2\mathbf{1}_n & -\mathbf{1}_n & -\mathbf{1}_n \\ -\mathbf{1}_n & 2\mathbf{1}_n & -\mathbf{1}_n \\ -\mathbf{1}_n & -\mathbf{1}_n & 2\mathbf{1}_n \end{bmatrix},$$

giving

$$\mathbf{R}_{22}^T\mathbf{R}_{22} = \frac{4}{3}\begin{bmatrix} 2 & -1 & -1 \\ -1 & 2 & -1 \\ -1 & -1 & 2 \end{bmatrix} = an\left(\mathbf{I}_b - \frac{1}{b}\mathbf{1}_b\mathbf{1}_b^T\right),$$

so that

$$\mathbf{R}_{22} = \sqrt{\frac{4}{3}}\begin{bmatrix} \sqrt{2} & -1/\sqrt{2} & -1/\sqrt{2} \\ 0 & \sqrt{3/2} & -\sqrt{3/2} \end{bmatrix},$$

again showing the dependence in $\mathbf{X}_2$. The final step is the construction of $\mathbf{Q}_2$, which requires some algebra to get

$$\mathbf{Q}_2 = \begin{bmatrix} \frac{2}{\sqrt{24}}\mathbf{1}_2 & \mathbf{0} \\ \frac{-1}{\sqrt{24}}\mathbf{1}_2 & \frac{1}{\sqrt{8}}\mathbf{1}_2 \\ \frac{-1}{\sqrt{24}}\mathbf{1}_2 & \frac{-1}{\sqrt{8}}\mathbf{1}_2 \\ \frac{2}{\sqrt{24}}\mathbf{1}_2 & \mathbf{0} \\ \frac{-1}{\sqrt{24}}\mathbf{1}_2 & \frac{1}{\sqrt{8}}\mathbf{1}_2 \\ \frac{-1}{\sqrt{24}}\mathbf{1}_2 & \frac{-1}{\sqrt{8}}\mathbf{1}_2 \end{bmatrix}$$

so that

$$\mathbf{R} = \begin{bmatrix} \sqrt{12} & \sqrt{3} & \sqrt{3} & \sqrt{4/3} & \sqrt{4/3} & \sqrt{4/3} \\ 0 & \sqrt{3} & -\sqrt{3} & 0 & 0 & 0 \\ 0 & 0 & 0 & \sqrt{8/3} & -\sqrt{2/3} & -\sqrt{2/3} \\ 0 & 0 & 0 & 0 & \sqrt{2} & -\sqrt{2} \end{bmatrix}.$$

Note also that

$$\mathbf{X}^T\mathbf{X} = \mathbf{R}^T\mathbf{R} = \begin{bmatrix} 12 & 6 & 6 & 4 & 4 & 4 \\ 6 & 6 & 0 & 2 & 2 & 2 \\ 6 & 0 & 6 & 2 & 2 & 2 \\ 4 & 2 & 2 & 4 & 0 & 0 \\ 4 & 2 & 2 & 0 & 4 & 0 \\ 4 & 2 & 2 & 0 & 0 & 4 \end{bmatrix} = \begin{bmatrix} abn & bn\mathbf{1}_a^T & an\mathbf{1}_b^T \\ bn\mathbf{1}_a & bn\mathbf{I}_a & n\mathbf{1}_a\mathbf{1}_b^T \\ an\mathbf{1}_b & n\mathbf{1}_b\mathbf{1}_a^T & an\mathbf{I}_b \end{bmatrix}.$$

Returning now to the general problem, let's look more closely at $k = 2$ to see the effects of this structure and its relationship with sequential sums of squares (recall also Exercise 5.29). Consider the vector whose squared length we intend to minimize:

$$\mathbf{Q}^T(\mathbf{y} - \mathbf{X}\mathbf{b}) = \begin{bmatrix} \mathbf{Q}_0^T\mathbf{y} \\ \mathbf{Q}_1^T\mathbf{y} \\ \mathbf{Q}_2^T\mathbf{y} \\ \mathbf{Q}_3^T\mathbf{y} \end{bmatrix} - \begin{bmatrix} \mathbf{R}_{00} & \mathbf{R}_{01} & \mathbf{R}_{02} \\ \mathbf{0} & \mathbf{R}_{11} & \mathbf{R}_{12} \\ \mathbf{0} & \mathbf{0} & \mathbf{R}_{22} \\ \mathbf{0} & \mathbf{0} & \mathbf{0} \end{bmatrix} \begin{bmatrix} \mathbf{b}_0 \\ \mathbf{b}_1 \\ \mathbf{b}_2 \end{bmatrix}.$$

Notice that the last component is unaffected by $\mathbf{b}$, and so its contribution to the sum of squares is

$$\left\|\mathbf{Q}_3^T\mathbf{y}\right\|^2 = \mathbf{y}^T(\mathbf{I} - \mathbf{P}_\mathbf{X})\mathbf{y} = SSE.$$

To see this more clearly, the sum of squares to be minimized can be written as

$$\begin{aligned} \|\mathbf{Q}^T(\mathbf{y} - \mathbf{X}\mathbf{b})\|^2 = {} & \left\|\mathbf{Q}_0^T\mathbf{y} - \mathbf{R}_{00}\mathbf{b}_0 - \mathbf{R}_{01}\mathbf{b}_1 - \mathbf{R}_{02}\mathbf{b}_2\right\|^2 \\ & + \left\|\mathbf{Q}_1^T\mathbf{y} - \mathbf{R}_{11}\mathbf{b}_1 - \mathbf{R}_{12}\mathbf{b}_2\right\|^2 + \left\|\mathbf{Q}_2^T\mathbf{y} - \mathbf{R}_{22}\mathbf{b}_2\right\|^2 + \left\|\mathbf{Q}_3^T\mathbf{y}\right\|^2. \end{aligned} \tag{7.3}$$

The first three block components can be made zero by solving the 3×3 block system of equations:

$$\begin{bmatrix} \mathbf{Q}_0^T \mathbf{y} \\ \mathbf{Q}_1^T \mathbf{y} \\ \mathbf{Q}_2^T \mathbf{y} \end{bmatrix} = \begin{bmatrix} \mathbf{R}_{00} & \mathbf{R}_{01} & \mathbf{R}_{02} \\ \mathbf{0} & \mathbf{R}_{11} & \mathbf{R}_{12} \\ \mathbf{0} & \mathbf{0} & \mathbf{R}_{22} \end{bmatrix} \begin{bmatrix} \mathbf{b}_0 \\ \mathbf{b}_1 \\ \mathbf{b}_2 \end{bmatrix}, \tag{7.4}$$

solving from the bottom up, since each diagonal block has full-row rank; this also gives the least squares estimator $\hat{\mathbf{b}}$. While this has some computational advantages, look at what happens if we sequentially add columns to the design matrix. If we want $R(\mathbf{b}_0)$, which corresponds to the regression problem with just $\mathbf{X}_0$ as the design matrix, or, in view of the problem above, $\mathbf{b}_1 = \mathbf{b}_2 = \mathbf{0}$, then (7.3) becomes just

$$\|\mathbf{Q}^T(\mathbf{y} - \mathbf{X}\mathbf{b})\|^2 = \left\|\mathbf{Q}_0^T\mathbf{y} - \mathbf{R}_{00}\mathbf{b}_0\right\|^2 + \left\|\mathbf{Q}_1^T\mathbf{y}\right\|^2 + \left\|\mathbf{Q}_2^T\mathbf{y}\right\|^2 + \left\|\mathbf{Q}_3^T\mathbf{y}\right\|^2, \tag{7.5}$$

and so only the small system $\mathbf{Q}_0^T\mathbf{y} = \mathbf{R}_{00}\mathbf{b}_0$ needs to be solved. The two pieces $\mathbf{Q}_1^T\mathbf{y}$ and $\mathbf{Q}_2^T\mathbf{y}$ shift to join $\mathbf{Q}_3^T\mathbf{y}$, contributing to the error sum of squares, and $R(\mathbf{b}_0) = \|\mathbf{Q}_0^T\mathbf{y}\|^2$. Now for $R(\mathbf{b}_0, \mathbf{b}_1)$, set $\mathbf{b}_2 = \mathbf{0}$ and solve the remaining block 2×2 system, and so

$$R(\mathbf{b}_0, \mathbf{b}_1) = \left\|\mathbf{Q}_0^T\mathbf{y}\right\|^2 + \left\|\mathbf{Q}_1^T\mathbf{y}\right\|^2,$$

so that the difference is just

$$R(\mathbf{b}_0, \mathbf{b}_1) - R(\mathbf{b}_0) = \left\|\mathbf{Q}_1^T\mathbf{y}\right\|^2.$$

Following similar calculations,

$$\begin{aligned} R(\mathbf{b}_0, \mathbf{b}_1, \mathbf{b}_2) - \mathrm{R}(\mathbf{b}_0, \mathbf{b}_1) &= \left\|\mathbf{Q}_0^T\mathbf{y}\right\|^2 + \left\|\mathbf{Q}_1^T\mathbf{y}\right\|^2 + \left\|\mathbf{Q}_2^T\mathbf{y}\right\|^2 \\ &\quad - \left\|\mathbf{Q}_0^T\mathbf{y}\right\|^2 - \left\|\mathbf{Q}_1^T\mathbf{y}\right\|^2 = \left\|\mathbf{Q}_2^T\mathbf{y}\right\|^2. \end{aligned}$$

The matrix $\mathbf{Q}$ can be viewed as rotating space so that the incremental dimensions that each block $\mathbf{X}_j$ contributes to the column space of $\mathbf{X}$ get rotated into the next coordinates in sequence. The column space of $\mathbf{X}_0$ is rotated into the first r_0 coordinates. Then the part of the column space of $\mathbf{X}_1^*$ that is orthogonal to $\mathbf{X}_0$ is rotated into the next r_1 coordinates, and so on. The remaining $N - r$ coordinates, not part of the column space of $\mathbf{X}$, are rotated to be orthogonal to $\mathcal{C}(\mathbf{X})$, and contribute to the error sum of squares.

As a final note, Gram–Schmidt orthonormalization from Section 2.6 could be used to construct $\mathbf{X} = \mathbf{QR}$ to form $\mathbf{Q}^T\mathbf{X} = \mathbf{R}$ in (7.3). However, the Gram–Schmidt procedure does not construct the last block of columns $\mathbf{Q}_{k+1}$, corresponding to $\mathcal{N}(\mathbf{X}^T)$, or $\mathbf{I} - \mathbf{P}_\mathbf{X}$.

Example 7.3: continued
Now add some responses to the multiple regression problem $y_i = \beta_0 + \beta_1 x_{i1} + \beta_2 x_{i2} + e_i$, for $N = 4$, with $\mathbf{y}$ and the design matrix $\mathbf{X}$ given below:

$$\mathbf{y} = \begin{bmatrix} 8 \\ 7 \\ 2 \\ 3 \end{bmatrix}, \quad \mathbf{X} = \begin{bmatrix} 1 & 0 & 2 \\ 1 & 3 & 3 \\ 1 & 2 & 2 \\ 1 & 3 & 1 \end{bmatrix}, \quad \mathbf{y}^T\mathbf{y} = 126,$$

so that (7.4) becomes

$$\begin{bmatrix} \mathbf{Q}_0^T\mathbf{y} \\ \mathbf{Q}_1^T\mathbf{y} \\ \mathbf{Q}_2^T\mathbf{y} \end{bmatrix} = \begin{bmatrix} 10 \\ -\sqrt{6} \\ 2\sqrt{2} \end{bmatrix} = \mathbf{Rb} = \begin{bmatrix} 2 & 4 & 4 \\ 0 & \sqrt{6} & 0 \\ 0 & 0 & \sqrt{2} \end{bmatrix} \begin{bmatrix} \beta_0 \\ \beta_1 \\ \beta_2 \end{bmatrix}.$$

Solving for $\mathbf{b}$ gives

$$\hat{\mathbf{b}} = \begin{bmatrix} 3 \\ -1 \\ 2 \end{bmatrix}.$$

Now we have $SSE = \mathbf{y}^T\mathbf{y} - \mathbf{y}^T\mathbf{Q}\mathbf{Q}^T\mathbf{y} = 126 - 114 = 12$. If we want to restrict $\beta_2 = 0$, then we increase SSE by $(2\sqrt{2})^2 = 8$, and while the estimate for β_1 remains -1, the estimate for β_0 changes to 7. Dropping the other covariate x_{i1} from the model leads to an increase of 6 in SSE. The two type I (sequential) sums of squares are $(\sqrt{6})^2 = 6$ for X_1 and $(2\sqrt{2})^2 = 8$ for X_2.

7.4 Orthogonal Polynomials and Contrasts

Orthogonal polynomials provide a method for fitting simple models for an experiment that usually would be analyzed as a one-way ANOVA, but where the group designations arise as levels of some continuous variable, such as dosages. The usual parameterization for the one-way analysis of variance problem follows the form

$$y_{ij} = \mu + \alpha_i + e_{ij}$$

where the α_i represent the group effects. Another reparameterization from Section 3.6, barely mentioned among the many in Example 3.2, uses a polynomial regression model:

$$y_{ij} = \beta_0 + \beta_1 x_i + \beta_2 x_i^2 + \cdots + \beta_{a-1} x_i^{a-1} + e_{ij},$$

where the a different levels of the covariate x_i designate the groups. Most commonly, the levels are equally spaced, so that the group index could be used as the covariate, that is, $x_i = i$. The equivalence of these two parameterizations follows from the equivalence of the group means model design matrix $\mathbf{Z}$ and the Vandermonde matrix $\mathbf{B}$:

$$\mathbf{V}\gamma = \begin{bmatrix} \mathbf{1}_{n_1} & x_1\mathbf{1}_{n_1} & x_1^2\mathbf{1}_{n_1} & \dots & x_1^{a-1}\mathbf{1}_{n_1} \\ \mathbf{1}_{n_2} & x_2\mathbf{1}_{n_2} & x_2^2\mathbf{1}_{n_2} & \dots & x_2^{a-1}\mathbf{1}_{n_2} \\ \mathbf{1}_{n_3} & x_3\mathbf{1}_{n_3} & x_3^2\mathbf{1}_{n_3} & \dots & x_3^{a-1}\mathbf{1}_{n_3} \\ \dots & \dots & \dots & \dots & \dots \\ \mathbf{1}_{n_a} & x_a\mathbf{1}_{n_a} & x_a^2\mathbf{1}_{n_a} & \dots & x_a^{a-1}\mathbf{1}_{n_a} \end{bmatrix} \begin{bmatrix} \beta_0 \\ \beta_1 \\ \beta_2 \\ \dots \\ \beta_{a-1} \end{bmatrix} = \mathbf{ZB}\gamma \tag{7.6}$$

$$= \begin{bmatrix} \mathbf{1}_{n_1} & \mathbf{0} & \mathbf{0} & \dots & \mathbf{0} \\ \mathbf{0} & \mathbf{1}_{n_2} & \mathbf{0} & \dots & \mathbf{0} \\ \mathbf{0} & \mathbf{0} & \mathbf{1}_{n_3} & \dots & \mathbf{0} \\ \dots & \dots & \dots & \dots & \dots \\ \mathbf{0} & \mathbf{0} & \dots & \mathbf{0} & \mathbf{1}_{n_a} \end{bmatrix} \begin{bmatrix} 1 & x_1 & x_1^2 & \dots & x_1^{a-1} \\ 1 & x_2 & x_2^2 & \dots & x_2^{a-1} \\ 1 & x_3 & x_3^2 & \dots & x_3^{a-1} \\ \dots & \dots & \dots & \dots & \dots \\ 1 & x_a & x_a^2 & \dots & x_a^{a-1} \end{bmatrix} \begin{bmatrix} \beta_0 \\ \beta_1 \\ \beta_2 \\ \dots \\ \beta_{a-1} \end{bmatrix}. \tag{7.7}$$

The $a \times a$ Vandermonde matrix $\mathbf{B}$ has the well-known determinant

$$|\mathbf{B}| = \prod_{j>i}(x_j - x_i),$$

which is nonzero as long as the design points $\{x_i, i = 1, \dots, a\}$ are distinct. As a result, the matrix $\mathbf{B}$ is nonsingular, and the two parameterizations are equivalent.

The mathematical theory behind orthogonal polynomials is just the application of Gram–Schmidt to the design matrix $\mathbf{V}$ to create $\mathbf{V} = \mathbf{QR}$, where the $N \times a$ matrix $\mathbf{Q}$ has orthonormal columns and $\mathbf{R}$ is $a \times a$ and upper triangular. Certain modifications are employed to make the arithmetic more convenient but also make the algebra a little more complicated. Beginning with the parameterization above, $\mathbf{y} = \mathbf{V}\gamma + \mathbf{e}$, write $\mathbf{V}$ in terms of its pieces:

$$\mathbf{V}\gamma = \mathbf{ZB}\gamma = \mathbf{QR}\gamma$$

where $\mathbf{R}$ is the upper triangular matrix arising from Gram–Schmidt. But the matrix with orthonormal columns $\mathbf{Q}$ can be written as $\mathbf{ZCD}$ where $\mathbf{Z}$ is the cell means model design matrix, as above, and the diagonal matrix $\mathbf{D}$ constructed so that the elements of the $a \times a$ matrix $\mathbf{C}$ are integers. Then one more step has

$$\mathbf{V}\gamma = \mathbf{ZB}\gamma = \mathbf{QR}\gamma = \mathbf{ZC}(\mathbf{DR}\gamma) = \mathbf{ZC}\phi$$

where $\phi = \mathbf{DR}\gamma$, so that ϕ is the parameter vector for the $\mathbf{ZC}$ parameterization. The upper triangular shape of $\mathbf{DR}$ means that the first k components of ϕ involve only the first k components of γ. Often we wish to test the hypothesis that the mean function is a low-order polynomial in x_i, that is, H: $\beta_{k+1} = \cdots = \beta_{a-1} = 0$. In the **ZC** parameterization, this is equivalent to H: $\phi_{k+1} = \cdots = \phi_{a-1} = 0$ (counting components of ϕ from 0 to $a-1$), which, using the sequential orthonormalization via Gram–Schmidt, is very easy to test.

To illustrate, consider the simpler, but common balanced case where $n_i = n$, and use $x_i = i$. Given the structure of **V** in terms of **B**, the structure of **Q** can be seen from the Gram–Schmidt orthonormalization of **B**:

$$a = 3, \mathbf{B} = \begin{bmatrix} 1 & 1 & 1 \\ 1 & 2 & 4 \\ 1 & 3 & 9 \end{bmatrix}, \quad \mathbf{C} = \begin{bmatrix} 1 & -1 & 1 \\ 1 & 0 & -2 \\ 1 & 1 & 1 \end{bmatrix},$$

$$\mathbf{Q} = \begin{bmatrix} \frac{1}{\sqrt{3n}}\mathbf{1}_n & \frac{-1}{\sqrt{2n}}\mathbf{1}_n & \frac{1}{\sqrt{6n}}\mathbf{1}_n \\ \frac{1}{\sqrt{3n}}\mathbf{1}_n & \mathbf{0}_n & \frac{-2}{\sqrt{6n}}\mathbf{1}_n \\ \frac{1}{\sqrt{3n}}\mathbf{1}_n & \frac{1}{\sqrt{2n}}\mathbf{1}_n & \frac{1}{\sqrt{6n}}\mathbf{1}_n \end{bmatrix} = (\mathbf{ZC})\mathbf{D} = \begin{bmatrix} \mathbf{1}_n & -\mathbf{1}_n & \mathbf{1}_n \\ \mathbf{1}_n & \mathbf{0}_n & -2\mathbf{1}_n \\ \mathbf{1}_n & \mathbf{1}_n & \mathbf{1}_n \end{bmatrix} \begin{bmatrix} \frac{1}{\sqrt{3n}} & 0 & 0 \\ 0 & \frac{1}{\sqrt{2n}} & 0 \\ 0 & 0 & \frac{1}{\sqrt{6n}} \end{bmatrix}$$

$$a = 4, \mathbf{B} = \begin{bmatrix} 1 & 1 & 1 & 1 \\ 1 & 2 & 4 & 8 \\ 1 & 3 & 9 & 27 \\ 1 & 4 & 16 & 64 \end{bmatrix}, \quad \mathbf{C} = \begin{bmatrix} 1 & -3 & 1 & -1 \\ 1 & -1 & -1 & 3 \\ 1 & 1 & -1 & -3 \\ 1 & 3 & 1 & 1 \end{bmatrix}$$

$$\mathbf{Q} = \mathbf{ZCD} = \begin{bmatrix} \mathbf{1}_n & \mathbf{0}_n & \mathbf{0}_n & \mathbf{0}_n \\ \mathbf{0}_n & \mathbf{1}_n & \mathbf{0}_n & \mathbf{0}_n \\ \mathbf{0}_n & \mathbf{0}_n & \mathbf{1}_n & \mathbf{0}_n \\ \mathbf{0}_n & \mathbf{0}_n & \mathbf{0}_n & \mathbf{1}_n \end{bmatrix} \begin{bmatrix} 1 & -3 & 1 & -1 \\ 1 & -1 & -1 & 3 \\ 1 & 1 & -1 & -3 \\ 1 & 3 & 1 & 1 \end{bmatrix} \begin{bmatrix} \frac{1}{2\sqrt{n}} & 0 & 0 & 0 \\ 0 & \frac{1}{\sqrt{20n}} & 0 & 0 \\ 0 & 0 & \frac{1}{2\sqrt{n}} & 0 \\ 0 & 0 & 0 & \frac{1}{\sqrt{20n}} \end{bmatrix}.$$

The matrix **C** is the result of Gram–Schmidt on the columns of **B**, but with the columns rescaled so that the elements are integers—this is the way orthogonal polynomials are usually tabled—and without the first column of ones (and indexed 0 to $a-1$). The columns of **C** then form a set of coefficients named by the polynomial—linear, quadratic, cubic, etc.

Testing under this reparameterization is easy. First, the normal equations are diagonal since $\mathbf{Z}^T\mathbf{Z} = n\mathbf{I}_a$:

$$(\mathbf{ZC})^T(\mathbf{ZC})\phi = (\mathbf{ZC})^T\mathbf{y}$$
$$n\mathbf{C}^T\mathbf{C}\phi = \mathbf{C}^T\mathbf{Z}^T\mathbf{y} = n\mathbf{C}^T\overline{\mathbf{y}}$$

where $\overline{\mathbf{y}}$ is the vector of group means $(\overline{\mathbf{y}})_i = \overline{y}_{i.}$, $i = 1, \ldots, a$. A little more algebra leads to simple coefficient estimates:

$$\hat{\phi}_i = \sum_{j=1}^{a} C_{ji}\overline{y}_{j.} \Bigg/ \left(\sum_{j=1}^{a} C_{ji}^2\right), \quad \text{for} \quad i = 0, \ldots, a-1,$$

since $\mathbf{C}^T\mathbf{C}$ is diagonal. Following from the usual normality assumptions, we have

$$\hat{\phi}_i \sim N\left(\phi_i, \sigma^2 \Big/ \left(n\sum_{j=1}^{a} C_{ji}^2\right)\right),$$

and $\hat{\phi}_i$ are independent. Testing the hypothesis that the mean function is a polynomial of degree k in $x_i = i$, rewritten as H:$\phi_{k+1} = \cdots = \phi_{a-1} = 0$, leads to the numerator sum of squares of

$$n\sum_{i=k+1}^{a-1}\left[\left(\sum_j C_{ji}\overline{y}_{j.}\right)^2 \Big/ \left(\sum_j C_{ji}^2\right)\right],$$

with degrees of freedom $(a-1)-k$ matching the length of the sum on i.

Example 7.5: Orthogonal Polynomials
For an experiment with $a = 4$ equally spaced levels, let us test for a linear effect. More specifically, we want to test H: $\beta_2 = \beta_3 = 0$, which is equivalent to H: $\phi_2 = \phi_3 = 0$, against the alternative that they are not both zero. From the matrix $\mathbf{C}$ above we have the following orthogonal contrasts for $a = 4$ levels:

Linear: $-3, -1, 1, 3$

Quadratic: $1, -1, -1, 1$

Cubic: $-1, 3, -3, 1$

For testing the absence of any quadratic or cubic effects, the numerator sum of squares becomes

$$n\sum_{i=k+1}^{a-1}\left[\left(\sum_j C_{ji}\overline{y}_{j.}\right)^2 \Big/ \left(\sum_j C_{ji}^2\right)\right],$$

which in this case takes the form

$$n(\overline{y}_{1.} - \overline{y}_{2.} - \overline{y}_{3.} + \overline{y}_{4.})^2/4 + n(-\overline{y}_{1.} + 3\overline{y}_{2.} - 3\overline{y}_{3.} + \overline{y}_{4.})^2/20.$$

Dividing the expression above by $2\hat{\sigma}^2$ yields the F-statistic with 2 and $a(n-1)$ df.

All of this analysis relies on two common, but critical assumptions: that the design is balanced, that is, $n_i = n$, and that the covariate values are equally spaced—at least on some scale. With the departure from either of these two assumptions, the decomposition loses its simplicity, and testing for a simple polynomial structure can just as easily be done as a regular full model versus the reduced model likelihood ratio test from Section 6.3.

7.5 Pure Error and the Lack of Fit Test

Although we regularly pose complicated models for observed phenomena, rarely can we verify the assumption that the model we pose is correct. In certain situations, that is, when we have replicates, we can test whether a regression model is correct. By replicates, we mean observations with exactly the same values for the covariates, and hence have the same mean. In multiple regression with two covariates x and z, we may use two indices for the observations,

$$y_{ij} = \beta_0 + \beta_1 x_i + \beta_2 z_i + e_{ij}, \tag{7.8}$$

with $i = 1, \ldots, a$ indexing the group of observations with the same value of the covariates—a group of observations with the same mean—and $j = 1, \ldots, n_i$ indexing the replicates within those groups. This suggests the following structure for the design matrix with two covariates:

$$E(\mathbf{y}) = \mathbf{Xb} = \begin{bmatrix} \mathbf{1}_{n_1} & x_1\mathbf{1}_{n_1} & z_1\mathbf{1}_{n_1} \\ \mathbf{1}_{n_2} & x_2\mathbf{1}_{n_2} & z_2\mathbf{1}_{n_2} \\ \ldots & \ldots & \ldots \\ \mathbf{1}_{n_a} & x_a\mathbf{1}_{n_a} & z_a\mathbf{1}_{n_a} \end{bmatrix} \begin{bmatrix} \beta_0 \\ \beta_1 \\ \beta_2 \end{bmatrix}. \tag{7.9}$$

Now to evaluate the adequacy of this model, we might pose another model that merely has different means for each group—a one-way ANOVA model with a different mean for each design point:

$$E(\mathbf{y}) = \mathbf{Z}\mu = \begin{bmatrix} \mathbf{1}_{n_1} & \mathbf{0} & & \mathbf{0} \\ \mathbf{0} & \mathbf{1}_{n_2} & \ldots & \mathbf{0} \\ \ldots & & \ldots & \\ \mathbf{0} & \mathbf{0} & \ldots & \mathbf{1}_{n_a} \end{bmatrix} \begin{bmatrix} \mu_1 \\ \mu_2 \\ \ldots \\ \mu_a \end{bmatrix}. \tag{7.10}$$

Testing the adequacy of the model (7.8) can then be viewed in terms of the more general model (7.10) as testing the hypothesis H: $\mathbf{Z}\mu = \mathbf{Xb}$, or in view of Example 6.1, H: $\mathbf{Z}\mu \in \mathcal{C}(\mathbf{X})$. Writing this hypothesis in the general linear hypothesis form, H: $\mathbf{K}^T\mu = \mathbf{0}$, is often a challenge (see Exercise 7.11); however, testing this hypothesis as a full versus reduced LRT is easy: just fit the two models (7.9) and (7.10). The likelihood ratio test reveals the geometry of the problem from the test statistic

$$F = \frac{\mathbf{y}^T(\mathbf{P_Z} - \mathbf{P_X})\mathbf{y}/(r(\mathbf{Z}) - r(\mathbf{X})}{\mathbf{y}^T(\mathbf{I} - \mathbf{P_X})\mathbf{y}/(N - r(\mathbf{Z})}.$$

If $\mathbf{Z}\mu \in \mathcal{C}(\mathbf{X})$, then $(\mathbf{P_Z} - \mathbf{P_X})\mathbf{Z}\mu = \mathbf{0}$ and the noncentrality parameter will be zero. We can also see how to write the hypothesis as $\mathbf{K}^T\mu = \mathbf{0}$ using the basis vectors of $(\mathbf{P_Z} - \mathbf{P_X})\mathbf{Z}$.

The error sum of squares arising from fitting (7.10), namely, $\mathbf{y}^T(\mathbf{I} - \mathbf{P_Z})\mathbf{y}$, is often known as *pure error* since it will have a central χ^2 distribution as long as the observations within a replicate group have the same mean—a highly reliable assumption. The test of this hypothesis also goes by the name of *lack of fit test*, since it tests the adequacy of a model assumption—often only available in distributional goodness-of-fit tests. In observational studies, the presence of replicates is unusual. In some designed experiments, the cost of an observation may preclude any replicates.

Example 7.6: Pure Error/Lack of Fit
Consider a simple linear regression model with $N = 5$ and covariate values $x_1 = x_2 = 1, x_3 = 2, x_4 = x_5 = 3$. Then we have the difference in projection matrices:

$$\mathbf{P_Z} - \mathbf{P_X} = \begin{bmatrix} 1/2 & 1/2 & 0 & 0 & 0 \\ 1/2 & 1/2 & 0 & 0 & 0 \\ 0 & 0 & 1 & 0 & 0 \\ 0 & 0 & 0 & 1/2 & 1/2 \\ 0 & 0 & 0 & 1/2 & 1/2 \end{bmatrix} - \begin{bmatrix} .45 & .45 & .2 & -.05 & -.05 \\ .45 & .45 & .2 & -.05 & -.05 \\ .2 & .2 & .2 & .2 & .2 \\ -.05 & -.05 & .2 & .45 & .45 \\ -.05 & -.05 & .2 & .45 & .45 \end{bmatrix}$$

$$= \begin{bmatrix} .05 & .05 & -.2 & .05 & .05 \\ .05 & .05 & -.2 & .05 & .05 \\ -.2 & -.2 & .8 & -.2 & -.2 \\ .05 & .05 & -.2 & .05 & .05 \\ .05 & .05 & -.2 & .05 & .05 \end{bmatrix} = \frac{1}{20}\begin{bmatrix} 1 \\ 1 \\ -4 \\ 1 \\ 1 \end{bmatrix} [1 \quad 1 \quad -4 \quad 1 \quad 1].$$

Now this gives the hypothesis matrix $\mathbf{K}$ from

$$(\mathbf{P_Z} - \mathbf{P_X})\mathbf{Z}\mu = \frac{1}{20}\begin{bmatrix} 1 \\ 1 \\ -4 \\ 1 \\ 1 \end{bmatrix} [1 \quad 1 \quad -4 \quad 1 \quad 1] \begin{bmatrix} 1 & 0 & 0 \\ 1 & 0 & 0 \\ 0 & 1 & 0 \\ 0 & 0 & 1 \\ 0 & 0 & 1 \end{bmatrix} \begin{bmatrix} \mu_1 \\ \mu_2 \\ \mu_3 \end{bmatrix}$$

$$= \frac{1}{20}\begin{bmatrix} 1 \\ 1 \\ -4 \\ 1 \\ 1 \end{bmatrix} [2 \quad -4 \quad 2] \begin{bmatrix} \mu_1 \\ \mu_2 \\ \mu_3 \end{bmatrix} = \mathbf{0},$$

so that we can find $\mathbf{K}^T = (2, -4, 2)^T$ and write $\mathbf{K}^T\mu = \mathbf{0}$ more simply as $\mu_1 - 2\mu_2 + \mu_3 = 0$. See Exercise 7.10.

7.6 Heresy: Testing Nontestable Hypotheses

In writing the general linear hypothesis, we used the form H: $\mathbf{K}^T\mathbf{b} = \mathbf{m}$ and insisted that (1) $\mathbf{K}$ had full-column rank (no redundancies, consistent equations), and (2) each component of $\mathbf{K}^T\mathbf{b}$ was estimable. These restrictions allowed us to construct the first principles test statistic and derive its distribution. Later in Section 6.5, we showed that this first principles test was also a likelihood ratio test. However, there are certain hypotheses that do not easily fall into the $\mathbf{K}^T\mathbf{b}$ formulation, for example, the case of testing whether the interaction is zero in the two-way crossed model (Example 6.2). Nonetheless, we found that we can easily construct the LRT test statistic for this case and derive its distribution without much difficulty. The issue arises, then, whether we can use the LRT for testing nontestable hypotheses.

The likelihood ratio test requires the solution of least squares problem over a restricted parameter space. We can solve such a problem and find $\hat{\mathbf{b}}_H$ using the restricted normal equations for the parameter space $\mathcal{T} = \{\mathbf{b} : \mathbf{P}^T\mathbf{b} = \delta\}$. We can then compute an equivalent form of the likelihood ratio statistic as $[Q(\hat{\mathbf{b}}_H) - Q(\hat{\mathbf{b}})]/Q(\hat{\mathbf{b}})$. As typical with the LRT, however, we are left with the problem of deriving its distribution in order to find an appropriate critical value.

Result 7.1 *Let $\mathcal{C}(\mathbf{C}) = \mathcal{N}(\mathbf{P}^T)$ and let $\hat{\mathbf{b}}_H$ form part of the solution to the restricted normal equations arising from $\mathbf{P}^T\mathbf{b} = \delta$. If r = rank ($\mathbf{X}$), $r*$ = rank ($\mathbf{XC}$), and $\mathbf{y} \sim N_N(\mathbf{Xb}, \sigma^2\mathbf{I}_N)$, then*

$$F = \frac{Q(\hat{\mathbf{b}}_H) - Q(\hat{\mathbf{b}})}{Q(\hat{\mathbf{b}})} \times \frac{N-r}{r-r*} \sim F_{r-r*,N-r}(\lambda_H)$$

where $\lambda_H = (\mathbf{Xb} - \mathbf{Xb}_*)^T(\mathbf{P_X} - \mathbf{P_{XC}})(\mathbf{Xb} - \mathbf{Xb}_*)/(2\sigma^2)$ and $\mathbf{b}_*$ is any particular solution to the equations $\mathbf{P}^T\mathbf{b} = \delta$, so $\mathbf{b}_* \in \mathcal{T}$. If $\mathbf{b} \in \mathcal{T}$, then $\lambda_H = 0$.

Proof: Following Results A.13 and A.15, we can write any vector $\mathbf{b} \in \mathcal{T}$ as $\mathbf{b}(z) = \mathbf{b}_* + \mathbf{Cz}$, and clearly $\mathbf{b}_* + \mathbf{Cz} \in \mathcal{T}$. Then the least squares problem over $\mathcal{T}$ can be rewritten

$$Q(\hat{\mathbf{b}}_H) = \min_{\mathbf{b}\in\mathcal{T}} \|\mathbf{y} - \mathbf{Xb}\|^2 = \min_{\mathbf{z}} \|\mathbf{y} - \mathbf{Xb}(\mathbf{z})\|^2 = \min_{\mathbf{z}} \|\mathbf{y} - \mathbf{Xb}_* - \mathbf{XCz}\|^2$$

which is a least squares problem with response $\mathbf{y} - \mathbf{Xb}_*$ and design matrix $\mathbf{XC}$. As a result

$$Q(\hat{\mathbf{b}}_H) = (\mathbf{y} - \mathbf{Xb}_*)^T(\mathbf{I} - \mathbf{P_{XC}})(\mathbf{y} - \mathbf{Xb}_*).$$

With a slight rewriting, we also have

$$Q(\hat{\mathbf{b}}) = \mathbf{y}^T(\mathbf{I} - \mathbf{P_X})\mathbf{y} = (\mathbf{y} - \mathbf{Xb}_*)^T(\mathbf{I} - \mathbf{P_X})(\mathbf{y} - \mathbf{Xb}_*)$$

since $(\mathbf{I} - \mathbf{P_X})\mathbf{X}\mathbf{b}_* = \mathbf{0}$. Taking the difference of the two sums of squares, we have

$$[Q(\hat{\mathbf{b}}_H) - Q(\hat{\mathbf{b}})] = (\mathbf{y} - \mathbf{X}\mathbf{b}_*)^T(\mathbf{P_X} - \mathbf{P_{XC}})(\mathbf{y} - \mathbf{X}\mathbf{b}_*).$$

Note as well that $(\mathbf{P_X} - \mathbf{P_{XC}})$ is symmetric, idempotent with rank $r - r*$, since $\mathcal{C}(\mathbf{XC}) \subseteq \mathcal{C}(\mathbf{X})$. Employing Cochran's Theorem, we have $[Q(\hat{\mathbf{b}}_H) - Q(\hat{\mathbf{b}})]/\sigma^2 \sim \chi^2_{r-r*}(\lambda_H)$, and independent of $Q(\hat{\mathbf{b}})/\sigma^2 \sim \chi^2_{N-r}$ so that we have the noncentral F-distribution as described. The noncentrality parameter is

$$\lambda_H = (\mathbf{Xb} - \mathbf{X}\mathbf{b}_*)^T(\mathbf{P_X} - \mathbf{P_{XC}})(\mathbf{Xb} - \mathbf{X}\mathbf{b}_*)/(2\sigma^2).$$

When $\mathbf{b} \in \mathcal{T}$, then $\mathbf{b} - \mathbf{b}_* \in \mathcal{N}(\mathbf{P}^T)$, $(\mathbf{Xb} - \mathbf{X}\mathbf{b}_*) \in \mathcal{C}(\mathbf{XC})$, and so $\lambda_H = 0$.

After the insistence in Chapter 3 on estimability, and later in Chapter 6 on testability, this result appears iconoclastic and prompts us to question our previous beliefs. Estimability and its more general form of identifiability (Section 6.7) address our ability to infer on certain linear combinations of parameters, based on the observations from an experiment. If we were to ignore estimability, then two individuals using different solutions to the normal equations could construct different estimates of the same quantity from the same experiment. Restricting to estimability leads to uniqueness of least squares estimators and their properties following from Gauss–Markov. This carries over into testable hypothesis, so that from the same experiment, different solutions to the normal equations would lead to the same test statistic and inference. However, in using the likelihood ratio approach, different solutions to the normal equations are no longer involved, and that permits us to construct such a test as above.

So what is so wrong with testing a supposedly nontestable hypothesis? The drawback is that this test may have power equal to its level far into the alternative. This test may be no better than ignoring the data and flipping a coin to decide whether to reject or not.

Example 7.6: Testing in One-Way ANOVA

Consider the one-way ANOVA model, $y_{ij} = \mu + \alpha_i + e_{ij}$ with $a = 3$. If we test the nontestable H: $\alpha_1 = \alpha_2 = \alpha_3 = 0$, then $\mathcal{C}(\mathbf{C}) = \mathcal{N}(\mathbf{P}^T) = span\{\mathbf{e}^{(2)}, \mathbf{e}^{(3)}, \mathbf{e}^{(4)}\}$ and $\mathbf{XC} = \mathbf{1}_N$. The resulting LRT is just the usual test for the hypothesis that the α's are equal.

Considering the consequence of Example 7.6, the LRT of the nontestable hypothesis would have power equal to its level for points in the alternative of the form $\mathbf{b} = (\mu, c, c, c)^T$. In Example 6.2, testing for no interaction in a two-way crossed model, writing the hypothesis as H: $\gamma_{ij} = 0$ again leads to the familiar test statistic, which would have power equal to its level to points in the alternative where $\gamma_{ij} = a + b_i + c_j$. The argument that, in principle, these points are not really in the alternative that we really meant to test can be viewed as a rationalization for sloppy work that could lead to getting the wrong degrees of freedom. But in cases such as the two-way crossed model without interaction discussed in Section 3.5, blindly testing a nontestable hypothesis can lead to power equal to its level far in an important alternative.

7.7 Summary

1. Reparameterization is the most common method for fitting the reduced model in a likelihood ratio test. The supporting theory is outlined.

2. Sequential partitioning of the design matrix leads to sequential sums of squares for conveniently testing subset models. The underlying theory is explained by applying Cochran's Theorem.

3. Orthogonal polynomials are used for testing low-order polynomial responses' one-way ANOVA models.

4. When replicate design points are available, the lack of fit of a model can be tested. This test can be viewed as testing whether a mean vector in an ANOVA model lies in a subspace.

5. The consequence of testing a nontestable hypothesis is discussed.

7.8 Exercises

7.1. * Show algebraically that the two reparameterizations of the reduced model in Example 7.1 are equivalent. One parameterization uses $y_i - x_{i2}$ as the response and $x_{i1} - x_{i2}$ as the covariate. The other parameterization uses $y_i - x_{i1}$ as the response and $x_{i2} - x_{i1}$ as the covariate.

7.2. Find the generalized inverse of $\mathbf{K}^T$ that gives the second reparameterization (in terms of b_2) for Example 7.1.

7.3. Taking Example 7.2 one step larger, with $a = 4$, leads to two equations in the hypothesis, H: $3\alpha_1 + \alpha_2 - \alpha_3 - 3\alpha_4 = 0$ and $\alpha_1 - \alpha_2 + \alpha_3 - \alpha_4 = 0$. Repeat the same analysis and find $[(\mathbf{XQ})_2 - (\mathbf{XQ})_1\mathbf{C}^{-1}\mathbf{D}]\mathbf{b}_2^*$ so that the new design matrix gives columns corresponding to an intercept and linear covariate $x_i = i$.

7.4. In Example 7.3, find the SSE and $\hat{\mathbf{b}}$ for all $2^3 - 1 = 7$ possible subset regression models.

7.5. In Example 7.3, find $\mathbf{Q}_3$, which is associated with the residuals.

7.6. In Section 7.4, show that as long as the x_i's are equally spaced, say, $x_i = a + bi$, then the $\mathbf{C}$ matrix remains unchanged.

7.7. Using a computer, or by hand, compute the contrast matrix $\mathbf{C}$ for $a = 8$.

7.8. For $\mathbf{X}$ given in (7.9) and $\mathbf{W}$ in (7.10), show that $\mathcal{C}(\mathbf{X}) \subset \mathcal{C}(\mathbf{W})$ by finding $\mathbf{B}$ such that $\mathbf{X} = \mathbf{WB}$.

7.9. Show that the lack of fit test applied to a simple linear regression model,

$$y_{ij} = \beta_0 + \beta_1 x_i + e_{ij},$$

for $i = 1, \ldots, a$; $j = 1, \ldots, n_i$, is equivalent to testing H: $\beta_2 = \ldots = \beta_{a-1} = 0$ in the model

$$y_{ij} = \beta_0 + \beta_1 x_i + \beta_2 x_i^2 + \ldots + \beta_{a-1} x_i^{a-1} + e_{ij}.$$

7.10. Show that in Example 7.6 the numerator in the F-statistic for testing H:$\mu_1 - 2\mu_2 + \mu_3 = 0$ with $\mathbf{K}^T = (2, -4, 2)^T$ is the same as $\mathbf{y}^T(\mathbf{P_Z} - \mathbf{P_X})\mathbf{y}$ from the likelihood ratio test.

7.11. In many situations, we assume that a subject responds in a linear way to the dosage level. Consider measuring a response y_{ij} to dosage level i on subject $j, j = 1, 2$ (balanced, $n_i = 2$), $i = 1, \ldots, a = 4$, leading to the familiar one-way ANOVA model:

$$y_{ij} = \mu + \alpha_i + e_{ij} \text{ (model 1).}$$

Assume, as usual, that e_{ij} are iid $N(0, \sigma^2)$. We are interested in testing the hypothesis that the response is linear in the dosage, where this reduced model might be written as

$$y_{ij} = \beta_0 + \beta_1 x_i + e_{ij} \text{ (model 2),}$$

where $x_i = i, i = 1, \ldots, 4$.

a. Write out $E(\mathbf{y}) = \mathbf{Xb}$ for the *full* model (model 1) by giving $\mathbf{X}$ and $\mathbf{b}$. Also give *rank* $(\mathbf{X})$.

b. Write out $E(\mathbf{y}) = \mathbf{Wc}$ for the *reduced* model (model 2) by giving $\mathbf{W}$ and $\mathbf{c}$. Also give *rank* $(\mathbf{W})$.

c. Show $\mathcal{C}(\mathbf{W}) \subseteq \mathcal{C}(\mathbf{X})$.

d. Let $\mathbf{v} = \begin{bmatrix} \overline{y}_{2.} - \overline{y}_{1.} \\ \overline{y}_{3.} - \overline{y}_{2.} \end{bmatrix}$. Give $E(\mathbf{v})$ and *Cov* $(\mathbf{v})$ under the reduced model 2.

e. Write the hypothesis that the mean of the response is linear in the dosage i in the form H: $\mathbf{K}^T\mathbf{b} = \mathbf{m}$ by giving $\mathbf{K}$ and $\mathbf{m}$. (Notice the hypothesis is stated in terms of the full-model parameter vector $\mathbf{b}$.)

f. Show that each component of $\mathbf{K}^T\mathbf{b}$ is estimable (so that the hypothesis is testable).

g. For the following data, compute $(\mathbf{K}^T\hat{\mathbf{b}} - \mathbf{m})^T[\mathbf{K}^T(\mathbf{X}^T\mathbf{X})^g\mathbf{K}]^{-1}(\mathbf{K}^T\hat{\mathbf{b}} - \mathbf{m})$.

h. Give the distribution of $\mathbf{y}^T(\mathbf{P_X} - \mathbf{P_W})\mathbf{y}/\sigma^2$ under both model 1 and model 2.

i. For the same data, compute *SSE* (*Full*) and *SSE* (*Reduced*), the error sums of squares under the full and reduced models.

j. Test the hypothesis that the mean response is linear at level $\alpha = 0.05$. Be sure to give the test statistic, rejection region, and conclusion.

i	1	1	2	2	3	3	4	4
j	1	2	1	2	1	2	1	2
y_{ij}	4	2	5	9	9	13	8	6

$\sum_i \sum_j (y_{ij} - \bar{y}_{i.})^2 = 20$, $\sum_i \sum_j (x_i - \bar{x})(y_{ij} - \bar{y}_{i.}) = 16$, $\sum_i \sum_j (x_i - \bar{x})^2 = 10$, $\sum_i \sum_j (y_{ij} - \bar{y}_{..})^2 = 84$, and recall $x_i = i$.

7.12. Professors Lewis Sadweather and Clark Williams are ethnobiologists working with new species. They performed an experiment that can be simplified to two factors: a treatment at three levels, and four blocks. Working under extreme hardship, some of the treatment combinations were lost due to bear attacks, and so we can consider the resulting trial in terms of a two-way crossed model with missing cells $y_{ij} = \mu + \alpha_i + \beta_j + e_{ij}$, $E(\mathbf{e}) = \mathbf{0}$, $Cov(\mathbf{e}) = \sigma^2 \mathbf{I}_N$, where **y**, **X**, and **b** are as follows:

$$\mathbf{y} = \begin{bmatrix} y_{12} \\ y_{13} \\ y_{22} \\ y_{23} \\ y_{31} \\ y_{34} \end{bmatrix}, \quad \mathbf{Xb} = \begin{bmatrix} 1 & 1 & 0 & 0 & 0 & 1 & 0 & 0 \\ 1 & 1 & 0 & 0 & 0 & 0 & 1 & 0 \\ 1 & 0 & 1 & 0 & 0 & 1 & 0 & 0 \\ 1 & 0 & 1 & 0 & 0 & 0 & 1 & 0 \\ 1 & 0 & 0 & 1 & 1 & 0 & 0 & 0 \\ 1 & 0 & 0 & 1 & 0 & 0 & 0 & 1 \end{bmatrix} \begin{bmatrix} \mu \\ \alpha_1 \\ \alpha_2 \\ \alpha_3 \\ \beta_1 \\ \beta_2 \\ \beta_3 \\ \beta_4 \end{bmatrix}$$

Note that $r = rank(\mathbf{X}) = 5$.

a. Give a set of basis vectors for $\mathcal{N}(\mathbf{X})$.

b. Professor Williams questions whether any of the following are estimable? Briefly explain.

   i. $\beta_1 - \beta_3$
   ii. $\alpha_2 - \alpha_3$
   iii. $\beta_1 - \beta_4$
   iv. $\mu + \alpha_1$

c. Professor Lewis Sadweather is interested in testing whether there is any block effect, that is, testing the hypothesis H: $\beta_1 = \beta_2 = \beta_3 = \beta_4$ versus A: not all equal. Is H a testable hypothesis? Briefly explain.

   Professor Sadweather has faced many obstacles in working in the wilderness without PROC GLM (or even a computer), and decides to go ahead and construct a test of the hypothesis H using his own instincts. He takes the differences of the pairs of observations with the same treatment:

$$z_1 = y_{12} - y_{13}$$

$$z_2 = y_{22} - y_{23}$$

$$z_3 = y_{31} - y_{34}$$

and constructs the statistic $G = \frac{(z_1+z_2)^2/2+z_3^2}{(z_1-z_2)^2}$, intending to reject H if G is too large.

d. Find $E(\mathbf{z})$ and $Cov(\mathbf{z})$ where $\mathbf{z} = (z_1, z_2, z_3)^T$.
e. Regardless of whether the hypothesis is true, $(z_1 - z_2)^2$ can be used to estimate the variance. Prove that $c_1(z_1 - z_2)^2$ has a (central) χ^2 distribution with 1 df, for the appropriate value of c_1 (and give the value of c_1), whether H or A is true.
f. Similarly, find an appropriate constant c_2, so that $c_2[(z_1 + z_2)^2/2 + z_3^2]$ has a χ^2 distribution, where the noncentrality parameter is zero when H is true.
g. Find c_3 such that $Pr(G > c_3|H) = 0.05$.
h. Find values of $\beta_* = (\beta_1, \beta_2, \beta_3, \beta_4)^T$ where H is not true, although $Pr(G > c_3|\beta_*)$ is still 0.05.

7.13. Following the design in Exercise 7.13 above, derive the LRT test for a non-testable hypothesis via Result 7.1. Is your test the same as Professor Sadweather's G?

Chapter 8

Variance Components and Mixed Models

8.1 Introduction

This discussion of variance components begins with the usual one-way ANOVA model. But before launching into that discourse, the reader should notice that this problem looks very easy from a purely theoretical viewpoint. There are just three parameters to estimate: an overall mean, a variance parameter for groups/treatments/varieties, and the usual noise variance. This simplicity belies the fact that variance component problems are indeed quite difficult. As the analysis of the one-way model will indicate, while the balanced case remains somewhat straightforward, the unbalanced case becomes surprisingly complicated, permitting only limited analysis before deferring to the general case. Moreover, the focus of our attention has changed from a mean or linear parameter to a variance—and variances are so much harder to estimate than means. As a result, the presentation here consists of much of the analysis that can be done with the one-way and two-way models, before giving up and deferring to the general case in Section 8.4. The important special case of the split plot is undertaken in Section 8.5, followed by estimation of the random effects in Section 8.6.

8.2 Variance Components: One Way

Consider the one-way ANOVA model, $y_{ij} = \mu + \alpha_i + e_{ij}$, where $i = 1, \ldots, a$; $j = 1, \ldots, n_i$, with $\sum n_i = N$; and we will assume e_{ij} are iid $N(0, \sigma^2)$. The ANOVA table looks like:

Source	**df**	**Projection**	**Sum of Squares**	**Noncentrality**
Mean	1	$\mathbf{P}_1$	$SSM = \mathbf{y}^T\mathbf{P}_1\mathbf{y} = N\overline{y}^2$	ϕ_M
Treatment	$a-1$	$\mathbf{P}_\mathbf{X} - \mathbf{P}_1$	$SSA = \mathbf{y}^T(\mathbf{P}_\mathbf{X} - \mathbf{P}_1)\mathbf{y} = \sum_{i=1}^a n_i(\overline{y}_{i.} - \overline{y}_{..})^2$	ϕ_a
Error	$N-a$	$\mathbf{I} - \mathbf{P}_\mathbf{X}$	$SSE = \mathbf{y}^T(\mathbf{I} - \mathbf{P}_\mathbf{X})\mathbf{y} = \sum_{i=1}^a \sum_{j=1}^{n_i}(\overline{y}_{ij} - \overline{y}_{i.})^2$	0

Note that the three sums of squares are mutually independent, with the first two, when divided by σ^2, noncentral χ^2 with the degrees of freedom and noncentralities as given in the table. In particular, these are

$$\begin{aligned} 2\sigma^2\phi_M = E(\mathbf{y})^T\mathbf{P}_1E(\mathbf{y}) &= \frac{1}{N}\left[\sum\sum E(y_{ij})\right]^2 \\ &= \frac{1}{N}\left[N\mu + \sum n_i\alpha_i\right]^2 = N(\mu+\overline{\alpha})^2 \end{aligned}$$

and

$$2\sigma^2\phi_a = E(\mathbf{y})^T(\mathbf{P_X}-\mathbf{P}_1)E(\mathbf{y}) = \sum n_i(\mu+\alpha_i)^2 - N(\mu+\overline{\alpha})^2 = \sum n_i(\alpha_i-\overline{\alpha})^2$$

where $\overline{\alpha} = \sum n_i\alpha_i/N$. The last sum of squares SSE, scaled by σ^2 of course, has a central chi-square distribution, as will SSA if there are no treatment effects, that is, the α_i's are equal.

All of these distributions are conditional on fixed α_i. But what if these effects α_i are random, arising from sampling groups from a population? For example, school effects α_i sampled from a population of schools with individual students j sampled within school i. The common distributional assumptions are then

$$\begin{aligned} \alpha_i &\sim N(0,\sigma_a^2) \\ e_{ij} &\sim N(0,\sigma^2) \end{aligned}$$

with all mutually independent. The consequence of moving from fixed effects to random is that (unconditionally) SSM, SSA, and SSE may no longer be mutually independent (see Exercise 5.21). We are interested in obtaining all of the information about these three quantities that we can, perhaps being satisfied with just means if that's the best we can do. We're mostly interested in estimating σ_a^2 and σ^2, the *variance components*.

One result is rather easy, that is, since the distribution of SSE conditional on the α_i's does not depend on the α_i's, SSE is independent of the α_i's, and hence its distribution (when divided by σ^2) remains central chi-square.

Another route is to try to get the mgf for SSA from our conditional result, and then unconditional on α_i's. Let $U = SSA$, then $U/\sigma^2 \sim \chi^2_{a-1}(\phi_a = \frac{1}{2}\sum n_i(\alpha_i-\overline{\alpha})^2/\sigma^2)$, or in terms of its mgf,

$$E(e^{Ut/\sigma^2}|\alpha_i' s) = (1-2t)^{-(a-1)/2}e^{2t\phi_a/(1-2t)} \tag{8.1}$$

so the mgf of U with argument $s = t/\sigma^2$ is

$$E(e^{Us}|\alpha_i' s) = (1-2s\sigma^2)^{-(a-1)/2}e^{s\sigma^2/(1-2s\sigma^2)\times 2\phi_a} \tag{8.2}$$

so that the effect of the α_i's is isolated in ϕ_a. So can we get the distribution of, say, $2\phi_a$? And does $2\phi_a$ have a chi-square distribution? In general, the dissappointing answer is negative. But let's see how far some simple analysis can take us.

The key to the noncentrality parameter ϕ_a is

$$\sum n_i(\alpha_i-\overline{\alpha})^2 = \sum n_i\alpha_i^2 - N\overline{\alpha}^2;$$

so since α_i are $N(0, \sigma_a^2)$, can we show that $\sum n_i(\alpha_i - \overline{\alpha})^2/\sigma_a^2 \sim \chi^2_{a-1}$? It looks promising! Stack the α_i's in a vector $\alpha \sim N_a(\mathbf{0}, \sigma_a^2 \mathbf{I}_a)$, then examine

$$2\phi_a \sigma^2/N = \frac{1}{n}\sum n_i \alpha_i^2 - \overline{\alpha}^2 = \alpha^T \frac{1}{n}\mathbf{D}\alpha - \frac{1}{N^2}\alpha^T \mathbf{D}\mathbf{1}\mathbf{1}^T\mathbf{D}\alpha$$
$$= \alpha^T \left[\frac{1}{N}\mathbf{D} - \frac{1}{N^2}\mathbf{D}\mathbf{1}\mathbf{1}^T\mathbf{D}\right]\alpha$$

where $\mathbf{D} = diag\,(n_1, n_2, \ldots, n_a)$ and noting that $\overline{\alpha} = N^{-1}\mathbf{1}^T\mathbf{D}\alpha$. Now the expression above (properly scaled) would have a chi-square distribution if the matrix in brackets [.] (or a proper scaling) were idempotent. Computing its square leads to

$$\left[\frac{1}{n}\mathbf{D} - \frac{1}{N^2}\mathbf{D}\mathbf{1}\mathbf{1}^T\mathbf{D}\right]^2 = \frac{1}{N^2}\mathbf{D}^2 - \frac{1}{N^3}\mathbf{D}^2\mathbf{1}\mathbf{1}^T\mathbf{D} - \frac{1}{N^3}\mathbf{D}\mathbf{1}\mathbf{1}^T\mathbf{D}^2 + \frac{1}{N^4}\mathbf{D}\mathbf{1}\mathbf{1}^T\mathbf{D}^2\mathbf{1}\mathbf{1}^T\mathbf{D},$$

and further continuation appears futile. In the general case, this matrix will not be idempotent, no matter what scaling we use. However, we can get an idempotent matrix in the balanced case, that is, when $n_i \equiv n$. In the balanced case, then, $\mathbf{D} = n\mathbf{I}$, and $N = na$, so

$$\frac{1}{N}\mathbf{D} - \frac{1}{N^2}\mathbf{D}\mathbf{1}\mathbf{1}^T\mathbf{D} = \frac{n}{na}\mathbf{I} - \frac{n^2}{n^2a^2}\mathbf{I}\mathbf{1}\mathbf{1}^T\mathbf{I} = \frac{1}{a}\mathbf{I} - \frac{1}{a^2}\mathbf{1}\mathbf{1}^T = \frac{1}{a}\left[\mathbf{I} - \mathbf{P}_1\right]$$

which is clearly a scaled idempotent matrix. Putting together all of the right scale factors, we have

$$\frac{2\phi_a}{n} \times \frac{\sigma^2}{\sigma_a^2} = \sum(\alpha_i - \overline{\alpha})^2/\sigma_a^2 \sim \chi^2_{a-1}. \tag{8.3}$$

This result can be pursued further. Using the Expression (8.2) from above, the (unconditional) mgf for U is

$$E\left(E\left(e^{Us}|\alpha_i's\right)\right) = (1 - 2s\sigma^2)^{-(a-1)/2} E(e^s \sigma^2/(1 - 2s\sigma^2) \times 2\phi_a) \tag{8.4}$$

whose second piece is the mgf of $\frac{2\phi_a}{n} \times \frac{\sigma^2}{\sigma_a^2}$, a central χ^2, evaluated at $t = n\sigma_a^2 s/(1 - s\sigma^2)$, leading to

$$E(e^{Us}) = (1 - 2s\sigma^2)^{-(a-1)/2}\left(1 - 2n\sigma_a^2 s/(1 - 2s\sigma^2)\right)^{-(a-1)/2} \tag{8.5}$$
$$= \left((1 - 2s\sigma^2)\left(1 - \frac{2n\sigma_a^2 s}{(1 - 2s\sigma^2)}\right)\right)^{-(a-1)/2} = \left(1 - 2s\left(\sigma^2 + n\sigma_a^2\right)\right)^{-(a-1)/2} \tag{8.6}$$

which is the mgf of a χ^2_{a-1} random variable, scaled by $(\sigma^2 + n\sigma_a^2)$. The result is that unconditionally, $SSA/(\sigma^2 + n\sigma_a^2) \sim \chi^2_{a-1}$.

An alternative route to this same result starts with the unconditional distribution of the group means:

$$\overline{y}_{i.}|\alpha_i's \sim N(\mu + \alpha_i, \sigma^2/n). \tag{8.7}$$

Then unconditionally (and jointly independent),

$$\bar{y}_{i.} \sim N\left(\mu, \sigma_a^2 + \sigma^2/n\right) \tag{8.8}$$

or

$$\sqrt{n}\bar{y}_{i.} \sim N\left(\mu, \sigma^2 + n\sigma_a^2\right),$$

so rescaling by the variance, the sum of squares around the mean is χ^2_{a-1}, hence

$$SSA/\left(\sigma^2 + n\sigma_a^2\right) = n\sum(\bar{y}_{i.} - \bar{y}_{..})^2/\left(\sigma^2 + n\sigma_a^2\right) \sim \chi^2_{a-1}.$$

So in the *balanced* case, $n_i \equiv n$, when we do the F-test with random α_i's, under the alternative A: $\sigma_a^2 > 0$ we have

$$F = \frac{\frac{SSA}{(\sigma^2+n\sigma_a^2)}/(a-1)}{\frac{SSE}{\sigma^2/[a(n-1)]}} = \frac{\frac{SSA}{(a-1)}}{\frac{SSE}{a(n-1)}} \times \frac{\sigma^2}{(\sigma^2 + n\sigma_a^2)} \sim F_{a-1,a(n-1)}. \tag{8.9}$$

Under the alternative A, F is stochastically larger (we had to multiply by the ratio that is less than 1 to get the F-distribution). When the treatment effects were fixed, the distribution under the alternative was a noncentral F-distribution; when the effects are random, the distribution under the alternative is a scaled F-distribution. Again for emphasis, in the case of random effects α_i, we have the same test statistic and rejection region, but a different distribution under the alternative, but only available in the *balanced* case. Figures 8.1 and 8.2 show densities of the scaled F-distribution where the scaling parameters were chosen to match the means of the densities shown in Figures 5.3 and 5.4. Only a close comparison reveals differences between Figures 8.1 and 5.3, and Figures 8.2 and 5.4.

In the unbalanced case, we can use the result from Equation (8.2) to obtain the expectations at least. We will find it convenient to express expectations in terms of *expected mean squares*, by dividing the sums of squares by their degrees of freedom. Now using $\alpha_i \sim N(0, \sigma_a^2)$ and $\bar{\alpha} \sim N(0, \sigma_a^2(\sum n_i^2)/N^2)$, we get

$$\begin{aligned}
E(SSA)/(a-1) &= E(E(SSA|\alpha_i' s))/(a-1) \\
&= E((a-1)\sigma^2 + \sum n_i(\alpha_i - \bar{\alpha})^2)/(a-1) \\
&= \sigma^2 + E\left[\sum n_i\alpha_i^2 - N\bar{\alpha}^2\right]/(a-1) \\
&= \sigma^2 + \sigma_a^2\left[N - \sum n_i^2/N\right]/(a-1),
\end{aligned}$$

and, of course, $E(SSE)/(N-a) = \sigma^2$.

One route for estimation employs these expectations, and just solves for the estimators in terms of these expectations—known as *ANOVA estimators*:

$$\hat{\sigma}^2 = SSE/(N-a) \tag{8.10}$$

$$\hat{\sigma}_a^2 = \frac{SSA - \frac{(a-1)}{(N-a)}SSE}{\left[N - \sum n_i^2/N\right]}. \tag{8.11}$$

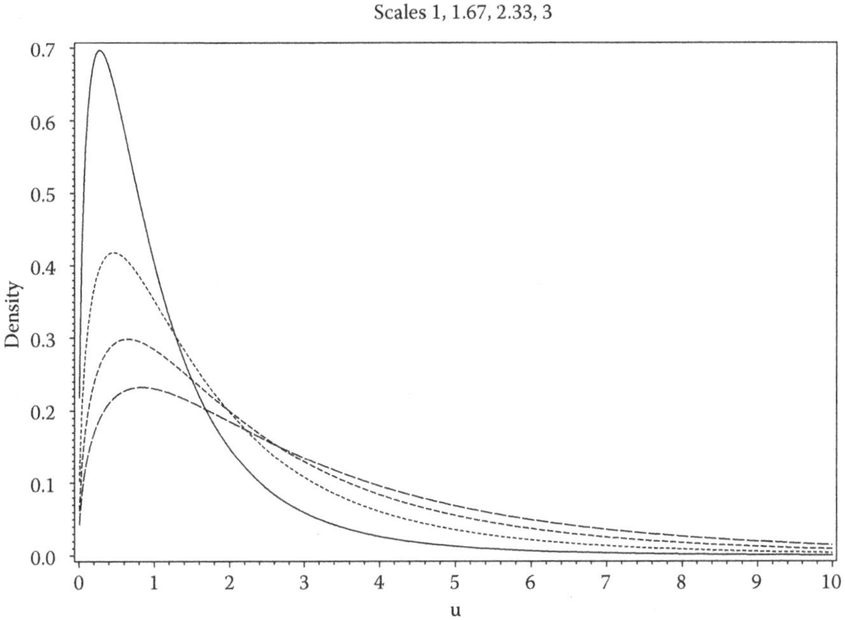

Figure 8.1: Scaled F densities with 3 and 10 degrees of freedom. Scaling chosen to match the means of noncentral F random variables with same degrees of freedom and noncentralities 0, 1, 2, 3 as in Figure 5.3.

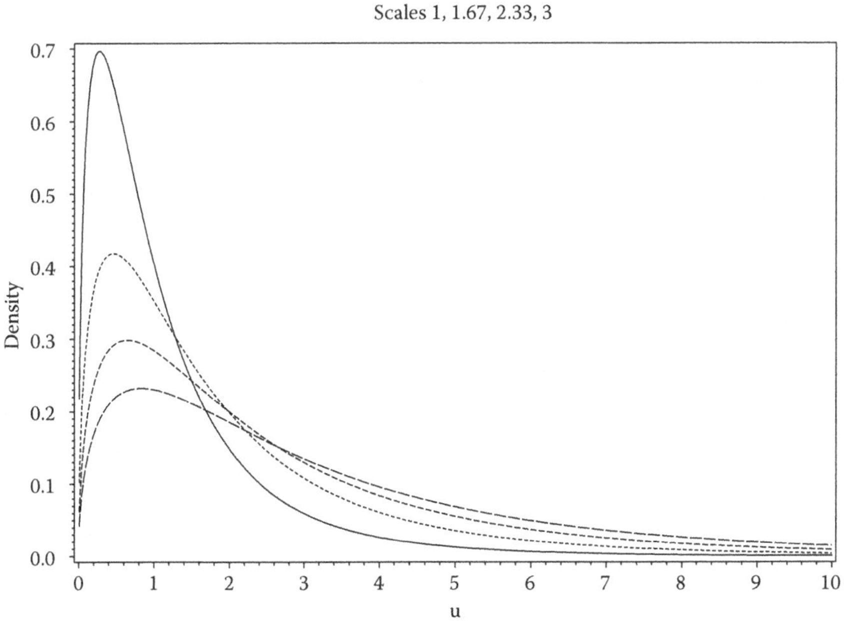

Figure 8.2: Scaled F densities with 6 and 10 degrees of freedom. Scaling chosen to match the means of noncentral F random variables with same degrees of freedom and noncentralities 0, 1, 2, 3 as in Figure 5.4.

These estimators offer simplicity and unbiasedness, but not much more. A common fault is that the estimator of σ_a^2, a positive quantity, can be negative (see Exercise 8.1). And even more unsettling, if we adjust $\hat{\sigma}_a^2$ in this case by setting it to zero instead of a negative value, this estimator loses the property of unbiasedness.

8.3 Variance Components: Two-Way Mixed ANOVA

Consider the next most complicated case, a two-way model with interaction $y_{ijk} = \mu + \alpha_i + \beta_j + \gamma_{ij} + e_{ijk}$, where $i = 1, \ldots, a$; $j = 1, \ldots, b$; $k = 1, \ldots, n$, and first we will deal with the case with all random factors:

- α_i's random: $\alpha_i \sim N(0, \sigma_a^2)$
- β_j's random: $\beta_j \sim N(0, \sigma_b^2)$
- γ_{ij}'s random: $\gamma_{ij} \sim N(0, \sigma_c^2)$
- $e_{ijk} \sim N(0, \sigma^2)$

Given the difficulties without balance in the one-way problem, consider all departures from the balanced case deferred to the general case in Section 8.3. A view of the usual ANOVA table suggests an algebraic route, although some comments on the geometry will be included.

Source	**df**	**SS**
Mean	1	$SSM = N\overline{y}_{...}^2$
Factor A	$a-1$	$SSA = bn\sum_i(\overline{y}_{i..} - \overline{y}_{...})^2$
Factor B	$b-1$	$SSB = an\sum_i(\overline{y}_{.j.} - \overline{y}_{...})^2$
Interaction	$(a-1)(b-1)$	$SSC = n\sum_i\sum_j(\overline{y}_{ij.} - \overline{y}_{i..} - \overline{y}_{.j.} + \overline{y}_{...})^2$
Error	$ab(n-1)$	$SSE = \sum_i\sum_j\sum_k(y_{ijk} - \overline{y}_{ij.})^2$

Examining each quantity that is being squared and summed is enlightening; first construct

$$\begin{aligned}
\overline{y}_{ij.} &= \mu + \alpha_i + \beta_j + \gamma_{ij} + \overline{e}_{ij.} \\
\overline{y}_{i..} &= \mu + \alpha_i + \overline{\beta}_{.} + \overline{\gamma}_{i.} + \overline{e}_{i..} \\
\overline{y}_{.j.} &= \mu + \overline{\alpha}_{.} + \beta_j + \overline{\gamma}_{.j} + \overline{e}_{.j.} \\
\overline{y}_{...} &= \mu + \overline{\alpha}_{.} + \overline{\beta}_{.} + \overline{\gamma}_{..} + \overline{e}_{...},
\end{aligned}$$

then for each quantity we find

$$\begin{aligned}
\overline{y}_{ijk} - \overline{y}_{ij.} &= e_{ijk} - \overline{e}_{ij.} \\
\overline{y}_{ij.} - \overline{y}_{i..} - \overline{y}_{.j.} + \overline{y}_{...} &= (\gamma_{ij} - \overline{\gamma}_{i.} - \overline{\gamma}_{.j} + \overline{\gamma}_{..}) + (\overline{e}_{ij.} - \overline{e}_{i..} - \overline{e}_{.j.} + \overline{e}_{...}) \\
\overline{y}_{i..} - \overline{y}_{...} &= (\alpha_i - \overline{\alpha}_.) + (\overline{\gamma}_{i.} - \overline{\gamma}_{..}) + (\overline{e}_{i..} - \overline{e}_{...}) \\
\overline{y}_{.j.} - \overline{y}_{...} &= (\beta_j - \overline{\beta}_.) + (\overline{\gamma}_{.j} - \overline{\gamma}_{..}) + (\overline{e}_{.j.} - \overline{e}_{...}).
\end{aligned}$$

From here we can compute the expected mean squares

$$\begin{aligned}
E(SSE)/\,[ab(n-1)] &= \sigma^2 \\
E(SSC)/\,[(a-1)(b-1)] &= n\sigma_c^2 + \sigma^2 \\
E(SSA)/\,[(a-1)] &= bn\sigma_a^2 + n\sigma_c^2 + \sigma^2 \\
E(SSB)/\,[(b-1)] &= an\sigma_b^2 + n\sigma_c^2 + \sigma^2.
\end{aligned}$$

Moreover, we can get the distribution of each as an independent central χ^2:

$$\begin{aligned}
SSE/\sigma^2 &\sim \chi^2_{ab(n-1)} \\
SSC/\left[n\sigma_c^2 + \sigma^2\right] &\sim \chi^2_{(a-1)(b-1)} \\
SSA/\left[bn\sigma_a^2 + n\sigma_c^2 + \sigma^2\right] &\sim \chi^2_{(a-1)} \\
SSB/\left[an\sigma_b^2 + n\sigma_c^2 + \sigma^2\right] &\sim \chi^2_{(b-1)}.
\end{aligned}$$

Establishing these last distributions is not so difficult, as each is the sum of squares of components (with a mean subtracted) that are jointly multivariate normal. Establishing independence is more difficult (Exercise 8.3).

The usual route for the two-way model is the ANOVA route: equate the usual sums of squares with their expectations and solve for the unknown parameters—the components of variance: $\sigma_a^2, \sigma_b^2, \sigma_c^2$, and σ^2. In this case we have

$$\begin{aligned}
\hat{\sigma}^2 &= SSE/ab(n-1) \\
\hat{\sigma}_c^2 &= \{SSC/\,[(a-1)(b-1)] - \hat{\sigma}^2\}/n \\
\hat{\sigma}_a^2 &= \{SSA/(a-1) - SSC/\,[(a-1)(b-1)]\}/(bn) \\
\hat{\sigma}_b^2 &= \{SSB/(b-1) - SSC/\,[(a-1)(b-1)]\}/(an).
\end{aligned} \tag{8.12}$$

These estimators are unbiased, and their variances can be computed using moments of χ^2 random variables. See Exercise 8.11.

Now the more difficult case is the mixed model, where β_j's are fixed, while the α_i's and γ_{ij}'s are random, satisfying the same distributional assumptions as before.

The expectations of the usual sums of squares can be computed:

$$\begin{aligned} E(SSE)/[ab(n-1)] &= \sigma^2 \\ E(SSC)/[(a-1)(b-1)] &= n\sigma_c^2 + \sigma^2 \\ E(SSA)/[(a-1)] &= bn\sigma_a^2 + n\sigma_c^2 + \sigma^2 \\ E(SSB)/[(b-1)] &= an\sum_j (\beta_j - \overline{\beta}_.)^2/(b-1) + n\sigma_c^2 + \sigma^2, \end{aligned} \tag{8.13}$$

and following the same route as before, estimates of the variance components can be computed by solving these equations with the observed sums of squares in place of their expectations. Note that if we wanted to test the hypothesis $H : \sigma_a^2 = 0$, then we would use the F-statistic

$$F = \frac{SSA/(a-1)}{SSC/((a-1)(b-1))},$$

whose distribution under the alternative would be an F-distribution scaled by the ratio of mean squares

$$\frac{E(SSA)/[(a-1)]}{E(SSC)/[(a-1)(b-1)]} = \frac{bn\sigma_a^2 + n\sigma_c^2 + \sigma^2}{n\sigma_c^2 + \sigma^2} = 1 + \frac{bn\sigma_a^2}{n\sigma_c^2 + \sigma^2}.$$

A variant on this way of thinking has substantial, but not overwhelming, support in the statistical community. Some folks assume that the interactions sum to zero over the fixed subscript, that is, $\sum_j \gamma_{ij} = 0$ for all $i = 1, \ldots, a$. This assumption is motivated by the following interpretation of the cell means parameter μ_{ij} in the two-way model:

$$\begin{aligned} y_{ijk} &= \mu_{ij} + e_{ijk} \\ \mu_{ij} &= \overline{\mu}_{..} + (\overline{\mu}_{i.} - \mu_{..}) + (\overline{\mu}_{.j} - \overline{\mu}_{..}) + (\mu_{ij} - \overline{\mu}_{i.} - \overline{\mu}_{.j} + \overline{\mu}_{..}) \\ &= \mu + \alpha_i + \beta_j + \gamma_{ij}. \end{aligned}$$

When we say the α_i's are random, we are saying that $\alpha_1, \ldots, \alpha_a$ are a random sample from an infinite population. When we say the β_j's are fixed, we have all levels of the B-effect in the experiments, and hence β_j can be considered at the deviation from the mean and thus $\sum_j \beta_j = 0$. Similarly, $\sum_j \gamma_{ij} = 0$ since we are summing over all B-effects. We can incorporate this condition, $\sum_j \gamma_{ij} = 0$, by writing $\tau_{ij} = v_{ij} - \overline{v}_{i.}$ where v_{ij} are iid $N(0, \sigma_v^2)$. Then we get the following:

$$\overline{\gamma}_{i.} = \overline{v}_{..}\overline{\gamma}_{.j} - \overline{\gamma}_{..} = \overline{v}_{.j} - \overline{v}_{..}\gamma_{ij} - \overline{\gamma}_{i.} - \overline{\gamma}_{.j} + \overline{\gamma}_{..} = v_{ij} - \overline{v}_{i.} - \overline{v}_{.j} + \overline{v}_{..}$$

We get different expected mean squares as a result:

$$\begin{aligned} E(SSE)/[ab(n-1)] &= \sigma^2 \\ E(SSC)/[(a-1)(b-1)] &= n\sigma_v^2 + \sigma^2 \\ E(SSA)/[(a-1)] &= bn\sigma_a^2 + \sigma^2 \\ E(SSB)/[(a-1)(b-1)] &= an\sum_j (\beta_j - \overline{\beta}_.)^2/(b-1) + n\sigma_v^2 + \sigma^2. \end{aligned} \tag{8.14}$$

Here the relationship between γ's and υ's is $\sigma_\upsilon^2 = \frac{b}{b-1}\sigma_c^2$. With this different formulation, if we again wanted to test the hypothesis $H : \sigma_a^2 = 0$ we would now use the F-statistic

$$F = \frac{SSA/(a-1)}{SSE/(ab(n-1))},$$

whose distribution under the alternative would be an F-distribution scaled by the ratio of mean squares:

$$\frac{E(SSA)/\,[(a-1)]}{E(SSE)/\,[ab(n-1)]} = \frac{bn\sigma_a^2 + \sigma^2}{\sigma^2} = 1 + \frac{bn\sigma_a^2}{\sigma^2}.$$

These two approaches to parameterizing the two-way mixed-model interpretations lead to two different interpretations of the parameters and to two different ways of testing what appears to be a simple hypothesis. Hocking ([22], Section 10.4) explains the differences as differences in interpretation of the parameters. Voss [48] provides a common framework for comparison and an excellent entrance to this discussion.

8.4 Variance Components: General Case

For all but the special cases presented so far, that is, unbalanced cases and most mixed models, the problem has to be treated as a general case. The general form for the mixed model is

$$\mathbf{y} = \mathbf{X}\mathbf{b} + \mathbf{Z}\mathbf{u} + \mathbf{e} \tag{8.15}$$

where $\mathbf{X}$ is an $N \times p$ design matrix for fixed effects $\mathbf{b}$, $\mathbf{Z}$ is an $N \times q$ design matrix for random effects $\mathbf{u}$, and $\mathbf{e}$ is the usual additive error. The most general assumptions are $\mathbf{e} \sim N_n(\mathbf{0}, \mathbf{R})$ and, independently, $\mathbf{u} \sim N_q(\mathbf{0}, \mathbf{G})$.

For the sake of simplicity, let's assume that the fixed-effects design matrix $\mathbf{X}$ has full-column rank; the reader should be able to make any further adjustments to handle the non-full-rank case that we are more accustomed to. The more common situation has $\mathbf{Z}$ partitioned,

$$\mathbf{Z} = [\mathbf{Z}_1 \,|\, \mathbf{Z}_2 \,|\, \ldots \,|\, \mathbf{Z}_m],$$

where each piece $\mathbf{Z}_i$ is $N \times q_i$ corresponding to a random factor, with $\mathbf{u}$ partitioned in the corresponding vertical manner, so that the model can be rewritten as

$$\mathbf{y} = \mathbf{X}\mathbf{b} + \mathbf{Z}_1\mathbf{u}_1 + \mathbf{Z}_2\mathbf{u}_2 + \ldots + \mathbf{Z}_m\mathbf{u}_m + \mathbf{e}. \tag{8.16}$$

In this more common case, the distributional assumptions simplify to $\mathbf{e} \sim N_n(\mathbf{0}, \sigma^2\mathbf{I}_n)$ and $\mathbf{u}_j \sim N_{q_j}(\mathbf{0}, \sigma_j^2\mathbf{I}_{q_j})$, with each random effect independent. The covariance matrix of $\mathbf{y}$ then takes the form

$$\mathbf{V} = \sigma^2\mathbf{I}_n + \sum_{j=1}^{m} \sigma_j^2\mathbf{Z}_j\mathbf{Z}_j^T = \mathbf{V}(\theta) \tag{8.17}$$

to emphasize that $\mathbf{V}$ is a function of the variance components $\theta^T = (\sigma^2, \sigma_1^2, \ldots, \sigma_m^2)$.

Example 8.1: Two-Way Mixed Model
Consider the two-way crossed, mixed model with no interaction, $y_{ijk} = \mu + \alpha_i + \beta_j + e_{ijk}$, with $i = 1, \ldots, a$; $j = 1, \ldots, b$; $k = 1, \ldots, n_{ij}$, where α_i's are fixed and β_j's random. For the special case of $a = 2, b = 4$, and $n_{ij} = 2$, we have

$$\mathbf{y} = \begin{bmatrix} y_{111} \\ y_{112} \\ y_{121} \\ y_{122} \\ y_{131} \\ y_{132} \\ y_{141} \\ y_{142} \\ y_{211} \\ y_{212} \\ y_{221} \\ y_{222} \\ y_{231} \\ y_{232} \\ y_{241} \\ y_{242} \end{bmatrix}, \quad \mathbf{Xb} = \begin{bmatrix} 1 & 1 & 0 \\ 1 & 1 & 0 \\ 1 & 1 & 0 \\ 1 & 1 & 0 \\ 1 & 1 & 0 \\ 1 & 1 & 0 \\ 1 & 1 & 0 \\ 1 & 1 & 0 \\ 1 & 0 & 1 \\ 1 & 0 & 1 \\ 1 & 0 & 1 \\ 1 & 0 & 1 \\ 1 & 0 & 1 \\ 1 & 0 & 1 \\ 1 & 0 & 1 \\ 1 & 0 & 1 \end{bmatrix}, \begin{bmatrix} \mu \\ \alpha_1 \\ \alpha_2 \end{bmatrix}, \quad \mathbf{Zu} = \begin{bmatrix} 1 & 0 & 0 & 0 \\ 1 & 0 & 0 & 0 \\ 0 & 1 & 0 & 0 \\ 0 & 1 & 0 & 0 \\ 0 & 0 & 1 & 0 \\ 0 & 0 & 1 & 0 \\ 0 & 0 & 0 & 1 \\ 0 & 0 & 0 & 1 \\ 1 & 0 & 0 & 0 \\ 1 & 0 & 0 & 0 \\ 0 & 1 & 0 & 0 \\ 0 & 1 & 0 & 0 \\ 0 & 0 & 1 & 0 \\ 0 & 0 & 1 & 0 \\ 0 & 0 & 0 & 1 \\ 0 & 0 & 0 & 1 \end{bmatrix} \begin{bmatrix} v_1 \\ v_2 \\ v_3 \\ v_4 \end{bmatrix}$$

and $Cov\,(\mathbf{y}) = \sigma^2\mathbf{I}_n + \mathbf{Z}(\sigma_\beta^2\mathbf{I}_b)\mathbf{Z}^T = \mathbf{V}(\theta) = \mathbf{V}(\sigma^2, \sigma_\beta^2)$.

8.4.1 Maximum Likelihood

The first general approach to be discussed is the usual maximum likelihood approach, where the parameters are the fixed effects $\mathbf{b}$, and the variance components θ. The log-likelihood for the data then is

$$\ell_N(\mathbf{b}, \theta) = \text{constant} - \frac{1}{2}\log|\mathbf{V}| - \frac{1}{2}(\mathbf{y} - \mathbf{Xb})^T\mathbf{V}^{-1}(\mathbf{y} - \mathbf{Xb}). \tag{8.18}$$

Maximizing with respect to $\mathbf{b}$ employs the generalized least squares estimator $\tilde{\mathbf{b}}$,

$$\tilde{\mathbf{b}} = (\mathbf{X}^T\mathbf{V}^{-1}\mathbf{X})^{-1}\mathbf{X}^T\mathbf{V}^{-1}\mathbf{y},$$

and the concentrated or profile likelihood function is

$$\ell_*(\theta) = \ell_N(\tilde{\mathbf{b}}, \theta) = \text{constant} - \frac{1}{2}\log|\mathbf{V}| - \frac{1}{2}(\mathbf{y} - \mathbf{X}\tilde{\mathbf{b}})^T\mathbf{V}^{-1}(\mathbf{y} - \mathbf{X}\tilde{\mathbf{b}}). \tag{8.19}$$

The maximum likelihood approach proceeds by numerically maximizing $\ell_*(\theta)$, whose further discussion is beyond the scope of this course. (See Tobias, Wolfinger, and Sall [49] for computational details underlying SAS's PROC MIXED.)

The main appeal of both ML and REML estimates of variance components is that they can be restricted to the proper parameter space; that is, we cannot have negative estimates of positive quantities. Their computation is a bit more complicated, and the only hope for a standard error must rely on asymptotics, whose assumptions some argue are specious at best. One need only take a small step toward a Bayesian route, but this approach using Gibbs' sampling for computation is fraught with problems (see Hobert and Casella [21]) .

Example 8.2: ML for Unbalanced One-Way Problem
Consider our usual one-way model $y_{ij} = \mu + \alpha_i + e_{ij}$, where $i = 1, \ldots, a$; $j = 1, \ldots, n_i$, with $\sum_{i=1}^{a} n_i = N$. Here with random effects α_i, we have α_i iid $N(0, \sigma_a^2)$ and the errors e_{ij} iid $N(0, \sigma^2)$. Here the fixed effect design matrix is $\mathbf{X} = \mathbf{1}_N$ and the random effect design matrix is

$$\mathbf{Z} = \begin{bmatrix} \mathbf{1}_{n_1} & \mathbf{0} & \mathbf{0} & \ldots & \mathbf{0} \\ \mathbf{0} & \mathbf{1}_{n_2} & \mathbf{0} & \ldots & \mathbf{0} \\ \mathbf{0} & \mathbf{0} & \mathbf{1}_{n_3} & \ldots & \mathbf{0} \\ \ldots & \ldots & \ldots & \ldots & \ldots \\ \mathbf{0} & \mathbf{0} & \ldots & \mathbf{0} & \mathbf{1}_{n_a} \end{bmatrix}$$

so that

$$\mathbf{V} = \begin{bmatrix} \mathbf{V}_1 & \mathbf{0} & \mathbf{0} & \ldots & \mathbf{0} \\ \mathbf{0} & \mathbf{V}_2 & \mathbf{0} & \ldots & \mathbf{0} \\ \mathbf{0} & \mathbf{0} & \mathbf{V}_3 & \ldots & \mathbf{0} \\ \ldots & \ldots & \ldots & \ldots & \ldots \\ \mathbf{0} & \mathbf{0} & \ldots & \mathbf{0} & \mathbf{V}_a \end{bmatrix}$$

where $\mathbf{V}_i = \sigma^2 \mathbf{I}_{n_i} + \sigma_a^2 \mathbf{1}_{n_i} \mathbf{1}_{n_i}^T$. Using some exercises from the appendix, we have

$$|\mathbf{V}| = \prod_{i=1}^{a} |\mathbf{V}_i| = (\sigma^2)^N \prod_{i=1}^{a} |\mathbf{I}_{n_i} + \rho \mathbf{1}_{n_i} \mathbf{1}_{n_i}^T| = (\sigma^2)^N \prod_{i=1}^{a} (1 + \rho n_i)$$
$$= (\sigma^2)^{N-a} \prod_{i=1}^{a} \left(\sigma^2 + n_i \sigma_a^2\right)$$

and

$$(\mathbf{V}_i)^{-1} = \sigma^{-2} (\mathbf{I}_{n_i} + \rho \mathbf{1}_{n_i} \mathbf{1}_{n_i}^T)^{-1} = \sigma^{-2} \left(\mathbf{I}_{n_i} - \frac{\sigma_a^2}{\sigma^2 + n_i \sigma_a^2} \mathbf{1}_{n_i} \mathbf{1}_{n_i}^T \right)$$

using $\rho = \sigma_a^2 / \sigma^2$. After considerable algebra, the expressions become impenetrable, although the likelihood equations resemble the form for the ANOVA estimators below. See Exercise 8.4.

8.4.2 Restricted Maximum Likelihood (REML)

One of the annoying things about maximum likelihood is its estimation of variances does not take into account the loss in degrees of freedom for estimating the mean, or any fixed effects (see Exercises 8.6–8.8). This is one of the motivations behind REML, which uses statistics that are unaffected by the fixed effects: $(\mathbf{I}-\mathbf{P_X})\mathbf{y}$. Now to employ this principle in constructing a likelihood, since this vector has a singular normal distribution, we find a matrix $\mathbf{A}$ (its shape is $N \times (N-r)$) with orthonormal columns ($\mathbf{A}^T\mathbf{A} = \mathbf{I}_{N-r}$) such that $\mathbf{I}-\mathbf{P_X} = \mathbf{A}\mathbf{A}^T$ and seek the distribution of $\mathbf{A}^T\mathbf{y} \equiv \mathbf{w}$, which is $\mathbf{w} \sim N_{N-r}(\mathbf{0}, \mathbf{A}^T\mathbf{V}\mathbf{A})$. Now $\mathbf{w}$ has a density, so the likelihood based on only $\mathbf{w}$ takes the form

$$L_R(\theta) = (2\pi)^{-(N-r)/2}|\mathbf{A}^T\mathbf{V}\mathbf{A}|^{-1/2} exp\left\{-\frac{1}{2}\mathbf{w}^T(\mathbf{A}^T\mathbf{V}\mathbf{A})^{-1}\mathbf{w}\right\}.$$

After some clever computations and a slick argument, Harville [18] shows that we can compute the likelihood based on just $\mathbf{w}$ as the restricted log-likelihood function

$$\ell_R(\theta) = \text{constant} - \frac{1}{2}\log|\mathbf{V}| - \frac{1}{2}\log|\mathbf{X}^T\mathbf{V}^{-1}\mathbf{X}| - \frac{1}{2}(\mathbf{y}-\mathbf{X}\tilde{\mathbf{b}})^T\mathbf{V}^{-1}(\mathbf{y}-\mathbf{X}\tilde{\mathbf{b}}), \tag{8.20}$$

without having to construct $\mathbf{w}$ or $\mathbf{A}$. Here $\tilde{\mathbf{b}}$ is the GLS estimator $\tilde{\mathbf{b}} = (\mathbf{X}^T\mathbf{V}^{-1}\mathbf{X})^{-1}$ $\mathbf{X}^T\mathbf{V}^{-1}\mathbf{y}$. To simplify matters, assume that $\mathbf{X}$ has full-column rank, so $r = p$. The argument begins by constructing the $N \times p$ matrix $\mathbf{G} = \mathbf{V}^{-1}\mathbf{X}(\mathbf{X}^T\mathbf{V}^{-1}\mathbf{X})^{-1}$ and the $N \times N$ matrix $\mathbf{B} = [\mathbf{A} \mid \mathbf{G}]$ where the column's partitioning is $N-p$ and p. First the product,

$$\mathbf{B}^T\mathbf{B} = \begin{bmatrix}\mathbf{A}^T\\ \mathbf{G}^T\end{bmatrix}\begin{bmatrix}\mathbf{A} & \mathbf{G}\end{bmatrix} = \begin{bmatrix}\mathbf{A}^T\mathbf{A} & \mathbf{A}^T\mathbf{G}\\ \mathbf{G}^T\mathbf{A} & \mathbf{G}^T\mathbf{G}\end{bmatrix},$$

and then its determinant can be rewritten, using the formula (Result A.7(e)) for a partitioned matrix, as

$$\begin{aligned}|\mathbf{B}|^2 &= |\mathbf{B}^T\mathbf{B}| = |\mathbf{A}^T\mathbf{A}| \times |\mathbf{G}^T\mathbf{G} - \mathbf{G}^T\mathbf{A}(\mathbf{A}^T\mathbf{A})^{-1}\mathbf{A}^T\mathbf{G}|\\ &= |\mathbf{I}| \times |\mathbf{G}^T\mathbf{G} - \mathbf{G}^T(\mathbf{I}-\mathbf{P_X})\mathbf{G}| = |\mathbf{G}^T\mathbf{P_X}\mathbf{G}|\\ &= |(\mathbf{X}^T\mathbf{V}^{-1}\mathbf{X})^{-1}\mathbf{X}^T\mathbf{V}^{-1}\mathbf{X}(\mathbf{X}^T\mathbf{X})^{-1}\mathbf{X}^T\mathbf{V}^{-1}\mathbf{X}(\mathbf{X}^T\mathbf{V}^{-1}\mathbf{X})^{-1}| = |(\mathbf{X}^T\mathbf{X})^{-1}|,\end{aligned}$$

which also says that $\mathbf{B}$ is nonsingular. Now following similar steps, we find

$$\mathbf{B}^T\mathbf{V}\mathbf{B} = \begin{bmatrix}\mathbf{A}^T\\ \mathbf{G}^T\end{bmatrix}\mathbf{V}\begin{bmatrix}\mathbf{A} & \mathbf{G}\end{bmatrix} = \begin{bmatrix}\mathbf{A}^T\mathbf{V}\mathbf{A} & \mathbf{A}^T\mathbf{V}\mathbf{G}\\ \mathbf{G}^T\mathbf{V}\mathbf{A} & \mathbf{G}^T\mathbf{V}\mathbf{G}\end{bmatrix} = \begin{bmatrix}\mathbf{A}^T\mathbf{V}\mathbf{A} & \mathbf{0}\\ \mathbf{0} & (\mathbf{X}^T\mathbf{V}^{-1}\mathbf{X})^{-1}\end{bmatrix}$$

which yields only $|\mathbf{B}^T\mathbf{VB}| = |\mathbf{A}^T\mathbf{VA}|/|\mathbf{X}^T\mathbf{V}^{-1}\mathbf{X}|$. Some more manipulation puts all of the determinant results together:

$$|\mathbf{A}^T\mathbf{VA}| = |\mathbf{X}^T\mathbf{V}^{-1}\mathbf{X}| \times |\mathbf{B}^T\mathbf{VB}| = |\mathbf{X}^T\mathbf{V}^{-1}\mathbf{X}| \times |\mathbf{B}|^2 \times |\mathbf{V}|$$
$$|\mathbf{X}^T\mathbf{V}^{-1}\mathbf{X}| \times |\mathbf{B}^T\mathbf{B}| \times |\mathbf{V}| = |\mathbf{X}^T\mathbf{V}^{-1}\mathbf{X}| \times |\mathbf{V}|/|\mathbf{X}^T\mathbf{X}|.$$

Now we turn our attention to the quadratic form, where the key is constructing the GLS estimator $\tilde{\mathbf{b}} = \mathbf{G}^T\mathbf{y} = (\mathbf{X}^T\mathbf{V}^{-1}\mathbf{X})^{-1}\mathbf{X}^T\mathbf{V}^{-1}\mathbf{y}$. Since $\mathbf{B}$ is nonsingular, we have

$$\begin{aligned}\mathbf{y}^T\mathbf{V}^{-1}\mathbf{y} = \mathbf{y}^T\mathbf{B}(\mathbf{B}^T\mathbf{VB})^{-1}\mathbf{B}^T\mathbf{y} &= \begin{bmatrix}\mathbf{w}^T & \tilde{\mathbf{b}}^T\end{bmatrix}\begin{bmatrix}\mathbf{A}^T\mathbf{VA} & \mathbf{0} \\ \mathbf{0} & (\mathbf{X}^T\mathbf{V}^{-1}\mathbf{X})^{-1}\end{bmatrix}^{-1}\begin{bmatrix}\mathbf{w} \\ \tilde{\mathbf{b}}\end{bmatrix} \\ &= \mathbf{w}^T(\mathbf{A}^T\mathbf{VA})^{-1}\mathbf{w} + \tilde{\mathbf{b}}^T(\mathbf{X}^T\mathbf{V}^{-1}\mathbf{X})\tilde{\mathbf{b}},\end{aligned}$$

so that the quadratic form of interest can be rewritten

$$\mathbf{w}^T(\mathbf{A}^T\mathbf{VA})^{-1}\mathbf{w} = \mathbf{y}^T\mathbf{V}^{-1}\mathbf{y} - \tilde{\mathbf{b}}^T(\mathbf{X}^T\mathbf{V}^1\mathbf{X})\tilde{\mathbf{b}} = (\mathbf{y} - \mathbf{X}\tilde{\mathbf{b}})^T\mathbf{V}^{-1}(\mathbf{y} - \mathbf{X}\tilde{\mathbf{b}}).$$

The function $\ell_R(\theta)$ must then be maximized numerically.

Example 8.2: REML for Unbalanced One-Way Problem
Here the analysis proceeds similarly to the ML method, with one more piece, $|\mathbf{X}^T\mathbf{V}^{-1}\mathbf{X}| = \sum_{i=1}^{a}(n_i\sigma^2 + n_i^2\sigma_a^2)$. See Exercise 8.5.

8.4.3 The ANOVA Approach

The ANOVA approach employed in the previous two sections can also be generalized in the following manner. Construct matrices $\mathbf{A}_i$, $i = 0, 1, \ldots, m$, and corresponding quadratic forms in the response vector $\mathbf{y}$, that is, $\mathbf{y}^T\mathbf{A}_i\mathbf{y}$. Now employ Lemma 4.1:

$$\begin{aligned}E(\mathbf{y}^T\mathbf{A}\mathbf{y}) &= tr(\mathbf{A}Cov(\mathbf{y})) + [E(\mathbf{y})]^T\mathbf{A}[E(\mathbf{y})] \\ &= \sigma^2 tr(\mathbf{A}) + \sum_{j=1}^{m}\sigma_j^2 tr(\mathbf{Z}_j^T\mathbf{A}\mathbf{Z}_j) + (\mathbf{Xb})^T\mathbf{A}(\mathbf{Xb}).\end{aligned} \tag{8.21}$$

Choosing a set of matrices $\mathbf{A}_i$ such that $\mathbf{X}^T\mathbf{A}_i\mathbf{X} = \mathbf{0}$ for $i = 0, 1, \ldots, m$ leads to a system of equations by setting $E(\mathbf{y}^T\mathbf{A}_i\mathbf{y})$ equal to the observed $\mathbf{y}^T\mathbf{A}_i\mathbf{y}$. This system of equations looks like

$$\begin{bmatrix} tr(\mathbf{A}_0) & tr(\mathbf{Z}_1^T\mathbf{A}_0\mathbf{Z}_1) & \ldots & tr(\mathbf{Z}_m^T\mathbf{A}_0\mathbf{Z}_m) \\ tr(\mathbf{A}_1) & tr(\mathbf{Z}_1^T\mathbf{A}_1\mathbf{Z}_1) & \ldots & tr(\mathbf{Z}_m^T\mathbf{A}_1\mathbf{Z}_m) \\ \ldots & \ldots & \ldots & \ldots \\ tr(\mathbf{A}_m) & tr(\mathbf{Z}_1^T\mathbf{A}_m\mathbf{Z}_1) & \ldots & tr(\mathbf{Z}_m^T\mathbf{A}_m\mathbf{Z}_m)\end{bmatrix}\begin{bmatrix}\sigma^2 \\ \sigma_1^2 \\ \ldots \\ \sigma_m^2\end{bmatrix} = \begin{bmatrix}\mathbf{y}^T\mathbf{A}_0\mathbf{y} \\ \mathbf{y}^T\mathbf{A}_1\mathbf{y} \\ \ldots \\ \mathbf{y}^T\mathbf{A}_m\mathbf{y}\end{bmatrix}, \tag{8.22}$$

which can be solved for the variance components θ and lead to unbiased estimates.

Example 8.2: ANOVA Estimators continued further
With $\mathbf{X} = \mathbf{1}_n$ and $\mathbf{Z}$ as given previously, the key is the choice of the matrices $\mathbf{A}_0$ and $\mathbf{A}_1$. Choosing $\mathbf{A}_0 = \mathbf{P_Z} - \mathbf{P_1}$ and $\mathbf{A}_1 = \mathbf{I} - \mathbf{P_Z}$ yields the equations

$$\begin{bmatrix} tr(\mathbf{P_Z} - \mathbf{P_1}) & tr(\mathbf{Z}^T(\mathbf{P_Z} - \mathbf{P_1})\mathbf{Z}) \\ tr((\mathbf{I} - \mathbf{P_Z}) & tr(\mathbf{Z}^T(\mathbf{I} - \mathbf{P_Z})\mathbf{Z}) \end{bmatrix} \begin{bmatrix} \sigma^2 \\ \sigma_a^2 \end{bmatrix} = \begin{bmatrix} \mathbf{y}^T(\mathbf{P_Z} - \mathbf{P_1})\mathbf{y} \\ \mathbf{y}^T(\mathbf{I} - \mathbf{P_Z})\mathbf{y} \end{bmatrix},$$

which simplifies to the familiar

$$\begin{bmatrix} a-1 & N - \sum n_i^2/N \\ N-a & 0 \end{bmatrix} \begin{bmatrix} \sigma^2 \\ \sigma_a^2 \end{bmatrix} = \begin{bmatrix} SSA \\ SSE \end{bmatrix},$$

which reproduces Equations (8.10) and (8.11).

Example 8.1: continued
Consider again the two-way crossed, mixed model with no interaction, $y_{ijk} = \mu + \alpha_i + \beta_j + e_{ijk}$ with $i = 1, \ldots, a$; $j = 1, \ldots, b$; $k = 1, \ldots, n_{ij} = n$ (balanced), where α_i's are fixed and β_j's random. Choose again $\mathbf{A}_0 = \mathbf{P_Z} - \mathbf{P_1}$, $\mathbf{A}_1 = \mathbf{I} - \mathbf{P_Z}$ to obtain

$$\begin{bmatrix} tr(\mathbf{P_Z} - \mathbf{P_1}) & tr(\mathbf{Z}^T(\mathbf{P_Z} - \mathbf{P_1})\mathbf{Z}) \\ tr(\mathbf{I} - \mathbf{P_Z}) & tr(\mathbf{Z}^T(\mathbf{I} - \mathbf{P_Z})\mathbf{Z}) \end{bmatrix} \begin{bmatrix} \sigma^2 \\ \sigma_b^2 \end{bmatrix} = \begin{bmatrix} \mathbf{y}^T(\mathbf{P_Z} - \mathbf{P_1})\mathbf{y} \\ \mathbf{y}^T(\mathbf{I} - \mathbf{P_Z})\mathbf{y} \end{bmatrix},$$

which simplifies to

$$\begin{bmatrix} b-1 & an(b-1) \\ N-b & 0 \end{bmatrix} \begin{bmatrix} \sigma^2 \\ \sigma_b^2 \end{bmatrix} = \begin{bmatrix} SSB \\ SSE \end{bmatrix}$$

where $SSB = an \sum_i (\overline{y}_{.j.} - \overline{y}_{...})^2$. See also Exercise 8.11.

8.5 The Split Plot

The split plot arises from an experimental design with two (or more) fixed factors and experimental units of different sizes. More commonly, the experimental unit at one level (whole plot) is composed of experimental units of the other factor (subplots). Moreover, the random assignment of these experimental units to treatments is done in a nested or conditional fashion, inducing a special correlation structure.

For example, a scientist is measuring the sugar content of sweet basil leaves. One factor may be fertilizer, and the natural experimental unit may be a plant. However, a second factor may be the height of the leaf in the plant, and so at the second level, the experimental unit would be a leaf. In such an experiment, seedlings would be randomly assigned a fertilizer level, the whole plot factor. At harvest, leaves would be randomly selected at the top, middle, and bottom levels of the plant, as height is the split-plot factor. Denote the sugar content of plant j given fertilizer level i and a leaf selected at level k of the plant by y_{ijk}. The correlation among the leaves from the same plant should be the same across all leaves from the same plant (equicorrelated), so we might view

$$Cov\left(\begin{bmatrix} y_{i,j,\text{top}} \\ y_{i,j,\text{middle}} \\ y_{i,j,\text{bottom}} \end{bmatrix}\right) = \begin{bmatrix} a & b & b \\ b & a & b \\ b & b & a \end{bmatrix} = \sigma^2\mathbf{I}_3 + \sigma_s^2\mathbf{1}\mathbf{1}^T$$

where $a = \sigma^2 + \sigma_s^2$ and $b = \sigma_s^2$, and the correlation is $b/a = \sigma_s^2/(\sigma^2 + \sigma_s^2)$. Another viewpoint would be to take the mean response for each plant as the response from the whole plot experiment—a one-way ANOVA. Then within each whole plot are subplot experiments.

The following model expresses this correlation structure in a different form:

$$y_{ijk} = \mu + \alpha_i + \delta_{ij} + \beta_k + \gamma_{ik} + e_{ijk},$$

with random effects δ_{ij} iid $N(0, \sigma_s^2)$ arising from the random assignment of treatments to the whole plots (e.g., fertilizer level i), independent of e_{ijk} iid $N(0, \sigma^2)$, which includes the random selection subplots (leaves). As a result, we have correlation among leaves within a plant, with independence across leaves from different plants.

The decomposition of the sums of squares does not follow the usual pattern for a two-factor model with interaction for two reasons: first, because of the correlation structure, and second, this is really a mixed model with two variance components. For example, consider the balanced case where $i = 1, \ldots, I = 2$; $j = 1, \ldots, J = 2$; and $k = 1, \ldots, K = 3$; the mixed model form can be written as $\mathbf{y} = \mathbf{Xb} + \mathbf{Zu} + \mathbf{e}$ where

$$\mathbf{y} = \begin{bmatrix} y_{111} \\ y_{112} \\ y_{113} \\ y_{121} \\ y_{122} \\ y_{123} \\ y_{211} \\ y_{212} \\ y_{213} \\ y_{221} \\ y_{222} \\ y_{223} \end{bmatrix}, \quad \mathbf{Zu} = \begin{bmatrix} 1 & 0 & 0 & 0 \\ 1 & 0 & 0 & 0 \\ 1 & 0 & 0 & 0 \\ 0 & 1 & 0 & 0 \\ 0 & 1 & 0 & 0 \\ 0 & 1 & 0 & 0 \\ 0 & 0 & 1 & 0 \\ 0 & 0 & 1 & 0 \\ 0 & 0 & 1 & 0 \\ 0 & 0 & 0 & 1 \\ 0 & 0 & 0 & 1 \\ 0 & 0 & 0 & 1 \end{bmatrix} \begin{bmatrix} \delta_{11} \\ \delta_{12} \\ \delta_{21} \\ \delta_{22} \end{bmatrix},$$

$$\mathbf{Xb} = \begin{bmatrix} 1 & 1 & 0 & 1 & 0 & 0 & 1 & 0 & 0 & 0 & 0 & 0 \\ 1 & 1 & 0 & 0 & 1 & 0 & 0 & 1 & 0 & 0 & 0 & 0 \\ 1 & 1 & 0 & 0 & 0 & 1 & 0 & 0 & 1 & 0 & 0 & 0 \\ 1 & 1 & 0 & 1 & 0 & 0 & 1 & 0 & 0 & 0 & 0 & 0 \\ 1 & 1 & 0 & 0 & 1 & 0 & 0 & 1 & 0 & 0 & 0 & 0 \\ 1 & 1 & 0 & 0 & 0 & 1 & 0 & 0 & 1 & 0 & 0 & 0 \\ 1 & 0 & 1 & 1 & 0 & 0 & 0 & 0 & 0 & 1 & 0 & 0 \\ 1 & 0 & 1 & 0 & 1 & 0 & 0 & 0 & 0 & 0 & 1 & 0 \\ 1 & 0 & 1 & 0 & 0 & 1 & 0 & 0 & 0 & 0 & 0 & 1 \\ 1 & 0 & 1 & 1 & 0 & 0 & 0 & 0 & 0 & 1 & 0 & 0 \\ 1 & 0 & 1 & 0 & 1 & 0 & 0 & 0 & 0 & 0 & 1 & 0 \\ 1 & 0 & 1 & 0 & 0 & 1 & 0 & 0 & 0 & 0 & 0 & 1 \end{bmatrix} \begin{bmatrix} \mu \\ \alpha_1 \\ \alpha_2 \\ \beta_1 \\ \beta_2 \\ \beta_3 \\ \gamma_{11} \\ \gamma_{12} \\ \gamma_{13} \\ \gamma_{21} \\ \gamma_{22} \\ \gamma_{23} \end{bmatrix}$$

$$= [\mathbf{1}_n \quad \mathbf{X}_A \quad \mathbf{X}_B \quad \mathbf{X}_{AB}] \begin{bmatrix} \mu \\ \alpha' s \\ \beta' s \\ \gamma' s \end{bmatrix}$$

where the partitioning of $\mathbf{X}$ into columns for μ, α's, β's, and γ's will be convenient later. Note that the covariance matrix of $\mathbf{y}$ takes the form $Cov\,(\mathbf{y}) = \sigma_s^2\mathbf{Z}\mathbf{Z}^T + \sigma^2\mathbf{I}_n$, or

$$\begin{bmatrix} \sigma_s^2+\sigma^2 & \sigma_s^2 & \sigma_s^2 & 0 & 0 & 0 & 0 & 0 & 0 & 0 & 0 & 0 \\ \sigma_s^2 & \sigma_s^2+\sigma^2 & \sigma_s^2 & 0 & 0 & 0 & 0 & 0 & 0 & 0 & 0 & 0 \\ \sigma_s^2 & \sigma_s^2 & \sigma_s^2+\sigma^2 & 0 & 0 & 0 & 0 & 0 & 0 & 0 & 0 & 0 \\ 0 & 0 & 0 & \sigma_s^2+\sigma^2 & \sigma_s^2 & \sigma_s^2 & 0 & 0 & 0 & 0 & 0 & 0 \\ 0 & 0 & 0 & \sigma_s^2 & \sigma_s^2+\sigma^2 & \sigma_s^2 & 0 & 0 & 0 & 0 & 0 & 0 \\ 0 & 0 & 0 & \sigma_s^2 & \sigma_s^2 & \sigma_s^2+\sigma^2 & 0 & 0 & 0 & 0 & 0 & 0 \\ 0 & 0 & 0 & 0 & 0 & 0 & \sigma_s^2+\sigma^2 & \sigma_s^2 & \sigma_s^2 & 0 & 0 & 0 \\ 0 & 0 & 0 & 0 & 0 & 0 & \sigma_s^2 & \sigma_s^2+\sigma^2 & \sigma_s^2 & 0 & 0 & 0 \\ 0 & 0 & 0 & 0 & 0 & 0 & \sigma_s^2 & \sigma_s^2 & \sigma_s^2+\sigma^2 & 0 & 0 & 0 \\ 0 & 0 & 0 & 0 & 0 & 0 & 0 & 0 & 0 & \sigma_s^2+\sigma^2 & \sigma_s^2 & \sigma_s^2 \\ 0 & 0 & 0 & 0 & 0 & 0 & 0 & 0 & 0 & \sigma_s^2 & \sigma_s^2+\sigma^2 & \sigma_s^2 \\ 0 & 0 & 0 & 0 & 0 & 0 & 0 & 0 & 0 & \sigma_s^2 & \sigma_s^2 & \sigma_s^2+\sigma^2 \end{bmatrix},$$

which shows the block correlation structure described previously.

Often these experiments are analyzed in a nested fashion. The average response from the whole plot (average over k) can be viewed as

$$\overline{y}_{ij.} = \mu + \alpha_i + \delta_{ij} + \overline{\beta} + \overline{\gamma}_{i.} + \overline{e}_{ij.},$$

which can be rewritten as

$$\overline{y}_{ij.} = (\mu + \overline{\beta}) + (\alpha_i + \overline{\gamma}_{i.}) + (\delta_{ij} + \overline{e}_{ij.}).$$

This can be viewed as following the usual one-way ANOVA model, since $\overline{y}_{ij.}$ are independent with mean $\mu + \alpha_i + \overline{\beta} + \overline{\gamma}_{i.}$ and variance $\sigma^2/K + \sigma_s^2$ where K is the number of subplots per whole plot. Within each whole plot i, j is another one-way experiment:

$$y_{ijk} = \mu + \alpha_i + \delta_{ij} + \beta_k + \gamma_{ik} + e_{ijk}$$

where, in this case, the second factor is replicated across whole plots.

The ANOVA table for a balanced simple split-plot experiment is given in Table 8.1. Some care must be taken in establishing the distribution of the usual statistics.

For the main plot analysis, one key step is that $Cov\,(\mathbf{y}) = \sigma_s^2\mathbf{Z}\mathbf{Z}^T + \sigma^2\mathbf{I}_N = \sigma_s^2 K\mathbf{P_Z} + \sigma^2\mathbf{I}_N$, due to balance within each whole plot. From Result 5.15, we will get χ^2distributions for $SSA = \mathbf{y}^T(\mathbf{P}_{\mathbf{X}_A} - \mathbf{P}_1)\mathbf{y}$ and $SSE(A) = \mathbf{y}^T(\mathbf{P_Z} - \mathbf{P}_{\mathbf{X}_A})\mathbf{y}$ if, properly scaled, $(\mathbf{P}_{\mathbf{X}_A} - \mathbf{P}_1)Cov\,(\mathbf{y})$ and $(\mathbf{P_Z} - \mathbf{P}_{\mathbf{X}_A})Cov\,(\mathbf{y})$ are idempotent. For the first,

$$\begin{aligned}(\mathbf{P}_{\mathbf{X}_A} - \mathbf{P}_1)Cov\,(\mathbf{y}) &= (\mathbf{P}_{\mathbf{X}_A} - \mathbf{P}_1)\left(\sigma_s^2 K\mathbf{P_Z} + \sigma^2\mathbf{I}_N\right)\\ &= \left(\sigma_s^2 K + \sigma^2\right)(\mathbf{P}_{\mathbf{X}_A} - \mathbf{P}_1),\end{aligned}$$

since $\mathcal{C}(\mathbf{X}_A) \subseteq \mathcal{C}(\mathbf{Z})$ and $\mathcal{C}(\mathbf{1}) \subseteq \mathcal{C}(\mathbf{Z})$. So $\mathbf{y}^T(\mathbf{P}_{\mathbf{X}_A} - \mathbf{P}_1)\mathbf{y}/(\sigma_s^2 K + \sigma^2) \sim \chi^2$ with $trace(\mathbf{P}_{\mathbf{X}_A} - \mathbf{P}_1) = I - 1$ as its degrees of freedom, and noncentrality that is zero when there is no factor A effect—α's are equal. Similarly for the second,

$$\begin{aligned}(\mathbf{P_Z} - \mathbf{P}_{\mathbf{X}_A})Cov\,(\mathbf{y}) &= (\mathbf{P_Z} - \mathbf{P}_{\mathbf{X}_A})\left(\sigma_s^2 K\mathbf{P_Z} + \sigma^2\mathbf{I}_N\right)\\ &= \left(\sigma_s^2 K + \sigma^2\right)(\mathbf{P_Z} - \mathbf{P}_{\mathbf{X}_a})\end{aligned}$$

since $\mathcal{C}(\mathbf{X}_A) \subseteq \mathcal{C}(\mathbf{Z})$. So $\mathbf{y}^T(\mathbf{P_Z} - \mathbf{P}_{\mathbf{X}_A})\mathbf{y}/(\sigma_s^2 K + \sigma^2) \sim \chi^2$ with $trace(\mathbf{P_Z} - \mathbf{P}_{\mathbf{X}_A}) = IJ - I$ as its degrees of freedom, and zero noncentrality since $(\mathbf{P_Z} - \mathbf{P}_{\mathbf{X}_A})\mathbf{X} = \mathbf{0}$ (see Exercise 8.14). Independence between these two quadratic forms requires

$$(\mathbf{P_Z} - \mathbf{P}_{\mathbf{X}_A})Cov\,(\mathbf{y})(\mathbf{P}_{\mathbf{X}_A} - \mathbf{P}_1) = \mathbf{0} \tag{8.23}$$

(also Exercise 8.14). Consequently, the ratio of mean squares,

$$F = [SSA/(I-1)] \Big/ [SSE(A)/(IJ - I)],$$

will have the F-distribution with $(I - 1)$ and $I(J - 1)$ degrees of freedom, and zero noncentrality when there is neither interaction nor factor A effect ($\alpha_i + \overline{\gamma}_{i.}$ equal for all i).

Similar issues are faced in the analysis at the split-plot level. Again, if properly scaled, $SSB = \mathbf{y}^T(\mathbf{P}_{\mathbf{X}_B} - \mathbf{P}_1)\mathbf{y}$ will lead to a χ^2 distribution if $(\mathbf{P}_{\mathbf{X}_B} - \mathbf{P}_1)Cov\,(\mathbf{y})$ is idempotent. The proper scaling follows from

$$(\mathbf{P}_{\mathbf{X}_B} - \mathbf{P}_1)\left(\sigma_s^2 K\mathbf{P_Z} + \sigma^2\mathbf{I}_n\right) = \sigma^2(\mathbf{P}_{\mathbf{X}_B} - \mathbf{P}_1),$$

since both $\mathbf{P}_{\mathbf{X}_B}\mathbf{P_Z} = \mathbf{P}_1$ and $\mathbf{P}_1\mathbf{P_Z} = \mathbf{P}_1$. So using Result 5.15, $\mathbf{y}^T(\mathbf{P}_{\mathbf{X}_B} - \mathbf{P}_1)\mathbf{y}/\sigma^2$ has a χ^2 distribution with $trace(\mathbf{P}_{\mathbf{X}_B} - \mathbf{P}_1) = K - 1$ df, and a noncentrality that is zero when the β's are equal. For the interaction term $SSAB = \mathbf{y}^T(\mathbf{P}_{\mathbf{X}_{AB}} - \mathbf{P}_{\mathbf{X}_A} - \mathbf{P}_{\mathbf{X}_B} + \mathbf{P}_1)\mathbf{y}$, we have

$$(\mathbf{P}_{\mathbf{X}_{AB}} - \mathbf{P}_{\mathbf{X}_A} - \mathbf{P}_{\mathbf{X}_B} + \mathbf{P}_1)\left(\sigma_s^2 K\mathbf{P_Z} + \sigma^2\mathbf{I}_n\right) = \sigma^2(\mathbf{P}_{\mathbf{X}_{AB}} - \mathbf{P}_{\mathbf{X}_A} - \mathbf{P}_{\mathbf{X}_B} + \mathbf{P}_1),$$

TABLE 8.1 ANOVA Table for Simple Split Plot

Source	df	Projection	Sum of Squares
Whole Plot Analysis			
Mean	1	$\mathbf{P}_1$	$N\overline{y}_{...}^2$
Factor A	$I-1$	$\mathbf{P}_{\mathbf{X}_A} - \mathbf{P}_1$	$SSA = JK\sum_i(\overline{y}_{i..} - \overline{y}_{...})^2 = \mathbf{y}^T(\mathbf{P}_{\mathbf{X}_A} - \mathbf{P}_1)\mathbf{y}$
Error A	$I(J-1)$	$\mathbf{P}_\mathbf{Z} - \mathbf{P}_{\mathbf{X}_A}$	$SSE(A) = K\sum_i\sum_j(\overline{y}_{ij.} - \overline{y}_{i..})^2 = \mathbf{y}^T(\mathbf{P}_\mathbf{Z} - \mathbf{P}_{\mathbf{X}_A})\mathbf{y}$
Split Plot Analysis			
Factor B	$K-1$	$\mathbf{P}_{\mathbf{X}_B} - \mathbf{P}_1$	$SSB = IJ\sum_k(\overline{y}_{..k} - \overline{y}_{...})^2 = \mathbf{y}^T(\mathbf{P}_{\mathbf{X}_B} - \mathbf{P}_1)\mathbf{y}$
Interaction	$(I-1)(K-1)$	$\mathbf{P}_{\mathbf{X}_{AB}} - \mathbf{P}_{\mathbf{X}_A} - \mathbf{P}_{\mathbf{X}_B} + \mathbf{P}_1$	$SSAB = J\sum_{i,k}(\overline{y}_{i.k} - \overline{y}_{i..} - \overline{y}_{..k} + \overline{y}_{...})^2 = \mathbf{y}^T(\mathbf{P}_{\mathbf{X}_{AB}} - \mathbf{P}_{\mathbf{X}_A} - \mathbf{P}_{\mathbf{X}_B} + \mathbf{P}_1)\mathbf{y}$
Error B	$I(J-1)(K-1)$	$\mathbf{I}_N - \mathbf{P}_\mathbf{Z} + \mathbf{P}_{\mathbf{X}_A} - \mathbf{P}_{\mathbf{X}_{AB}}$	$SSE(B) = \sum_{i,j,k}(y_{ijk} - \overline{y}_{ij.} + \overline{y}_{i..} - \overline{y}_{i.k})^2 = \mathbf{y}^T(\mathbf{I}_N - \mathbf{P}_Z + \mathbf{P}_{\mathbf{X}_A} - \mathbf{P}_{\mathbf{X}_{AB}})\mathbf{y}$

following from $\mathbf{P}_{\mathbf{X}_{AB}}\mathbf{P}_\mathbf{Z} = \mathbf{P}_{\mathbf{X}_A}$ and $\mathbf{P}_{\mathbf{X}_A}\mathbf{P}_\mathbf{Z} = \mathbf{P}_{\mathbf{X}_A}$ as well as the previous $\mathbf{P}_{\mathbf{X}_B}\mathbf{P}_\mathbf{Z} = \mathbf{P}_1$ and $\mathbf{P}_1\mathbf{P}_\mathbf{Z} = \mathbf{P}_1$. Lastly, for the error B term, again we have

$$(\mathbf{I}_N - \mathbf{P}_\mathbf{Z} + \mathbf{P}_{\mathbf{X}_A} - \mathbf{P}_{\mathbf{X}_{AB}})(\sigma_s^2 K\mathbf{P}_\mathbf{Z} + \sigma^2\mathbf{I}_n) = \sigma^2(\mathbf{I}_N - \mathbf{P}_\mathbf{Z} + \mathbf{P}_{\mathbf{X}_A} - \mathbf{P}_{\mathbf{X}_{AB}}),$$

following the previous results for the products of projection matrices. For its degrees of freedom, note that *trace* $(\mathbf{I}_N - \mathbf{P}_\mathbf{Z} + \mathbf{P}_{\mathbf{X}_A} - \mathbf{P}_{\mathbf{X}_{AB}}) = IJK - IJ + I - IK$.

Notice that while our model has two variance components, the two error sums of squares ($SSE(A)$ and $SSE(B)$) are used to estimate variances $(\sigma_s^2 K + \sigma^2)$ and σ^2:

$$E(SSE(A)/(IJ - I)) = E(\mathbf{y}^T(\mathbf{P}_\mathbf{Z} - \mathbf{P}_{\mathbf{X}_A})\mathbf{y}/(IJ - I) = (\sigma_s^2 K + \sigma^2)$$

$$E(SSE(B)/(I(J-1)(K-1))) = E(\mathbf{y}^T(\mathbf{I}_N - \mathbf{P}_\mathbf{Z} + \mathbf{P}_{\mathbf{X}_A} - \mathbf{P}_{\mathbf{X}_{AB}})\mathbf{y}/ (I(J-1)(K-1)) = \sigma^2.$$

No attempt is made to estimate σ_s^2 by itself, only the two terms in the covariance matrix are estimated.

Comparison of factor A requires using error A, that is, the difference of two factor A means gives

$$E(\overline{y}_{1..} - \overline{y}_{2..}) = (\alpha_1 + \overline{\gamma}_{1.}) - (\alpha_2 + \overline{\gamma}_{2.})$$

$$Var(\overline{y}_{1..} - \overline{y}_{2..}) = Var((\overline{\delta}_{1.} + \overline{e}_{1..}) - (\overline{\delta}_{2.} + \overline{e}_{2..})) = 2(\sigma_s^2 + \sigma^2/K)/J.$$

The standard error for the difference can be constructed from error A as $\sqrt{2\,[SSE(A)/df]\,/(JK)}$ since

$$E(SSE(A)/df) = E(\mathbf{y}^T(\mathbf{P}_\mathbf{Z} - \mathbf{P}_{\mathbf{X}_A})\mathbf{y})/(I(J-1)) = K\sigma_s^2 + \sigma^2.$$

A similar analysis for the comparison of factor B means leads to using error B:

$$E(\overline{y}_{..1} - \overline{y}_{..2}) = (\overline{\alpha} + \beta_1 + \overline{\gamma}_{.1}) - (\overline{\alpha} + \beta_2 + \overline{\gamma}_{.2})$$

$$Var(\overline{y}_{..1} - \overline{y}_{..2}) = Var((\overline{\delta}_{..} + \overline{e}_{..1}) - (\overline{\delta}_{..} + \overline{e}_{..2})) = 2\sigma^2/(IJ),$$

and the standard error for the difference is $\sqrt{2\,[SSE(B)/df]\,/(IJ)}$ since $E(SSE(B)/df) = \sigma^2$.

8.6 Predictions and BLUPs

When we were discussing constructing confidence intervals back in Chapter 6, the focus was on fixed effects of the form $\lambda^T\mathbf{b}$. A typical situation would be a linear combination of the unknown coefficients that corresponded to the mean of the response at a particular design point $\mathbf{x}_*$, namely,

$$E(y|\mathbf{x}*) = \mathbf{x}_*^T\mathbf{b}.$$

Merely following the previously established formulae for a confidence interval for $\lambda^T \mathbf{b}$ and setting $\lambda = \mathbf{x}_*$, we construct the interval

$$\mathbf{x}_*^T \hat{\mathbf{b}} \pm t_{N-r,\alpha/2} \hat{\sigma} \sqrt{\mathbf{x}_*^T (\mathbf{X}^T \mathbf{X})^g \mathbf{x}_*}.$$

Among the variety of simultaneous confidence interval methods are those (e.g., Working-Hotelling) designed exactly for these situations.

But what if we are not interested in the mean response at a design point, but in constructing a confidence interval for a yet unobserved response y_* at the design point $\mathbf{x}_*$? This is a single random event when the covariates take on specified values. Some authors emphasize with the description *prediction* that a confidence interval is being constructed for a random variable rather than a fixed quantity—a confidence interval for a *prediction*.

Instead of looking at constructing just any prediction, let's look for the best predictor, in particular, an unbiased linear predictor $\mathbf{a}^T \mathbf{y}$ that has the smallest variance. Beginning with the Gauss–Markov assumptions, we have

$$E \begin{bmatrix} \mathbf{y} \\ y_* \end{bmatrix} = \begin{bmatrix} \mathbf{X} \\ \mathbf{x}_* \end{bmatrix}, \; Cov \left(\begin{bmatrix} \mathbf{y} \\ y_* \end{bmatrix} \right) = \sigma^2 \begin{bmatrix} \mathbf{I}_N & \mathbf{0} \\ \mathbf{0} & 1 \end{bmatrix}$$

so that

$$E\left(\mathbf{a}^T \mathbf{y} - y_*\right) = \mathbf{a}^T \mathbf{X}\mathbf{b} - \mathbf{x}_*^T \mathbf{b}, \; Var\left(\mathbf{a}^T \mathbf{y} - y_*\right) = \sigma^2(\mathbf{a}^T \mathbf{a} + 1).$$

Unbiasedness here means that $\mathbf{a}^T \mathbf{X}\mathbf{b} - \mathbf{x}_*^T \mathbf{b} = \mathbf{0}$ for all $\mathbf{b}$, or $\mathbf{X}^T \mathbf{a} - \mathbf{x}_* = \mathbf{0}$. Now minimizing the variance looks almost the same as finding the BLUE, except that the variance is $\sigma^2(\mathbf{a}^T \mathbf{a} + 1)$ instead of $\sigma^2 \mathbf{a}^T \mathbf{a}$ in the Gauss–Markov Theorem. Applying this theorem, we can obtain the same result, that the variance is minimized at $\mathbf{a} = \mathbf{X}(\mathbf{X}^T \mathbf{X})^{gT} \mathbf{x}_*$, so that the *best linear unbiased predictor* (BLUP) is $\mathbf{a}^T \mathbf{y} = \mathbf{x}_*^T (\mathbf{X}^T \mathbf{X})^g \mathbf{X}^T \mathbf{y} = \mathbf{x}_*^T \hat{\mathbf{b}}$, whose variance is $\sigma^2(\mathbf{a}^T \mathbf{a} + 1) = \sigma^2(\mathbf{x}_*^T (\mathbf{X}^T \mathbf{X})^g \mathbf{x}^* + 1)$.

Adding the assumption of joint multivariate normality allows us to construct a *prediction* or *forecast* confidence interval following the familiar steps:

$$\mathbf{x}_*^T \hat{\mathbf{b}} - y_* \sim N\left(0, \sigma^2\left(\mathbf{x}_*^T (\mathbf{X}^T \mathbf{X})^g \mathbf{x}^* + 1\right)\right)$$

$$Pr\left(-t_{\alpha/2} < \frac{\mathbf{x}_*^T \hat{\mathbf{b}} - y_*}{\hat{\sigma}\,(\mathbf{x}_*^T (\mathbf{X}^T \mathbf{X})^g \mathbf{x}^* + 1)^{1/2}} < t_{\alpha/2}\right) = 1 - \alpha \tag{8.24}$$

$$= Pr\left(\mathbf{x}_*^T \hat{\mathbf{b}} - t_{\alpha/2} \hat{\sigma} \left(\mathbf{x}_*^T (\mathbf{X}^T \mathbf{X})^g \mathbf{x}^* + 1\right)^{1/2} < y_* < \mathbf{x}_*^T \hat{\mathbf{b}} + t_{\alpha/2} \hat{\sigma} \left(\mathbf{x}_*^T (\mathbf{X}^T \mathbf{X})^g \mathbf{x}^* + 1\right)^{1/2}\right) \tag{8.25}$$

so that the prediction or forecast confidence interval is

$$\mathbf{x}_*^T \hat{\mathbf{b}} \pm t_{\alpha/2} \hat{\sigma} \left(\mathbf{x}_*^T (\mathbf{X}^T \mathbf{X})^g \mathbf{x}^* + 1\right)^{1/2}.$$

This approach can be easily extended to the Aitken or GLS case, following

$$E \begin{bmatrix} \mathbf{y} \\ y_* \end{bmatrix} = \begin{bmatrix} \mathbf{X} \\ \mathbf{x}_* \end{bmatrix} \mathbf{b}, \; Cov \left(\begin{bmatrix} \mathbf{y} \\ y_* \end{bmatrix} \right) = \begin{bmatrix} \mathbf{\Omega} & \omega \\ \omega & \omega_{**} \end{bmatrix}$$

so that

$$E(\mathbf{a}^T\mathbf{y} - y_*) = \mathbf{a}^T\mathbf{X}\mathbf{b} - \mathbf{x}_*^T\mathbf{b},\ Var(\mathbf{a}^T\mathbf{y} - y_*) = \mathbf{a}^T\boldsymbol{\Omega}\mathbf{a} + 2\mathbf{a}^T\omega + \omega_{**}.$$

Unbiasedness again means that $\mathbf{a}^T\mathbf{X}\mathbf{b} - \mathbf{x}_*^T\mathbf{b}$ for all $\mathbf{b}$, or $\mathbf{X}^T\mathbf{a} - \mathbf{x}_*$. Minimizing the variance subject to the unbiasedness constraint suggests Lagrange multipliers, leading to the equation

$$\boldsymbol{\Omega}\mathbf{a} + \omega + \mathbf{X}\lambda = \mathbf{0} \tag{8.26}$$

where λ is the Lagrange multiplier. A few lines of algebra (Exercise 8.17) leads to the expression (Goldberger [12]) for the BLUP of y_*:

$$\mathbf{a}^T\mathbf{y} = \mathbf{x}_*^T\hat{\mathbf{b}}_{GLS} + \omega^T\boldsymbol{\Omega}^{-1}(\mathbf{y} - \mathbf{x}_*^T\hat{\mathbf{b}}_{GLS}) \tag{8.27}$$

Example 8.3: Autoregressive Errors
Consider the first-order autoregressive model, where $\boldsymbol{\Omega}_{ij} \propto \rho^{|i-j|}$. If we are forecasting k periods ahead, so that $\omega_i = \rho^{k+i-1}$, then the second part of Expression (8.27) is

$$\omega^T\boldsymbol{\Omega}^{-1}\hat{\mathbf{e}}_{GLS} = \rho^k\mathbf{e}^{(N)}\hat{\mathbf{e}}_{GLS} = \rho^k\hat{e}_N$$

where $\mathbf{e}^{(N)}$ is the N-th elementary vector and so $\hat{e}_N$ is the last residual from $\hat{e}_{GLS}$.

An alert reader is probably wondering if perhaps this result was misplaced from Chapter 6. Well, not really, since only now have we needed to deal with prediction. Next on the agenda, following the mixed-model scenario, is to construct predictors for the unobserved random effects $\mathbf{u}$.

There are several motivations for the same BLUP expressions; some of the motivations are strong, others dubious. The important thing is that a direct approach is long and tedious and not recommended. The first approach puts a simple twist on the Aitken forecast model.

$$E\begin{bmatrix}\mathbf{y}\\\mathbf{u}\end{bmatrix} = \begin{bmatrix}\mathbf{X}\mathbf{b}\\\mathbf{0}\end{bmatrix},\ Cov\left(\begin{bmatrix}\mathbf{y}\\\mathbf{u}\end{bmatrix}\right) = \begin{bmatrix}\boldsymbol{\Omega} & \omega\\ \omega & \omega_{**}\end{bmatrix}$$

so that here $\boldsymbol{\Omega} = \mathbf{R} + \mathbf{Z}\mathbf{G}\mathbf{Z}^T$, $\omega = \mathbf{Z}\mathbf{G}$, and $\omega_{**} = \mathbf{G}$. From this we can obtain

$$\hat{\mathbf{b}}_{GLS} = [\mathbf{X}^T(\mathbf{R} + \mathbf{Z}\mathbf{G}\mathbf{Z}^T)^{-1}\mathbf{X}]\mathbf{X}^T(\mathbf{R} + \mathbf{Z}\mathbf{G}\mathbf{Z}^T)^{-1}\mathbf{y}$$

and

$$\hat{\mathbf{u}} = \mathbf{G}\mathbf{Z}^T(\mathbf{R} + \mathbf{Z}\mathbf{G}\mathbf{Z}^T)^{-1}(\mathbf{y} - \mathbf{X}\hat{\mathbf{b}}_{GLS}).$$

A second motivation is to construct a likelihood from the densities of $\mathbf{y}|\mathbf{u}$ and $\mathbf{u}$, giving the joint density of $(\mathbf{y}, \mathbf{u})$—but by no means a likelihood. Maximizing with respect to $\mathbf{b}$ and $\mathbf{u}$ leads to the *mixed-model equations* (MMEs):

$$\begin{bmatrix}\mathbf{X}^T\mathbf{R}^{-1}\mathbf{X} & \mathbf{X}^T\mathbf{R}^{-1}\mathbf{Z}\\ \mathbf{Z}^T\mathbf{R}^{-1}\mathbf{X} & \mathbf{G}^{-1} + \mathbf{Z}^T\mathbf{R}^{-1}\mathbf{Z}\end{bmatrix}\begin{bmatrix}\mathbf{b}\\\mathbf{u}\end{bmatrix} = \begin{bmatrix}\mathbf{X}^T\mathbf{R}^{-1}\mathbf{y}\\\mathbf{Z}^T\mathbf{R}^{-1}\mathbf{y}\end{bmatrix}. \tag{8.28}$$

Now with a bit of algebra (Exercise 8.19), one can show that $\hat{\mathbf{b}}_{GLS}$ and $\hat{\mathbf{u}}$ satisfy the MMEs. A third and more substantive motivation is to show that a scalar estimator of the form $\mathbf{c}^T\hat{\mathbf{b}}_{GLS} + \mathbf{d}^T\hat{\mathbf{u}}$ is uncorrelated with all unbiased estimators of zero, by demonstrating

$$Cov\,(\hat{\mathbf{b}}_{GLS}, \mathbf{a}^T\mathbf{y}) = 0$$

and

$$Cov\,(\hat{\mathbf{u}}, \mathbf{a}^T\mathbf{y}) = 0$$

for $E(\mathbf{a}^T\mathbf{y}) = 0$. See Exercise 8.21.

A last motivation, following somewhat from the first, originates from a conditioning argument. Let $\mathbf{Ay}$ be any potential predictor for $\mathbf{u}$. Then the expression

$$\begin{aligned} Cov\,(\mathbf{Ay} - \mathbf{y}) &= E\,[Cov\,(\mathbf{Ay} - \mathbf{u}|\mathbf{y})] + Cov\,[E(\mathbf{Ay} - \mathbf{u}|\mathbf{y})] \\ &= E\,[Cov\,(\mathbf{Ay} - \mathbf{u}|\mathbf{y})] + Cov\,[\mathbf{Ay} - E(\mathbf{u}|\mathbf{y})], \end{aligned}$$

which is minimized by choosing $\mathbf{Ay} = E(\mathbf{u}|\mathbf{y})$, leads to

$$E(\mathbf{u}|\mathbf{y}) = \mathbf{GZ}^T(\mathbf{R} + \mathbf{ZGZ}^T)^{-1}(\mathbf{y} - \mathbf{Xb}),$$

which differs from our expression for $\hat{\mathbf{u}}$ only by replacing $\mathbf{b}$ with the best estimator $\hat{\mathbf{b}}_{GLS}$. Adding rigor to this argument merely repeats the same analysis as showing that $\hat{\mathbf{u}}$ is uncorrelated with unbiased estimators of zero.

8.7 Summary

1. Only in the balanced cases can we derive the exact χ^2-distribution of the sums of squares in the random effects and mixed models.

2. Three methods are outlined for the general case: ANOVA, maximum likelihood (ML), and restricted maximum likelihood (REML).

3. The details of the split-plot experimental design, with its two variance components, are elaborated.

4. Best linear unbiased predictors (BLUPs) are introduced for estimating random effects as well as constructing confidence intervals for prediction.

8.8 Notes

- For some parts of the statistical community, the R in REML represents *residual*. A reader interested in the origins should begin with papers by Patterson and Thompson [34] and Thompson [46], and a later one by Harville [17].

- Robinson [41] provides an excellent survey of the literature on BLUPs.

- In practice, BLUPs are often not possible since we have to estimate the parameters, leading to eBLUPs. These eBLUPs do not quite follow the advertised distribution, and confidence sets do not have the right coverage, so some recommend other approaches; see, for example, Harville and Carriquiry [19].

8.9 Exercises

8.1. Find the probability that the ANOVA estimator $\hat{\sigma}_a^2$ from (8.11) is negative when $n_i = n = 5$ and $a = 3$, with σ_a^2/σ^2 ranging from 1/10 to 2.

8.2. Consider the balanced one-way random effects model. Construct a confidence interval for σ_a^2/σ^2 using Equation (8.9).

8.3. Show that the four sums of squares in the two-way crossed problem in Section 8.4, SSA, SSB, SSC, and SSE, are mutually independent.

8.4. Continue the ML calculations in Example 8.2 for the balanced case $n_i = n$.

8.5. Continue the REML calculations in Example 8.2 for the balanced case $n_i = n$.

8.6. (Neyman and Scott [33]) Let Y_{ij} iid $N(\mu_i, \sigma^2)$, $j = 1, \ldots, n$ (balanced) and $i = 1, \ldots, k$. Find the MLEs for μ_i and σ^2. What is their behavior (mean, variance) when n is fixed and $k \to \infty$?

8.7. For the problem in Exercise 8.6, compute the REML estimators (here $\mathbf{V} = \sigma^2\mathbf{I}_{nk}$), following Section 8.4B, and discuss their behavior when n is fixed and $k \to \infty$.

8.8. (Neyman and Scott [33]) Let Y_{ij} iid $N(\mu, \sigma_i^2)$, $j = 1, \ldots, n$ (balanced) and $i = 1, \ldots, k$. Find the MLEs for μ and σ_i^2. What is their behavior when n is fixed and $k \to \infty$?

8.9. Repeat the analysis of the two-way crossed model in Section 8.3, but in the case of no replication, that is, $n = 1$.

8.10. For the ANOVA estimators in Equations (8.10) and (8.11), find their variances and covariances.

8.11. For the ANOVA estimators in Equation (8.12), find their variances and covariances.

8.12. In Example 8.1, especially the continuation, show that $\mathbf{X}^T(\mathbf{P_Z} - \mathbf{P_1})\mathbf{X} = \mathbf{0}$ and $\mathbf{X}^T(\mathbf{I} - \mathbf{P_Z})\mathbf{X} = \mathbf{0}$.

8.13. A study of student performance may follow a balanced two-way nested model

$$y_{ijk} = \mu + \alpha_i + \beta_{ij} + e_{ijk}, i = 1, \ldots, a;\ j = 1, \ldots, b;\ k = 1, \ldots, c$$

where i may represent schools, j teachers within a school, and k individual students. The standard sums of squares decomposition follows:

$$SSA = \sum_i bc(y_{i..} - \overline{y}_{...})^2,$$

$$SSB(A) = \sum_i \sum_j c(y_{ij.} - \overline{y}_{i..})^2, \text{ and}$$

$$SSE = \sum_i \sum_j \sum_k (y_{ijk} - \overline{y}_{ij.})^2.$$

Use the tools outlined in Section 8.2 for the following problems.

a. Derive (show the steps, fill in the degrees of freedom, df) the following expected mean squares where all three effects are random, that is, $\alpha_i iidN(0, \sigma_a^2)$, $\beta_{ij} iidN(0, \sigma_b^2)$, and $e_{ijk} iidN(0, \sigma^2)$, and all three α_i, β_{ij}, and e_{ijk} mutually independent:

$$E(SSA/df) = \sigma^2 + c\sigma_b^2 + bc\sigma_a^2$$

$$E(SSB(A)/df) = \sigma^2 + c\sigma_b^2$$

$$E(SSE/df) = \sigma^2.$$

b. Derive the distribution of an appropriate test statistic for testing the hypothesis $H : \sigma_a^2 = 0$.

c. If α_i were fixed, and the other two effects random, derive the expected mean squares for the same three sums of squares.

d. Following the mixed model in (c), can you compute the expected mean squares if the problem were unbalanced, say, $i = 1, \ldots, a;\ j = 1, \ldots, b_i;\ k = 1, \ldots, c_{ij}$?

8.14. In the split-plot experiment in Section 8.5, show $(\mathbf{P_Z} - \mathbf{P_{X_A}})\mathbf{X} = \mathbf{0}$, and $(\mathbf{P_Z} - \mathbf{P_{X_A}})Cov(\mathbf{y})(\mathbf{P_{X_A}} - \mathbf{P_1}) = \mathbf{0}$.

8.15. Recall Exercise 6.17 and let's make the usual linear model's distributional assumptions, $\mathbf{y} \sim N_6(\mathbf{Xb}, \sigma^2\mathbf{I}_6)$. Now suppose these four rods were really a sample from a large population of rods, whose lengths could be modeled as coming from a normal population with mean μ and variance γ^2. We want to estimate γ^2 using the ANOVA method for variance components (essentially method of moments).

a. First let's get the model right. We have $\mathbf{y} = \mathbf{Xb} + \mathbf{e}$, where $\mathbf{e} \sim N_6(\mathbf{0}, \sigma^2 I_6)$ and, independently, $\mathbf{b} \sim N_4(\mu\mathbf{1}, \gamma^2 I_4)$. The response vector $\mathbf{y}$ will have a multivariate normal distribution; find its mean vector and covariance matrix.

b. Find the expectations for the two sums of squares given below.

$$E(\mathbf{y}^T(\mathbf{P_X} - \mathbf{P_1})\mathbf{y}) =$$

$$E(\mathbf{y}^T(\mathbf{I} - \mathbf{P_X})\mathbf{y}) =$$

c. Find the following traces (three of these four should be easy):

$$tr((\mathbf{P_X} - \mathbf{P_1})) =$$

$$tr((\mathbf{I} - \mathbf{P_X})) =$$

$$tr(\mathbf{X}^T(\mathbf{P_X} - \mathbf{P_1})\mathbf{X}) =$$

$$tr(\mathbf{X}^T(\mathbf{I} - \mathbf{P_X})\mathbf{X}) =$$

d. Construct estimators for σ^2 and γ^2 by equating the expectations of the sums of squares in (b) with their observed values $\mathbf{y}^T(\mathbf{P_X} - \mathbf{P_1})\mathbf{y}$ and $\mathbf{y}^T(\mathbf{I} - \mathbf{P_X})\mathbf{y}$.

8.16. For modeling the growth of trees, the following model is often employed: $y_{it} = \beta_0 + \alpha_i + \beta_1 t + e_{ij}$, $i = 1, \ldots, a$; $t = 1, \ldots, n$, where y_{it} measures the girth (circumference) of tree i at time t. Here β_0 represents an average intercept, β_1 a common slope (growth rate), and α_i a random effect due to the timing of the emergence of the sprout and the initial tagging of the tree. The errors would be modeled as usual as e_{it} iid $N(0, \sigma^2)$, and the random tree effect as α_i iid $N(0, \sigma_a^2)$, with both e_{it} and α_i independent.

a. For the simple case of $a = 3$ trees and $n = 4$ time points, select appropriate matrices and vectors for $\mathbf{X}$, $\mathbf{b}$, $\mathbf{Z}$, and $\mathbf{u}$ (four of them) from the possible matrices and vectors below, in order to write the model above in the general mixed-model form:

$$\mathbf{y} = \mathbf{Xb} + \mathbf{Zu} + \mathbf{e},$$

where $\mathbf{Xb}$ represents the fixed effects and $\mathbf{u}$ the random effects.

$$\mathbf{y} = \begin{bmatrix} y_{11} \\ y_{12} \\ y_{13} \\ y_{14} \\ y_{21} \\ y_{22} \\ y_{23} \\ y_{24} \\ y_{31} \\ y_{32} \\ y_{33} \\ y_{34} \end{bmatrix}, \quad \mathbf{e} = \begin{bmatrix} e_{11} \\ e_{12} \\ e_{13} \\ e_{14} \\ e_{21} \\ e_{22} \\ e_{23} \\ e_{24} \\ e_{31} \\ e_{32} \\ e_{33} \\ e_{34} \end{bmatrix} \begin{bmatrix} 1 & 0 & 0 \\ 1 & 0 & 0 \\ 1 & 0 & 0 \\ 1 & 0 & 0 \\ 0 & 1 & 0 \\ 0 & 1 & 0 \\ 0 & 1 & 0 \\ 0 & 1 & 0 \\ 0 & 0 & 1 \\ 0 & 0 & 1 \\ 0 & 0 & 1 \\ 0 & 0 & 1 \end{bmatrix} \begin{bmatrix} 1 \\ 1 \\ 1 \\ 1 \\ 1 \\ 1 \\ 1 \\ 1 \\ 1 \\ 1 \\ 1 \\ 1 \end{bmatrix} \begin{bmatrix} 1 & 0 & 0 \\ 2 & 0 & 0 \\ 3 & 0 & 0 \\ 4 & 0 & 0 \\ 0 & 1 & 0 \\ 0 & 2 & 0 \\ 0 & 3 & 0 \\ 0 & 4 & 0 \\ 0 & 0 & 1 \\ 0 & 0 & 2 \\ 0 & 0 & 3 \\ 0 & 0 & 4 \end{bmatrix} \begin{bmatrix} 1 & 1 \\ 1 & 2 \\ 1 & 3 \\ 1 & 4 \\ 1 & 1 \\ 1 & 2 \\ 1 & 3 \\ 1 & 4 \\ 1 & 1 \\ 1 & 2 \\ 1 & 3 \\ 1 & 4 \end{bmatrix}$$

$$\begin{bmatrix} 1 & 1 & 0 & 0 \\ 1 & 2 & 0 & 0 \\ 1 & 3 & 0 & 0 \\ 1 & 4 & 0 & 0 \\ 1 & 0 & 1 & 0 \\ 1 & 0 & 2 & 0 \\ 1 & 0 & 3 & 0 \\ 1 & 0 & 4 & 0 \\ 1 & 0 & 0 & 1 \\ 1 & 0 & 0 & 2 \\ 1 & 0 & 0 & 3 \\ 1 & 0 & 0 & 4 \end{bmatrix} \begin{bmatrix} \beta_0 \\ \beta_1 \end{bmatrix} \begin{bmatrix} 1 \\ 2 \\ 3 \\ 4 \end{bmatrix} \begin{bmatrix} 1 \\ 1 \\ 1 \\ 1 \end{bmatrix} \begin{bmatrix} \alpha_1 \\ \alpha_2 \\ \alpha_3 \end{bmatrix} \begin{bmatrix} \mu \end{bmatrix} \begin{bmatrix} \beta_1 \end{bmatrix} \begin{bmatrix} \beta_1 \\ \beta_2 \\ \beta_3 \\ \beta_4 \end{bmatrix} \begin{bmatrix} \beta_0 \\ \alpha_1 \\ \alpha_2 \\ \alpha_3 \end{bmatrix}$$

b. Give the distribution of your random effect vector $\mathbf{u}$.
c. We know that $(\mathbf{I} - \mathbf{P_X})\mathbf{Xb} = \mathbf{0}$. Show that $(\mathbf{P_Z} - \mathbf{P_1})\mathbf{Xb} = \mathbf{0}$ in this problem.
d. Find the following:

$$tr(\mathbf{I} - \mathbf{P_X}) =$$

$$tr(\mathbf{P_Z} - \mathbf{P_1}) =$$

$$tr(\mathbf{Z}^T(\mathbf{I} - \mathbf{P_X})\mathbf{Z}) =$$

$$tr(\mathbf{Z}^T(\mathbf{P_Z} - \mathbf{P_1})\mathbf{Z}) =$$

8.17. Fill in the missing algebra from Equation (8.26) to finding the BLUP for forecasting y_* in Equation (8.27).

8.18. Show that variance of the BLUP in Equation (8.27)) is

$$\omega^T(\boldsymbol{\Omega}^{-1} - \boldsymbol{\Omega}^{-1}\mathbf{X}(\mathbf{X}^T\boldsymbol{\Omega}^{-1}\mathbf{X})^g\mathbf{X}^T\boldsymbol{\Omega}^{-1})\omega + \mathbf{x}_*^T(\mathbf{X}^T\boldsymbol{\Omega}^{-1}\mathbf{X})^g\mathbf{x}_*.$$

8.19. Show that $\hat{\mathbf{b}}_{GLS}$ and $\hat{\mathbf{u}}$ satisfy the MME (Equation (8.28)), using the alternative expression for

$$\hat{\mathbf{u}} = (\mathbf{Z}^T\mathbf{R}^{-1}\mathbf{Z} + \mathbf{G}^{-1})^{-1}\mathbf{Z}^T\mathbf{R}^{-1}(\mathbf{y} - \mathbf{X}\hat{\mathbf{b}}_{GLS}).$$

8.20. Prove the alternative expression above in Exercise 8.19 using the binomial inverse theorem (Exercise A.75).

8.21. Show that BLUPs are uncorrelated with unbiased estimators of zero.

Chapter 9

The Multivariate Linear Model

9.1 Introduction

Our analysis of the multivariate linear model will attempt to repeat all of the previous material for the univariate case in one fell swoop for the multivariate case. Since the foundation for many of the issues has already been laid, this attempt may not be as ambitious as it may appear. Often, it will just be necessary to discuss how the multivariate case is similar or different from the univariate.

The main theme of the multivariate linear model is the change from a single response y_i for individual i to many responses for a single individual. In general, the different responses will have different units, as we may be looking at height, weight, and so on from a plant in response to fertilizer, water, or soil factors. In other cases, we may have multiple responses in the same units, such as concentrations of lead in different tissues, or blood pressure at different time points, for the same individual.

We will leap into Gauss–Markov estimation with the goal of repeating Chapters 2–4 for the multivariate case. Then we will look at maximum likelihood estimation and distribution theory in Section 9.3. Hypothesis testing in the multivariate case will be discussed in Section 9.4. Repeated measures problems, where the multiple responses are all the same type, will be treated in Section 9.5.

9.2 The Multivariate Gauss–Markov Model

The multivariate linear model is characterized by multiple responses; here we consider q responses from each individual. The multivariate linear model can then be written in a quite simple form as

$$\mathbf{Y} = \mathbf{XB} + \mathbf{E} \tag{9.1}$$

where $\mathbf{Y}$ is an $N \times q$ matrix of responses, $\mathbf{X}$ is an $N \times p$ design matrix, as used previously in this book, $\mathbf{B}$ is a $p \times q$ matrix of unknown coefficients, and $\mathbf{E}$ is an $N \times q$ matrix of errors. Here we assume that the design matrix $\mathbf{X}$ is the same for all q responses; the case where the design matrix may differ across responses leads to the *seemingly unrelated regression* model (Zellner [50]). Recall this issue was briefly discussed in Example 4.11.

Example 9.1: Simple Multivariate Linear Model
Consider a simple linear regression model with two responses, say

$$Y_{i1} = \text{weight of patient } i$$
$$Y_{i2} = \text{blood pressure of patient } i,$$

for $i = 1, \ldots, N$ and covariate x_i = dosage, so that

$$Y_{ij} = \mathbf{B}_{0j} + \mathbf{B}_{1j}x_i + e_{ij}$$

where $j = 1, 2$. In matrix form, with $N = 4$ patients for simplicity, the multivariate linear model in (9.1) takes the form

$$\begin{bmatrix} Y_{11} & Y_{12} \\ Y_{21} & Y_{22} \\ Y_{31} & Y_{32} \\ Y_{41} & Y_{42} \end{bmatrix} = \begin{bmatrix} 1 & x_2 \\ 1 & x_2 \\ 1 & x_3 \\ 1 & x_4 \end{bmatrix} \begin{bmatrix} B_{01} & B_{02} \\ B_{11} & B_{12} \end{bmatrix} + \begin{bmatrix} E_{11} & E_{12} \\ E_{21} & E_{22} \\ E_{31} & E_{32} \\ E_{41} & E_{42} \end{bmatrix}.$$

Note that the columns of **Y** and **B** correspond to the two responses (weight = 1, blood pressure = 2). The rows of **B** correspond to explanatory variables, as in the univariate simple linear regression. The rows of **Y**, **X**, and **E** correspond to the four individuals.

Example 9.2: Iris Data/One-Way MANOVA
This famous example includes four measurements (sepal length and width, petal length and width; sl, sw, pl, pw) on three species of iris (Setosa, Versicolor, Virginica; S,C,V). The model commonly employed for these data has the vector of four measurements $\mathbf{Y}^{ij}$ of individual j from species i having a multivariate normal distribution:

$$\mathbf{Y}^{ij} \sim N(\mu^i, \Sigma), i = 1, \ldots, a = 3;\ j = 1, \ldots, n_i,$$

so that a single measurement, say, petal length, $Y_3^{i,j}$ follows the usual one-way ANOVA model. A cell means model formulation for petal length would look like $Y_3^{ij} = \mu_3^i + E_3^{ij}$. In the form (9.1) with some different indexing (the usual indexing for a one-way model), we would have

$$\mathbf{XB} = \begin{bmatrix} \mathbf{1}_{n_S} & \mathbf{0} & \mathbf{0} \\ \mathbf{0} & \mathbf{1}_{n_C} & \mathbf{0} \\ \mathbf{0} & \mathbf{0} & \mathbf{1}_{n_V} \end{bmatrix} \begin{bmatrix} \mu_{sl}^S & \mu_{sw}^S & \mu_{pl}^S & \mu_{pw}^S \\ \mu_{sl}^C & \mu_{sw}^C & \mu_{pl}^C & \mu_{pw}^C \\ \mu_{sl}^V & \mu_{sw}^V & \mu_{pl}^V & \mu_{pw}^V \end{bmatrix}.$$

We will often look at this model by breaking it apart by either column (response) or row (individual). Dissecting by rows, we have for the ith individual, $i = 1, \ldots, N$,

$$\mathbf{Y}_{i.} = \mathbf{X}_{i.}\mathbf{B} + \mathbf{E}_{i.}$$

where $\mathbf{Y}_{i.}$ and $\mathbf{E}_{i.}$ are row vectors. To discuss the Gauss–Markov form of the multivariate linear model, let us transpose these row vectors to their usual orientation,

$$\mathbf{Y}_{i.}^T = \mathbf{B}^T\mathbf{X}_{i.}^T + \mathbf{E}_{i.}^T.$$

Now the error vectors here are usually assumed to be uncorrelated (more commonly, independent) across individuals, but the responses are not uncorrelated within an individual; hence, $E(\mathbf{E}_{i.}^T) = \mathbf{0}$ and $Cov(\mathbf{E}_{i.}^T) = \boldsymbol{\Sigma}$. The covariance relationships across both rows and columns are captured by

$$cov(E_{ij}, E_{st}) = \begin{cases} 0 & \text{if } i \neq s \\ \Sigma_{jt} & \text{if } i = s \end{cases}.$$

The mean and covariance within an individual can be written directly in terms of the response as

$$E\left(\mathbf{Y}_{i.}^T\right) = \mathbf{B}^T\mathbf{X}_{i.}^T, \text{ and } Cov\left(\mathbf{Y}_{i.}^T\right) = \boldsymbol{\Sigma} \tag{9.2}$$

or

$$cov(Y_{ij}, Y_{st}) = \begin{cases} 0 & \text{if } i \neq s \\ \Sigma_{jt} & \text{if } i = s \end{cases}.$$

This unknown covariance matrix $\boldsymbol{\Sigma}$ corresponds to the unknown variance σ^2 in the univariate case. We will assume that $\boldsymbol{\Sigma}$ is positive definite. Breaking apart by columns gives another view of this multivariate linear model column, as we have for the jth response, $j = 1, \ldots, q$,

$$\mathbf{Y}_{.j} = \mathbf{X}\mathbf{B}_{.j} + \mathbf{E}_{.j}, \tag{9.3}$$

which looks like the linear model considered for most of this book, with $E(\mathbf{E}_{.j}) = \mathbf{0}$ and $Cov(\mathbf{E}_{.j}) = \Sigma_{jj}\mathbf{I}_N$.

The Aitken form of (9.2) suggests that the least squares criterion of Chapter 2 should be modified to take into account the covariance across responses, leading to the natural extension to the generalized least squares criterion function of Chapter 4 where

$$Q(\mathbf{B}) = \sum_{i=1}^{N}(\mathbf{Y}_{i.} - \mathbf{X}_{i.}\mathbf{B})\boldsymbol{\Sigma}^{-1}(\mathbf{Y}_{i.} - \mathbf{X}_{i.}\mathbf{B})^T = trace\,[(\mathbf{Y} - \mathbf{X}\mathbf{B})\boldsymbol{\Sigma}^{-1}(\mathbf{Y} - \mathbf{X}\mathbf{B})^T].$$

At first glance, two directions are suggested in minimizing $Q(\mathbf{B})$. One route, following the column dissection, suggests a series of q univariate least squares problems, solving normal equations of the form

$$\mathbf{X}^T\mathbf{X}\mathbf{B}_{.j} = \mathbf{X}^T\mathbf{Y}_{.j}$$

for $j = 1, \ldots, q$, or one for each response. A second route, suggested by the Aitken form of (9.2), appears rather complicated. However, solving the multivariate normal equations is supported by the following result, using a matrix $\hat{\mathbf{B}}$ that solves to the multivariate normal equations (MNEs):

$$\mathbf{X}^T\mathbf{X}\mathbf{B} = \mathbf{X}^T\mathbf{Y}. \tag{9.4}$$

Result 9.1 *If the matrix* $\hat{\mathbf{B}}$ *solves the multivariate normal Equations (9.4), then* $\hat{\mathbf{B}}$ *minimizes* $Q(\mathbf{B})$.

Proof: Consider the following identity:

$$\begin{aligned} Q(\mathbf{B}) &= trace\,[(\mathbf{Y}-\mathbf{X}\mathbf{B})\Sigma^{-1}(\mathbf{Y}-\mathbf{X}\mathbf{B})^T] \\ &= tr[\Sigma^{-1}(\mathbf{Y}-\mathbf{X}\mathbf{B})^T(\mathbf{Y}-\mathbf{X}\mathbf{B})] \\ &= tr[\Sigma^{-1}(\mathbf{Y}-\mathbf{X}\hat{\mathbf{B}}+\mathbf{X}(\hat{\mathbf{B}}-\mathbf{B}))^T(\mathbf{Y}-\mathbf{X}\hat{\mathbf{B}}+\mathbf{X}(\hat{\mathbf{B}}-\mathbf{B}))] \\ &= tr[\Sigma^{-1}(\mathbf{Y}-\mathbf{X}\hat{\mathbf{B}})^T(\mathbf{Y}-\mathbf{X}\hat{\mathbf{B}})] + tr[\Sigma^{-1}(\hat{\mathbf{B}}-\mathbf{B})^T\mathbf{X}^T\mathbf{X}(\hat{\mathbf{B}}-\mathbf{B})] \\ &= Q(\hat{\mathbf{B}}) + tr[\Sigma^{-1}(\hat{\mathbf{B}}-\mathbf{B})^T\mathbf{X}^T\mathbf{X}(\hat{\mathbf{B}}-\mathbf{B})] \end{aligned}$$

since the cross-product term $(\hat{\mathbf{B}}-\mathbf{B})^T\mathbf{X}^T(\mathbf{Y}-\mathbf{X}\hat{\mathbf{B}})$ is zero. Rewriting the second part above into the trace of a nonnegative definite matrix (Exercise 9.1) should make it clear that $Q(\mathbf{B})$ is minimized at $\mathbf{B}=\hat{\mathbf{B}}$. □

Estimation of the covariance matrix follows the steps of Lemma 4.1, using a generalization of SSE with the error sum of squares and cross-products matrix $\mathbf{F}$, defined by

$$\mathbf{F} = (\mathbf{Y}-\mathbf{X}\hat{\mathbf{B}})^T(\mathbf{Y}-\mathbf{X}\hat{\mathbf{B}}). \tag{9.5}$$

Result 9.2 *The covariance matrix estimator*

$$\hat{\Sigma} = \frac{1}{N-rank\,(\mathbf{X})}\mathbf{F} = \frac{1}{N-rank\,(\mathbf{X})}(\mathbf{Y}-\mathbf{X}\hat{\mathbf{B}})^T(\mathbf{Y}-\mathbf{X}\hat{\mathbf{B}})$$

is an unbiased estimator of Σ.

Proof: The expectation of one entry of the matrix $\mathbf{F}$ is found by

$$\begin{aligned} E(F_{jk}) &= E((\mathbf{Y}_{.j}-\mathbf{X}\hat{\mathbf{B}}_{.j})^T(\mathbf{Y}_{.k}-\mathbf{X}\hat{\mathbf{B}}_{.k})) = E\left(\mathbf{Y}_{.j}^T(\mathbf{I}_N-\mathbf{P}_\mathbf{X})\mathbf{Y}_{.k}\right) \\ &= E\left[tr\left(\mathbf{Y}_{.j}^T(\mathbf{I}_N-\mathbf{P}_\mathbf{X})\mathbf{Y}_{.k}\right)\right] = E\left[tr\left((\mathbf{I}_N-\mathbf{P}_\mathbf{X})\mathbf{Y}_{.k}\mathbf{Y}_{.j}^T\right)\right] \\ &= tr\left[(\mathbf{I}_N-\mathbf{P}_\mathbf{X})\left(\mathbf{X}\mathbf{B}_{.k}\mathbf{B}_{.j}^T\mathbf{X}^T + (\mathbf{I}_N-\mathbf{P}_\mathbf{X})\Sigma_{jk}\right)\right] = (N-rank\,(\mathbf{X}))\Sigma_{jk}. \end{aligned}$$

□

See also Exercise 9.2.

Estimability in the multivariate case follows the univariate case, but the step is a subtle one. Estimating the scalar quantity $\lambda^T\mathbf{B}\mathbf{m}$ suggests taking the linear combination across responses expressed by the vector $\mathbf{m}$:

$$\mathbf{Y}\mathbf{m} = \mathbf{X}\mathbf{B}\mathbf{m} + \mathbf{E}\mathbf{m}. \tag{9.6}$$

We now have a vector response $\mathbf{Y}\mathbf{m}$, with the same design matrix $\mathbf{X}$, but with a coefficient vector $\mathbf{B}\mathbf{m}$ and error $\mathbf{E}\mathbf{m}$, whose covariance matrix is $Cov\,(\mathbf{E}\mathbf{m}) = (\mathbf{m}^T\Sigma\mathbf{m})\mathbf{I}_N = Cov\,(\mathbf{Y}\mathbf{m})$. The model in (9.6) now appears as a standard (univariate) linear model, and all the usual results follow that representation.

Definition 9.1 *A linear estimator of a scalar quantity $\lambda^T \mathbf{Bm}$ in the multivariate case is an estimator of the form $t(\mathbf{Y}) = \mathbf{t}^T \mathbf{Ym}$.*

Definition 9.2 *The scalar quantity $\lambda^T \mathbf{Bm}$ in the multivariate case is estimable iff a linear unbiased estimator of it exists.*

Result 9.3 *The scalar quantity $\lambda^T \mathbf{Bm}$ is estimable iff $\lambda \in \mathcal{C}(\mathbf{X}^T)$.*

Proof: See Exercise 9.3. □

Result 9.4 *If $\lambda^T \mathbf{Bm}$ is estimable, then $\lambda^T \hat{\mathbf{B}}\mathbf{m}$ is constant for all solutions $\hat{\mathbf{B}}$ of the multivariate normal equations.*

Proof: Suppose $\hat{\mathbf{B}}^{(1)}$ and $\hat{\mathbf{B}}^{(2)}$ are two solutions to the MNE. Then each column of $\hat{\mathbf{B}}^{(1)} - \hat{\mathbf{B}}^{(2)}$ is in $\mathcal{N}(\mathbf{X}^T\mathbf{X}) = \mathcal{N}(\mathbf{X})$, and so is any linear combination, such as $(\hat{\mathbf{B}}^{(1)} - \hat{\mathbf{B}}^{(2)})\mathbf{m}$; if $\lambda^T \mathbf{Bm}$ is estimable, then λ is in $\mathcal{C}(\mathbf{X}^T)$ and orthogonal to any vector in $\mathcal{N}(\mathbf{X})$. □

Following the usual rules, we can compute the mean and variance of this generalization of the least squares estimator.

$$\begin{aligned} E(\lambda^T \hat{\mathbf{B}}\mathbf{m}) &= \lambda^T E(\hat{\mathbf{B}})\mathbf{m} = \lambda^T (\mathbf{X}^T\mathbf{X})^g \mathbf{X}^T E(\mathbf{Y})\mathbf{m} \\ &= \lambda^T (\mathbf{X}^T\mathbf{X})^g \mathbf{X}^T \mathbf{X}\mathbf{B}\mathbf{m} = \lambda^T \mathbf{Bm} \end{aligned} \tag{9.7}$$

$$\begin{aligned} Var(\lambda^T \hat{\mathbf{B}}\mathbf{m}) &= \lambda^T (\mathbf{X}^T\mathbf{X})^g \mathbf{X}^T Cov(\mathbf{Ym})\mathbf{X}(\mathbf{X}^T\mathbf{X})^{gT} \lambda \\ &= \lambda^T (\mathbf{X}^T\mathbf{X})^g \mathbf{X}^T (\mathbf{m}^T \Sigma \mathbf{m})\mathbf{I}_N \mathbf{X}(\mathbf{X}^T\mathbf{X})^{gT} \lambda \\ &= (\mathbf{m}^T \Sigma \mathbf{m}) \lambda^T (\mathbf{X}^T\mathbf{X})^g \lambda \end{aligned} \tag{9.8}$$

Also see Exercise 9.4 for generalizations.

Result 9.5 *If $\hat{\mathbf{B}}$ solves the multivariate normal equations (MNEs), then the best linear unbiased estimator of estimable $\lambda^T \mathbf{Bm}$ is $\lambda^T \hat{\mathbf{B}}\mathbf{m}$.*

Proof: See Exercise 9.5. □

9.3 Inference under Normality Assumptions

As in the univariate case, we extend the Gauss–Markov model to include the assumption that the error distribution follows the multivariate normal distribution. In the case of multiple responses, observations across individuals are usually assumed to be independent, leading to $\mathbf{Y}_{i.}$ independent and

$$\mathbf{Y}_{i.}^T \sim N_q(\mathbf{B}\mathbf{X}_{i.}^T, \boldsymbol{\Sigma}), i = 1, \ldots, N. \tag{9.9}$$

This model then easily leads to the joint density of the data, and a likelihood function of the form

$$\begin{aligned}
log L(\mathbf{B}, \boldsymbol{\Sigma}) &= \ell(\mathbf{B}, \boldsymbol{\Sigma}) \\
&= -(Nq/2)log(2\pi) - (N/2)log(|\boldsymbol{\Sigma}|) - \frac{1}{2}\sum_{i=1}^{N}(\mathbf{Y}_{i.} - \mathbf{X}_{i.}\mathbf{B})\boldsymbol{\Sigma}^{-1}(\mathbf{Y}_{i.} - \mathbf{X}_{i.}\mathbf{B})^T \\
&= -(Nq/2)log(2\pi) - (N/2)log(|\boldsymbol{\Sigma}|) - \frac{1}{2}trace\,[(\mathbf{Y} - \mathbf{XB})\boldsymbol{\Sigma}^{-1}(\mathbf{Y} - \mathbf{XB})^T] \\
&= -\frac{Nq}{2}log(2\pi) - \frac{N}{2}log(|\boldsymbol{\Sigma}|) - \frac{1}{2}tr[\boldsymbol{\Sigma}^{-1}(\mathbf{Y} - \mathbf{X}\hat{\mathbf{B}})^T(\mathbf{Y} - \mathbf{X}\hat{\mathbf{B}})] \\
&\quad - \frac{1}{2}tr[\boldsymbol{\Sigma}^{-1}(\hat{\mathbf{B}} - \mathbf{B})^T\mathbf{X}^T\mathbf{X}(\hat{\mathbf{B}} - \mathbf{B})].
\end{aligned} \tag{9.10}$$

From the expression above, it should be clear that the log-likelihood $\ell(\mathbf{B}, \boldsymbol{\Sigma})$ can be maximized as a function of $\mathbf{B}$ by taking $\mathbf{B} = \hat{\mathbf{B}}$, leading to the concentrated or *profile likelihood function*

$$\begin{aligned}
\ell_*(\boldsymbol{\Sigma}) &= \ell(\hat{\mathbf{B}}, \boldsymbol{\Sigma}) \\
&= -\frac{Nq}{2}log(2\pi) - \frac{N}{2}log(|\boldsymbol{\Sigma}|) - \frac{1}{2}tr[\boldsymbol{\Sigma}^{-1}(\mathbf{Y} - \mathbf{X}\hat{\mathbf{B}})^T(\mathbf{Y} - \mathbf{X}\hat{\mathbf{B}})] \\
&= -\frac{Nq}{2}log(2\pi) - \frac{N}{2}log(|\boldsymbol{\Sigma}|) - \frac{1}{2}tr[\boldsymbol{\Sigma}^{-1}\mathbf{F}]
\end{aligned} \tag{9.11}$$

where $\mathbf{F}$ is the error sum of squares matrix defined in (9.5). Maximizing $\ell*$ as function of $\boldsymbol{\Sigma}$ can be tricky, but the following route (see, e.g., Mardia, et al. [29]) is remarkably clever and avoids derivatives.

Lemma 9.1 *Let $f(\mathbf{A}) = |\mathbf{A}|^{-k/2}exp\{-\frac{1}{2}trace\,(\mathbf{A}^{-1}\mathbf{F})\}$, where $\mathbf{F}$ is a symmetric, positive definite matrix of order p. Then $f(\mathbf{A})$ is maximized over positive definite matrices at $\mathbf{A} = \frac{1}{k}\mathbf{F}$ where $f(\frac{1}{k}\mathbf{F}) = |\frac{1}{k}\mathbf{F}|^{-k/2}e^{-kp/2}$.*

Proof: Let $s_1, s_2, \ldots, s_p$ be the eigenvalues of $\frac{1}{k}\mathbf{A}^{-1}\mathbf{F}$. Then

$$\begin{aligned}
logf\left(\frac{1}{k}\mathbf{F}\right) - logf(\mathbf{A}) &= -\frac{k}{2}log\left|\frac{1}{k}\mathbf{F}\right| - kp/2 + \frac{k}{2}log|\mathbf{A}| + \frac{1}{2}trace\,(\mathbf{A}^{-1}\mathbf{F}) \\
&= -\frac{k}{2}log\left(\prod_{i=1}^{p} s_i\right) - kp/2 + \frac{k}{2}\sum_{i=1}^{p} s_i \\
&= -\frac{k}{2}\left[\sum_{i=1}^{p} logs_i + p - \sum_{i=1}^{p} s_i\right].
\end{aligned} \tag{9.12}$$

Now since $log(x)$ is concave, it lies below its tangent line at $x = 1$; hence $log(x) \leq x - 1$, or $log(x) + 1 - x \leq 0$ for all positive x. Since all of the eigenvalues of $\frac{1}{k}\mathbf{A}^{-1}\mathbf{F}$ are positive (Exercise 9.6), applying this result to s_i makes the expression in the square brackets above [.] negative, and the left-hand side positive, or $logf(\frac{1}{k}\mathbf{F}) \geq logf(\mathbf{A})$. ▯

Result 9.6 *The maximum likelihood estimators of* $\mathbf{B}$ *and* $\mathbf{\Sigma}$ *in the multivariate linear model are the solution to the multivariate normal equations* $\hat{\mathbf{B}}$ *and* $\hat{\mathbf{\Sigma}} = \frac{1}{N}\mathbf{F} = \frac{1}{N}(\mathbf{Y} - \mathbf{X}\hat{\mathbf{B}})^T(\mathbf{Y} - \mathbf{X}\hat{\mathbf{B}})$.

Proof: Apply Lemma 9.1 to $\ell_*(\mathbf{\Sigma})$. ▯

Result 9.7 *If* $\lambda^T\mathbf{Bm}$ *is estimable, then* $\lambda^T\hat{\mathbf{B}}\mathbf{m}$ *is the MVU estimator of* $\lambda^T\mathbf{Bm}$, *and its distribution is*

$$\lambda^T\hat{\mathbf{B}}\mathbf{m} \sim N(\lambda^T\mathbf{Bm}, (\mathbf{m}^T\mathbf{\Sigma}\mathbf{m})\lambda^T(\mathbf{X}^T\mathbf{X})^g\lambda).$$

Proof: The distribution follows from (9.7); MVU follows from complete, sufficient statistics, similar to Corollary 6.4. See also Exercise 9.7. ▯

The distribution of the error sums of squares and cross-products matrix $\mathbf{F}$, however, requires some new definitions, including the *Wishart distribution*, which can be considered a generalization of the chi-square distribution.

Definition 9.3 *Let* $\mathbf{Z}^{(i)}$, $i = 1, \ldots, m$, *be iid* $N_q(\mathbf{0}, \mathbf{\Sigma})$, *then the* $q \times q$ *nonnegative definite matrix* $\mathbf{F} = \sum_{i=1}^m \mathbf{Z}^{(i)}\mathbf{Z}^{(i)T}$ *has the Wishart distribution with* m *degrees of freedom and scale matrix* $\mathbf{\Sigma}$, *or* $\mathbf{F} \sim W_q(m, \mathbf{\Sigma})$ *on the set* $\mathcal{V}_q$, *the set of symmetric, nonnegative definite* $q \times q$ *matrices.*

Result 9.8 *If* $\mathbf{F} \sim W_q(m, \mathbf{\Sigma})$, *then for any* $q \times 1$ *vector* $\mathbf{a}$, $\mathbf{a}^T\mathbf{Fa}/\mathbf{a}^T\mathbf{\Sigma a}$ *has the (central)* χ^2*-distribution with* m *degrees of freedom.*

Proof: Write $\mathbf{F} = \sum_{i=1}^m \mathbf{Z}^{(i)}\mathbf{Z}^{(i)T}$ so that $\mathbf{a}^T\mathbf{Fa} = \sum_{i=1}^m(\mathbf{a}^T\mathbf{Z}^{(i)})^2$ where $\mathbf{a}^T\mathbf{Z}^{(i)}$ are iid $N(0, \mathbf{a}^T\mathbf{\Sigma a})$. ▯

Result 9.9 *If* $\mathbf{F}_1 \sim W_q(m_1, \mathbf{\Sigma})$ *and, independently,* $\mathbf{F}_2 \sim W_q(m_2, \mathbf{\Sigma})$, *then* $\mathbf{F}_1 + \mathbf{F}_2 \sim W_q(m_1 + m_2, \mathbf{\Sigma})$.

Proof: Follows directly from the characterization in Definition 9.3. ▯

Result 9.10 *If* $\mathbf{F} \sim W_q(m, \mathbf{\Sigma})$, *and* $\mathbf{A}$ *is a* $p \times q$ *matrix, then* $\mathbf{AFA}^T \sim W_p(m, \mathbf{A\Sigma A}^T)$.

Proof: Again begin with $\mathbf{F} = \sum_{i=1}^m \mathbf{Z}^{(i)}\mathbf{Z}^{(i)T}$, so that

$$\mathbf{AFA}^T = \sum_{i=1}^{m} \mathbf{AZ}^{(i)}\mathbf{Z}^{(i)T}\mathbf{A}^T = \sum_{i=1}^{m} (\mathbf{AZ}^{(i)})(\mathbf{AZ}^{(i)})^T$$

where $\mathbf{AZ}^{(i)} \sim N_p(\mathbf{0}, \mathbf{A\Sigma A}^T)$. (Note that the new matrix is $p \times p$.) □

Result 9.11 *(Bartlett decomposition) Let* $\mathbf{F} \sim Wishart_q(m, \mathbf{I}_q)$ *with* $m \geq q$. *Let* $\mathbf{L}$ *be the Cholesky factor of* $\mathbf{F}$, *that is,* $\mathbf{F} = \mathbf{LL}^T$ *where* $\mathbf{L}$ *is lower triangular. Then we have*

- L_{kk}^2 *indep* χ_{m-k+1}^2
- L_{kj} *iid N(0, 1)*

Proof: Use the characterization of the Wishart from Definition 9.3 and let $\mathbf{F} = \sum_{i=1}^m \mathbf{Z}^{(i)}\mathbf{Z}^{(i)T}$. Start the Cholesky factorization with $L_{11}^2 = \sum_{i=1}^m Z_1^{(i)2}$, which is the sum of squares of standard normals, and hence $L_{11}^2 \sim \chi_m^2$. The induction step constructs the new row k of $\mathbf{L}$ (without the diagonal), from the factor of the first $k-1$ rows and columns $\mathbf{L}_{[k-1]}$ as $\ell_k = \mathbf{L}_{[k-1]}^{-1}\mathbf{f}_k$. This follows from the partitioning

$$\mathbf{L}_{[k]}\mathbf{L}_{[k]}^T = \begin{bmatrix} \mathbf{L}_{[k-1]} & \mathbf{0} \\ \ell_k^T & L_{kk} \end{bmatrix} \begin{bmatrix} \mathbf{L}_{[k-1]}^T & \ell_k \\ \mathbf{0} & L_{kk} \end{bmatrix} = \mathbf{F}_{[k]} = \begin{bmatrix} \mathbf{F}_{[k-1]} & \mathbf{f}_k \\ \mathbf{f}_k^T & F_{kk} \end{bmatrix}$$
$$= \begin{bmatrix} \mathbf{X}^T\mathbf{X} & \mathbf{X}^T\mathbf{y} \\ \mathbf{y}^T\mathbf{X} & \mathbf{y}^T\mathbf{y} \end{bmatrix}$$

with a slight abuse of notation, where $y_i = Z_k^{(i)}$ and the design matrix $X_{ij} = Z_j^{(i)}$ for $j = 1, \cdots, k-1$. Conditional on $\mathbf{X}$ (really the first $k-1$ components of the vectors $\mathbf{Z}^{(i)}$), $(\mathbf{X}^T\mathbf{X})^{-1}\mathbf{X}^T\mathbf{y} = \mathbf{L}_{[k-1]}^{-T}\mathbf{L}_{[k-1]}^{-1}\mathbf{f}_k$ is a least squares coefficient vector, and owing from independence, should be $N_{k-1}(\mathbf{0}, (\mathbf{X}^T\mathbf{X})^{-1} = \mathbf{L}_{[k-1]}^{-T}\mathbf{L}_{[k-1]}^{-1})$, so that premultiplying by $\mathbf{L}_{[k-1]}^T$ gives $\ell_k = \mathbf{L}_{[k-1]}^{-1}\mathbf{f}_k \sim N_{k-1}(0, \mathbf{I}_{k-1})$. The error sum of squares at this step gives the diagonal element of $\mathbf{L}$ as $L_{kk}^2 = \mathbf{y}^T\mathbf{y} - \mathbf{y}^T\mathbf{X}(\mathbf{X}^T\mathbf{X})^{-1}\mathbf{X}^T\mathbf{y}$, which has the χ_{m-k+1}^2 distribution. □

Theorem 9.1 *(Generalization of Cochran's Theorem) Let* $\mathbf{E}$ *follow the error distribution in the multivariate linear regression model (9.9), that is,* $\mathbf{E}_{i.}^T$ *indep* $N_q(\mathbf{0}, \mathbf{\Sigma})$, $i = 1, \ldots, N$, *where* $\mathbf{\Sigma}$ *is nonsingular. Also let* $\mathbf{A}_i$, $i = 1, \ldots, k$ *be symmetric, idempotent matrices with ranks* s_i, *respectively. If* $\sum_{i=1}^k \mathbf{A}_i = \mathbf{I}_N$, *then* $\mathbf{E}^T\mathbf{A}_i\mathbf{E}$ *are independently distributed as* $W_p(s_i, \mathbf{\Sigma})$, *and* $\sum_{i=1}^k s_i = N$.

Proof: Following the proof of Cochran's Theorem (Theorem 5.1), since $\mathbf{A}_i$ are idempotent, then by Lemma 5.17, we can write each one as $\mathbf{A}_i = \mathbf{Q}_i\mathbf{Q}_i^T$ where $\mathbf{Q}_i$ is a matrix of size $N \times s_i$ with rank s_i with orthonormal columns, that is, $\mathbf{Q}_i^T\mathbf{Q}_i = \mathbf{I}_{s_i}$.

The matrix $\mathbf{Q}$ by stacking these $\mathbf{Q}_i$ side by side,

$$\mathbf{Q} = [\mathbf{Q}_1 \quad \mathbf{Q}_2 \quad \ldots \quad \mathbf{Q}_k],$$

is an orthogonal matrix and $\sum_{i=1}^{k} s_i = N$. Since $\mathbf{\Sigma}$ is nonsingular, we can find a nonsingular matrix $\mathbf{L}$ such that $\mathbf{L}\mathbf{L}^T = \mathbf{\Sigma}$. Now construct $\mathbf{Z} = \mathbf{E}\mathbf{L}^{-T}$ and note that Z_{ij} iid $N(0, 1)$. Since the columns of $\mathbf{Z}$ are independent and $\mathbf{Q}$ is orthogonal, columns of the matrix $\mathbf{Q}^T\mathbf{Z}$ are $N_N(\mathbf{0}, \mathbf{I}_N)$ and also independent, composed of $N(0, 1)$ entries. Hence the elements of the $s_i \times p$ submatrices $\mathbf{Q}_i^T\mathbf{Z}$, $i = 1, \ldots, k$ are also independent, with each element iid $N(0, 1)$. Applying Definition 9.3, the $p \times p$ matrices $\mathbf{Z}^T\mathbf{Q}_i\mathbf{Q}_i^T\mathbf{Z}$ are each independent, with Wishart distributions, $W_p(s_i, \mathbf{I}_p)$. For the final step, since

$$\mathbf{Z}^T\mathbf{Q}_i\mathbf{Q}_i^T\mathbf{Z} = \mathbf{L}^{-1}\mathbf{E}^T\mathbf{Q}_i\mathbf{Q}_i^T\mathbf{E}\mathbf{L}^{-T} \sim W_p(s_i, \mathbf{I}_p)$$

and independent, applying Result 9.9 we have

$$\mathbf{L}\mathbf{Z}^T\mathbf{Q}_i\mathbf{Q}_i^T\mathbf{Z}\mathbf{L}^T = \mathbf{E}^T\mathbf{Q}_i\mathbf{Q}_i^T\mathbf{E} \sim W_p(s_i, \mathbf{L}(\mathbf{I}_p)\mathbf{L}^T) \text{ or } W_p(s_i, \mathbf{\Sigma}).$$

□

Result 9.12 *Let* $\mathbf{F} \sim \text{Wishart}_q(m, \mathbf{\Sigma})$ *with* $\mathbf{\Sigma}$ *positive definite, and partition both* $\mathbf{F}$ *and* $\mathbf{\Sigma}$ *similarly:*

$$\mathbf{F} = \begin{bmatrix} \mathbf{F}_{11} & \mathbf{F}_{12} \\ \mathbf{F}_{21} & \mathbf{F}_{22} \end{bmatrix} \begin{matrix} q_1 \\ q_2 \end{matrix} \quad \textit{and} \quad \mathbf{\Sigma} = \begin{bmatrix} \mathbf{\Sigma}_{11} & \mathbf{\Sigma}_{12} \\ \mathbf{\Sigma}_{21} & \mathbf{\Sigma}_{22} \end{bmatrix} \begin{matrix} q_1 \\ q_2 \end{matrix};$$

then $\mathbf{F}_{11} - \mathbf{F}_{12}\mathbf{F}_{22}^{-1}\mathbf{F}_{21} \sim \text{Wishart}_{q_1}(m - q_2, \mathbf{\Sigma}_{11} - \mathbf{\Sigma}_{12}\mathbf{\Sigma}_{22}^{-1}\mathbf{\Sigma}_{21})$.

Proof: Following the characterization of the Wishart distribution, partition the constituent matrix $\mathbf{U} = \begin{bmatrix} \mathbf{U}_1 & \mathbf{U}_2 \end{bmatrix}$ in the same fashion, where rows of $\mathbf{U}$, transposed, have the independent distribution

$$\mathbf{U}_{i.}^T = \begin{bmatrix} \mathbf{U}_{i1}^T \\ \mathbf{U}_{i2}^T \end{bmatrix} \sim N_q\left(\mathbf{0}, \mathbf{\Sigma} = \begin{bmatrix} \mathbf{\Sigma}_{11} & \mathbf{\Sigma}_{12} \\ \mathbf{\Sigma}_{21} & \mathbf{\Sigma}_{22} \end{bmatrix}\right),$$

as well as

$$\mathbf{F} = \begin{bmatrix} \mathbf{F}_{11} & \mathbf{F}_{12} \\ \mathbf{F}_{21} & \mathbf{F}_{22} \end{bmatrix} = \mathbf{U}^T\mathbf{U} = [\mathbf{U}_1 \quad \mathbf{U}_2]^T[\mathbf{U}_1 \quad \mathbf{U}_2].$$

Now transform as follows:

$$\mathbf{V}_{i.}^T = \begin{bmatrix} \mathbf{V}_{i1}^T \\ \mathbf{V}_{i2}^T \end{bmatrix} = \begin{bmatrix} \mathbf{I} & -\mathbf{\Sigma}_{12}\mathbf{\Sigma}_{22}^{-1} \\ \mathbf{0} & \mathbf{I} \end{bmatrix} \begin{bmatrix} \mathbf{U}_{i1}^T \\ \mathbf{U}_{i2}^T \end{bmatrix} \sim N_q\left(\mathbf{0}, \begin{bmatrix} \mathbf{\Sigma}_{11} - \mathbf{\Sigma}_{12}\mathbf{\Sigma}_{22}^{-1}\mathbf{\Sigma}_{21} & \mathbf{0} \\ \mathbf{0} & \mathbf{\Sigma}_{22} \end{bmatrix}\right)$$

so that $\mathbf{V}_{i1}$ and $\mathbf{V}_{i2}$ are independent. Then by the extension of Cochran's Theorem, Theorem 9.1 above, $\mathbf{V}_1^T(\mathbf{I} - \mathbf{P}_{\mathbf{V}_2})\mathbf{V}_1$ has the desired distribution. Remaining is some algebra:

$$\mathbf{V}_1^T(\mathbf{I} - \mathbf{P}_{\mathbf{V}_2})\mathbf{V}_1 = \mathbf{V}_1^T(\mathbf{I} - \mathbf{P}_{\mathbf{U}_2})\mathbf{V}_1 = \mathbf{U}_1^T(\mathbf{I} - \mathbf{P}_{\mathbf{U}_2})\mathbf{U}_1 = \mathbf{F}_{11} - \mathbf{F}_{12}\mathbf{F}_{22}^{-1}\mathbf{F}_{21}$$

since $(\mathbf{I} - \mathbf{P}_{\mathbf{U}_2})\mathbf{V}_1 = \mathbf{U}_1$. □

Corollary 9.1 *Let* $\mathbf{F} \sim Wishart_q(m, \mathbf{\Sigma})$, *and denote the upper-left corner of the inverse matrix* $\mathbf{F}^{-1}$ *partitioned as above in Result 9.12 by* $(\mathbf{F}^{-1})_{11}$. *Then* $((\mathbf{F}^{-1})_{11})^{-1} \sim Wishart_{q_1}(m - q_2, \mathbf{\Sigma}_{11} - \mathbf{\Sigma}_{12}\mathbf{\Sigma}_{22}^{-1}\mathbf{\Sigma}_{21})$.

Proof: Using the partitioned inverse result from Exercise A.72, we have $(\mathbf{F}^{-1})_{11} = (\mathbf{F}_{11} - \mathbf{F}_{12}\mathbf{F}_{22}^{-1}\mathbf{F}_{21})^{-1}$. ∎

9.4 Testing

In the usual case of a univariate response, we took two approaches to testing the general linear hypothesis. The first approach followed first principles and we could show that the resulting test was unbiased. The second approach was the likelihood ratio test, and we were able to show that the two approaches were identical. In the multivariate case, the situation is much more complicated.

We begin with the basic model of

$$\mathbf{Y} = \mathbf{XB} + \mathbf{E} \tag{9.13}$$

where the rows of $\mathbf{E}$ are independent with a multivariate normal distribution, that is, $\mathbf{E}_{i.}^T \sim N_q(\mathbf{0}, \mathbf{\Sigma})$. As before, the coefficient matrix $\mathbf{B}$ is unknown, as well as the covariance matrix $\mathbf{\Sigma}$. And we will briefly consider the most general form of the general linear hypothesis as $H : \mathbf{K}^T\mathbf{BL} = \mathbf{M}$ where $\mathbf{K}$ is $s \times p$ and follows the same restrictions as in Chapter 6. The matrix $\mathbf{L}$ is $q \times t$ and takes linear combinations across response coefficients, and the matrix $\mathbf{M}$ is $s \times t$ the known matrix of right-hand sides, often zero. In this discussion of testing, we will consider this only briefly for the following reasons. If we postmultiply Equation (9.13) by the matrix $\mathbf{L}$ we get

$$\mathbf{YL} = \mathbf{XBL} + \mathbf{EL}, \tag{9.14}$$

which now looks like the same form as (9.13) but with a new response matrix $\mathbf{Y}^* = \mathbf{YL}$, new coefficient matrix $\mathbf{B}^* = \mathbf{BL}$, and an error matrix $\mathbf{E}^* = \mathbf{EL}$ whose rows now have the covariance matrix $\mathbf{L}^T\mathbf{\Sigma}\mathbf{L}$. In this transformed model, the hypothesis to be tested takes the form $H : \mathbf{K}^T\mathbf{B}^* = \mathbf{M}^* = \mathbf{ML}$ with a new right-hand side. So for this discussion, we will need to only consider testing a hypothesis of the form $H : \mathbf{K}^T\mathbf{B} = \mathbf{M}$ with $\mathbf{M}$ being $s \times q$, since the more complicated form of the hypothesis can be transformed into the simpler general linear hypothesis.

Example 9.3: Testing a Subset
Consider the simplest model of a one-sample problem with mean vector μ, so that $\mathbf{XB} = \mathbf{1}\mu^T$. Then to test the hypothesis that a subset of the mean vector was zero, say, begin by partitioning,

$$\mu = \begin{bmatrix} \mu_1 \\ \mu_2 \end{bmatrix} \begin{matrix} q_1 \\ q_2 \end{matrix}$$

so that the hypothesis could be written as $H : \mu_2 = 0$. Then the matrix $\mathbf{L}$ above takes the form

$$\mathbf{L} = [\mathbf{0} \quad \mathbf{I}_s].$$

Now we merely test whether $\mathbf{Y}^* = \mathbf{YL}$ has a zero mean vector.

9.4.1 First Principles Again

The next step is to replicate some of the results of Chapter 6. Let us begin with the $N \times s$ matrix $\mathbf{W} = \mathbf{X}(\mathbf{X}^T\mathbf{X})^g\mathbf{K}$ so that $\mathbf{W}^T\mathbf{X} = \mathbf{K}^T$ and $\mathbf{W}^T\mathbf{W} = \mathbf{K}^T(\mathbf{X}^T\mathbf{X})^g\mathbf{K}$ are a nonsingular matrix. Generalizing the previous analysis, we can measure the departure of the estimated coefficient matrix from the hypothesized value using the matrix

$$\mathbf{G} = (\mathbf{K}^T\hat{\mathbf{B}} - \mathbf{M})^T(\mathbf{K}^T(\mathbf{X}^T\mathbf{X})^g\mathbf{K})^{-1}(\mathbf{K}^T\hat{\mathbf{B}} - \mathbf{M}).$$

To find the distribution of $\mathbf{G}$ when the hypothesis is true, begin with Equation (9.1) and multiply by $\mathbf{W}^T$ to obtain

$$\mathbf{W}^T\mathbf{Y} = \mathbf{K}^T\hat{\mathbf{B}} = \mathbf{K}^T\mathbf{B} + \mathbf{W}^T\mathbf{E}. \tag{9.15}$$

From the following analysis,

$$\begin{aligned}\mathbf{E}^T\mathbf{P_W}\mathbf{E} &= \mathbf{E}^T(\mathbf{W}^T\mathbf{W})^{-1}\mathbf{W}^T\mathbf{E}\\ &= (\mathbf{W}^T\mathbf{Y} - \mathbf{K}^T\mathbf{B})^T(\mathbf{W}^T\mathbf{W})^{-1}(\mathbf{W}^T\mathbf{Y} - \mathbf{K}^T\mathbf{B}),\end{aligned}$$

we see that if the hypothesis H is true, then

$$\mathbf{E}^T\mathbf{P_W}\mathbf{E} = (\mathbf{K}^T\hat{\mathbf{B}} - \mathbf{M})^T(\mathbf{K}^T(\mathbf{X}^T\mathbf{X})^g\mathbf{K})^{-1}(\mathbf{K}^T\hat{\mathbf{B}} - \mathbf{M}) = \mathbf{G}. \tag{9.16}$$

Using Theorem 9.1, when H is true, we have that $\mathbf{E}^T\mathbf{P_W}\mathbf{E} = \mathbf{G}$, $\mathbf{E}^T(\mathbf{I} - \mathbf{P_X})\mathbf{E} = \mathbf{Y}^T(\mathbf{I} - \mathbf{P_X})\mathbf{Y} = \mathbf{F}$, and $\mathbf{E}^T(\mathbf{P_X} - \mathbf{P_W})\mathbf{E}$ are independent Wishart matrices. In particular, when H is true we have the distributions of these matrices: $\mathbf{F} \sim \mathit{Wishart}(N - r, \boldsymbol{\Sigma})$ and $\mathbf{G} \sim \mathit{Wishart}(s, \boldsymbol{\Sigma})$. If the alternative is true, we still have $\mathbf{E}^T\mathbf{P_W}\mathbf{E} \sim \mathit{Wishart}(s, \boldsymbol{\Sigma})$; however,

$$\mathbf{E}^T\mathbf{P_W}\mathbf{E} = (\mathbf{K}^T\hat{\mathbf{B}} - \mathbf{K}^T\mathbf{B})^T(\mathbf{K}^T(\mathbf{X}^T\mathbf{X})^g\mathbf{K})^{-1}(\mathbf{K}^T\hat{\mathbf{B}} - \mathbf{K}^T\mathbf{B}),$$

but $\mathbf{G} \neq \mathbf{E}^T\mathbf{P_w}\mathbf{E}$ since $\mathbf{K}^T\mathbf{B} \neq \mathbf{M}$. We would like to say that the matrix $\mathbf{G}$ is stochastically larger than a $\mathit{Wishart}(s, \boldsymbol{\Sigma})$ random matrix, but we can't make such a general statement in the multivariate case. If we take quadratic forms, say, from taking linear combinations across responses, then we can say that

$$\mathbf{v}^T\mathbf{G}\mathbf{v}/\mathbf{v}^T\boldsymbol{\Sigma}\mathbf{v} \sim \chi_s^2(\mathbf{v}^T(\mathbf{K}^T\mathbf{B} - \mathbf{M})^T(\mathbf{K}^T(\mathbf{X}^T\mathbf{X})^g\mathbf{K})^{-1}(\mathbf{K}^T\mathbf{B} - \mathbf{M})\mathbf{v}/(2\mathbf{v}^T\boldsymbol{\Sigma}\mathbf{v})),$$

and construct the same type of F-statistic as in Chapter 6 and claim that the resulting test is unbiased, but there is little we can do directly with $\mathbf{G}$, except in the $s = 1$ case. In a different sense of univariate, the matrix $\mathbf{K}$ is now just a vector $\mathbf{k}$, as is

$(\mathbf{K}^T\hat{\mathbf{B}} - \mathbf{M})^T = \mathbf{d}$ for simplicity, and we can measure distance from the hypothesis in a natural way using the inverse of the covariance matrix $\mathbf{\Sigma}$ or an estimate based on $\mathbf{F}$. That is, the squared distance from the sample $\mathbf{d}$ from the hypothesized $\mathbf{0}$ takes the form $\mathbf{d}^T [Cov(\mathbf{d})]^{-1} \mathbf{d}$. Since $Cov(\mathbf{d}) = c\mathbf{\Sigma}$ where the scalar $c = \mathbf{k}^T(\mathbf{X}^T\mathbf{X})^{-1}\mathbf{k}$, following this analogy leads to *Hotelling's* T^2 statistic:

$$T^2 = \frac{N-r}{c}(\mathbf{k}^T\hat{\mathbf{B}} - \mathbf{M})\mathbf{F}^{-1}(\mathbf{k}^T\hat{\mathbf{B}} - \mathbf{M})^T = \frac{N-r}{c}\mathbf{d}^T\mathbf{F}^{-1}\mathbf{d}.$$

Properly scaled, T^2 has an F-distribution.

Lemma 9.2 *Let* $\mathbf{F} \sim Wishart_q(k, \mathbf{\Sigma})$ *with* $k > q$*; then* $\frac{\mathbf{e}^{(1)T}\mathbf{\Sigma}^{-1}\mathbf{e}^{(1)}}{\mathbf{e}^{(1)T}\mathbf{F}^{-1}\mathbf{e}^{(1)}} = \frac{(\mathbf{\Sigma}^{-1})_{11}}{(\mathbf{F}^{-1})_{11}} \sim \chi^2_{k-q+1}$

Proof: Apply Corollary 9.1 to the $(1, 1)$ element of $\mathbf{F}$ and employ Result 9.9 for a one-dimensional problem. □

Lemma 9.3 *Let* $\mathbf{F} \sim Wishart_q(k, \mathbf{\Sigma})$*; then for any vector* $\mathbf{d}$, $\mathbf{d}^T\mathbf{\Sigma}^{-1}\mathbf{d}/\mathbf{d}^T\mathbf{F}^{-1}\mathbf{d} \sim \chi^2_{k-q+1}$.

Proof: Let $\mathbf{Q}$ be an orthogonal matrix with first row $\|\mathbf{d}\|^{-1}\mathbf{d}^T$. Then since $\mathbf{QFQ}^T \sim Wishart_q(k, \mathbf{Q\Sigma Q}^T)$, just apply the previous lemma since $(\mathbf{Q\Sigma Q}^T)^{-1} = \mathbf{Q\Sigma}^{-1}\mathbf{Q}^T$, $\mathbf{Q}^T\mathbf{e}^{(1)} = \|\mathbf{d}\|^{-1}\mathbf{d}$, and the $\|\mathbf{d}\|$'s cancel in the ratio. □

Result 9.13 *Let* $\mathbf{F} \sim Wishart_q(N-r, \mathbf{\Sigma})$ *and, independently,* $\mathbf{d} \sim N_q(\delta, c\mathbf{\Sigma})$. *Then*

$$\frac{N-r-q+1}{q}\frac{T^2}{N-r} \sim F_{q,N-r-q+1}(\delta^T\mathbf{\Sigma}^{-1}\delta/2).$$

Proof: Begin by writing Hotelling's T^2 as

$$T^2 = \frac{N-r}{c}\mathbf{d}^T\mathbf{F}^{-1}\mathbf{d} = (N-r)\frac{c^{-1}\mathbf{d}^T\mathbf{\Sigma}^{-1}\mathbf{d}}{\frac{\mathbf{d}^T\mathbf{\Sigma}^{-1}\mathbf{d}}{\mathbf{d}^T\mathbf{F}^{-1}\mathbf{d}}}.$$

First we have from the previous lemma that the denominator has a $\chi^2_{N-r-q+1}$ distribution for any $\mathbf{d}$. Hence, the distribution of the denominator does not depend on $\mathbf{d}$, and so is independent of $\mathbf{d}$ and independent of the numerator. From Result 5.10, we have under the alternative $c^{-1}\mathbf{d}^T\mathbf{\Sigma}^{-1}\mathbf{d} \sim \chi^2_{q-1}(\delta^T\mathbf{\Sigma}^{-1}\delta/2)$ (and central χ^2 under the hypothesis). Multiplying by the right constants, we get the desired F-distribution. □

Example 9.4: Hotelling's T^2 in the Two Sample Case
Consider the two-sample problem with a cell means parameterization, so that

$$\mathbf{XB} = \begin{bmatrix} \mathbf{1}_{n_1} & \mathbf{0} \\ \mathbf{0} & \mathbf{1}_{n_2} \end{bmatrix} \begin{bmatrix} \mu_1^T \\ \mu_2^T \end{bmatrix}, \hat{\mathbf{B}} = \begin{bmatrix} \overline{\mathbf{Y}}_1^T \\ \overline{\mathbf{Y}}_2^T \end{bmatrix}$$

where $\overline{\mathbf{Y}}_1 = n_1^{-1}\sum_{i=1}^{n_1}\mathbf{Y}_{i.}^T$ and $\overline{\mathbf{Y}}_2 = n_2^{-1}\sum_{i=n_1+1}^{n_1+n_2}\mathbf{Y}_{i.}^T$. The hypothesis can be written as $H : \mu_1 = \mu_2$ or as $\mathbf{k}^T\mathbf{B} = \mathbf{0}$ with $\mathbf{k}^T = (1, -1)$, so that $\mathbf{d} = \overline{\mathbf{Y}}_1 - \overline{\mathbf{Y}}_2$. The scalar is then $c = n_1^{-1} + n_2^{-1}$ and $\mathbf{F}$ is the pooled sum of squares matrix

$$\mathbf{F} = \sum_{i=1}^{n_1}(\mathbf{Y}_{i.} - \overline{\mathbf{Y}}_1^T)(\mathbf{Y}_{i.}^T - \overline{\mathbf{Y}}_1) + \sum_{i=n_1+1}^{n_1+n_2}(\mathbf{Y}_{i.} - \overline{\mathbf{Y}}_2^T)(\mathbf{Y}_{i.}^T - \overline{\mathbf{Y}}_2).$$

Using the result above with $r = 2$, $N = n_1 + n_2$, and $\delta = \mu_1 - \mu_2$, we have $T^2 = ((N-2)/c)\mathbf{d}^T\mathbf{F}^{-1}\mathbf{d}$ and

$$\frac{N-2-q+1}{q}\frac{T^2}{N-2} = \frac{n_1+n_2-q-1}{q}\frac{\mathbf{d}^T\mathbf{F}^{-1}\mathbf{d}}{c} \sim F_{q,n_1+n_2-q-1}(\delta^T\mathbf{\Sigma}^{-1}\delta/2).$$

The following sections address various approaches for constructing multivariate tests for the general linear hypothesis. The basic outline of the problem is that when the hypothesis is true, $\mathbf{F} \sim \mathit{Wishart}(N-r, \mathbf{\Sigma})$ and $\mathbf{G} \sim \mathit{Wishart}(s, \mathbf{\Sigma})$. When the alternative is true, $\mathbf{G}$ is in some ways larger. All of these approaches attempt to measure how big $\mathbf{G}$ is compared to what it should be, with $\mathbf{F}$ and the degrees of freedom providing the basis for the comparison. All of these approaches coincide in the univariate $q = 1$ case (Exercise 9.16), as they attempt to generalize the F-statistic from the univariate case. All of the test statistics that arise are functions of the nonzero eigenvalues of $\mathbf{F}^{-1}\mathbf{G}$. The following result can be employed with all of these approaches.

Result 9.14 *Let* $\mathbf{A} \sim \mathit{Wishart}_p(m, \mathbf{I})$, $\mathbf{B} \sim \mathit{Wishart}_p(n, \mathbf{I})$, *with* $\mathbf{A}$ *and* $\mathbf{B}$ *independent, with* $m \geq p$ *and* $n \leq p$. *Then the distribution of the nonzero eigenvalues of* $\mathbf{A}^{-1}\mathbf{B}$ *is the same as if* $\mathbf{A} \sim \mathit{Wishart}_n(m-p+n, \mathbf{I})$ *and* $\mathbf{B} \sim \mathit{Wishart}_n(p, \mathbf{I})$.

Proof: Let $\mathbf{Q}$ be an orthogonal matrix, partitioned into p_1 and p_2 columns as

$$\mathbf{Q} = [\mathbf{Q}_1 \quad \mathbf{Q}_2].$$

Then since $\mathbf{Q}^T\mathbf{A}\mathbf{Q} \sim \mathit{Wishart}_p(m, \mathbf{I})$, using Corollary 9.1 we have

$$(\mathbf{Q}^T\mathbf{A}\mathbf{Q})^{-1} = \mathbf{Q}^T\mathbf{A}^{-1}\mathbf{Q} = \begin{bmatrix}\mathbf{Q}_1^T \\ \mathbf{Q}_2^T\end{bmatrix}\mathbf{A}[\mathbf{Q}_1 \quad \mathbf{Q}_2],$$

so that $(\mathbf{Q}_1^T\mathbf{A}^{-1}\mathbf{Q}_1)^{-1} \sim \mathit{Wishart}_{p_1}(m - p_2, \mathbf{I})$. Now write $\mathbf{B} = \mathbf{X}\mathbf{X}^T$ where $\mathbf{X}$ is a $p \times n$ matrix with its entries iid $N(0, 1)$. Since $\mathbf{X}^T\mathbf{X}$ is positive definite, write $\mathbf{X}^T\mathbf{X} = \mathbf{V}^2$ using the symmetric square root $\mathbf{V}$. Then let $\mathbf{Q}_1 = \mathbf{X}\mathbf{V}^{-1}$ so that $\mathbf{Q}_1$ is $p \times n$ with orthonormal columns so $\mathbf{Q}_1^T\mathbf{Q}_1 = \mathbf{I}$ and $\mathbf{P}_\mathbf{X} = \mathbf{Q}_1\mathbf{Q}_1^T$. Then the nonzero eigenvalues of

$$\mathbf{A}^{-1}\mathbf{B} = \mathbf{A}^{-1}\mathbf{X}\mathbf{X}^T = \mathbf{A}^{-1}\mathbf{Q}_1\mathbf{Q}_1^T\mathbf{X}\mathbf{X}^T\mathbf{Q}_1\mathbf{Q}_1^T$$

are the same as the nonzero eigenvalues of

$$[\mathbf{Q}_1^T \mathbf{A}^{-1} \mathbf{Q}_1] \, \{\mathbf{Q}_1^T \mathbf{X}\mathbf{X}^T \mathbf{Q}_1\}.$$

Now we have established the distribution of the first matrix [.], with $p_1 = n$ and $p_2 = p - n$, irrespective of the orthogonal matrix. Even though $\mathbf{Q}$ depends on $\mathbf{X}$, the conditional distribution doesn't depend on $\mathbf{X}$, and hence the first matrix [.] is independent of $\mathbf{X}$. Since $\mathbf{X}^T \mathbf{Q}_1 = \mathbf{X}^T \mathbf{X} \mathbf{V}^{-1} = \mathbf{V}$, the second matrix {.} is

$$\{\mathbf{Q}_1^T \mathbf{X}\mathbf{X}^T \mathbf{Q}_1\} = \mathbf{V}^2 = \mathbf{X}^T \mathbf{X},$$

which has the $Wishart_n(p, \mathbf{I})$ distribution. □

9.4.2 Likelihood Ratio Test and Wilks' Lambda

As in Chapter 6, the likelihood ratio test is designed to look at the ratio λ of likelihoods, where the numerator is restricted to the hypothesis, the denominator is unrestricted, and the hypothesis is to be rejected when this ratio is too small. The analysis is straightforward in the unrestricted case, where the likelihood is maximized over $\mathbf{B}$ at the least squares matrix $\hat{\mathbf{B}}$ and, following Result 6, the log of the unrestricted likelihood takes the form

$$\ell\left(\hat{\mathbf{B}}, \frac{1}{N}\mathbf{F}\right) = -\frac{Nq}{2} log(2\pi) - \frac{N}{2} log|\mathbf{F}| + \frac{Nq}{2} log(N) - \frac{Nq}{2}$$

where $\mathbf{F} = (\mathbf{Y} - \mathbf{X}\hat{\mathbf{B}})^T(\mathbf{Y} - \mathbf{X}\hat{\mathbf{B}})$. For the restricted case, following similar analysis and posing the existence of $\hat{\mathbf{B}}_H$ that achieves the appropriate minimum of the log likelihood, we reach

$$\ell\left(\hat{\mathbf{B}}_H, \frac{1}{N}\mathbf{F}_H\right) = -\frac{Nq}{2} log(2\pi) - \frac{N}{2} log|\mathbf{F}_H| + \frac{Nq}{2} log(N) - \frac{Nq}{2}$$

where $\mathbf{F}_H = (\mathbf{Y} - \mathbf{X}\hat{\mathbf{B}}_H)^T(\mathbf{Y} - \mathbf{X}\hat{\mathbf{B}}_H)$.

Example 9.2: continued
Recall the iris data one-way ANOVA example. Following the same analysis as in Chapter 6, we have

$$\mathbf{F} = \sum_{i=1}^{3}\sum_{j=1}^{n_i}(\mathbf{Y}^{ij} - \overline{\mathbf{Y}}^{i\cdot})(\mathbf{Y}^{ij} - \overline{\mathbf{Y}}^{i\cdot})^T,$$

and the restricted error sum of squares and cross-products matrix is

$$\mathbf{F}_H = \sum_{i=1}^{3}\sum_{j=1}^{n_i}(\mathbf{Y}^{ij} - \overline{\mathbf{Y}}^{\cdot\cdot})(\mathbf{Y}^{ij} - \overline{\mathbf{Y}}^{\cdot\cdot})^T.$$

Now when this same problem was faced in the univariate case in Chapter 6, we posed the existence of the minimizing coefficient vector and then recalled the solution to the constrained parameter problem from Section 3.9 using the restricted normal equations (RNEs). We then showed that the change in SSE led to the same statistic as in the first principles approach. In the multivariate case, these steps are similar, but different enough to suggest great care in filling in the details. And, of course, we will get a different test statistic and different distributions when the hypothesis is true.

The first step is to show that maximizing the likelihood under the constraint of the hypothesis leads to the same form of RNE. Following the analysis from the previous section, we have in Equation (9.10)

$$\ell(\mathbf{B}, \Sigma) - \frac{Nq}{2} log(2\pi) - \frac{N}{2} log(|\Sigma|) - \frac{1}{2} tr[\Sigma^{-1}(\mathbf{Y} - \mathbf{XB})^T(\mathbf{Y} - \mathbf{XB})].$$

Denoting the elements of the inverse of Σ as $\Sigma^{ij} = (\Sigma^{-1})_{ij}$, we can rewrite the trace part of the likelihood above – the only part that is a function of $\mathbf{B}$ – as

$$\sum_i \sum_j \Sigma^{ij}(\mathbf{Y}_{.J} - \mathbf{XB}_{.j})^T(\mathbf{Y}_{.i} - \mathbf{XB}_{.i}).$$

Now minimizing this with respect to $\mathbf{B}$ and subject to the constraint $\mathbf{K}^T\mathbf{B} = \mathbf{M}$ leads to the Lagrangian function (see Appendix B)

$$G_*(\mathbf{B}, \mathbf{\Gamma}) = \sum_i \sum_j \Sigma^{ij}(\mathbf{Y}_{.j} - \mathbf{XB}_{.j})^T(\mathbf{Y}_{.i} - \mathbf{XB}_{.i}) + 2\sum_j \mathbf{\Gamma}_{.j}^T(\mathbf{K}^T\mathbf{B}_{.j} - \mathbf{M}_{.j}) \tag{9.17}$$

with Lagrange multiplier matrix $\mathbf{\Gamma}$. Taking partial derivatives with respect to column m of $\mathbf{B}$ or $\mathbf{B}_{.m}$ gives

$$\frac{\partial G_*}{\partial \mathbf{B}_{.m}} = -2\sum_i \Sigma^{im}\mathbf{X}^T(\mathbf{Y}_{.i} - \mathbf{XB}_{.i}) + 2\mathbf{K\Gamma}_{.m}$$
$$= -2\mathbf{X}^T(\mathbf{Y} - \mathbf{XB})(\Sigma^{-1})_{.m} + 2\mathbf{K\Gamma}_{.m} \tag{9.18}$$

Dividing by 2 and stacking the equations from (9.18) side-by-side we get

$$-(\mathbf{X}^T\mathbf{Y} - \mathbf{X}^T\mathbf{XB})\Sigma^{-1} + \mathbf{K\Gamma} = \mathbf{0}$$

or multiplying through by Σ on the right leads to

$$\mathbf{X}^T\mathbf{XB} + \mathbf{K\Gamma\Sigma} = \mathbf{X}^T\mathbf{Y}.$$

Now adding in the linear constraint directly (as the derivative with respect to the Lagrange multiplier) and stacking for convenience, we arrive at the more familiar form

$$\begin{bmatrix} \mathbf{X}^T\mathbf{X} & \mathbf{K} \\ \mathbf{K}^T & \mathbf{0} \end{bmatrix} \begin{bmatrix} \mathbf{B} \\ \Theta \end{bmatrix} = \begin{bmatrix} \mathbf{X}^T\mathbf{Y} \\ \mathbf{M} \end{bmatrix} \tag{9.19}$$

in terms of a new matrix of Lagrange multipliers $\Theta = \boldsymbol{\Gamma\Sigma}$. Leaning on Result 3.8, we can now claim that this multivariate version of the RNE also has a solution matrix $\hat{\mathbf{b}}_H$ (and Lagrange multiplier Θ_H) that minimizes the trace and thereby maximizing the likelihood. The next step is to reproduce Theorem 6.1 for the multivariate case, showing that

$$\begin{aligned}(\mathbf{Y} - \mathbf{X}\hat{\mathbf{B}}_H)^T(\mathbf{Y} - \mathbf{X}\hat{\mathbf{B}}_H) &- (\mathbf{Y} - \mathbf{X}\hat{\mathbf{B}})^T(\mathbf{Y} - \mathbf{X}\hat{\mathbf{B}}) \\ &= (\mathbf{K}^T\hat{\mathbf{B}} - \mathbf{M})^T(\mathbf{K}^T(\mathbf{X}^T\mathbf{X})^g\mathbf{K})^{-1}(\mathbf{K}^T\hat{\mathbf{B}} - \mathbf{M}) = \mathbf{G} \end{aligned} \tag{9.20}$$

(see Exercise 9.17).

Example 9.3: Iris/One-Way MANOVA continued
The usual steps of subtracting and adding $\overline{\mathbf{Y}}^{i\cdot}$ can be employed to show the multivariate version of the familiar result

$$\mathbf{F}_H - \mathbf{F} = \sum_{i=1}^{3} n_i(\overline{\mathbf{Y}}^{i\cdot} - \overline{\mathbf{Y}}^{\cdot\cdot})(\overline{\mathbf{Y}}^{i\cdot} - \overline{\mathbf{Y}}^{\cdot\cdot})^T.$$

However, showing that this same expression equals

$$(\mathbf{K}^T\hat{\mathbf{B}} - \mathbf{M})^T(\mathbf{K}^T(\mathbf{X}^T\mathbf{X})^g\mathbf{K})^{-1}(\mathbf{K}^T\hat{\mathbf{B}} - \mathbf{M}) = \mathbf{G}$$

requires the same several steps as Example 6.5.

Now we can return to the likelihood ratio to show

$$\phi(\mathbf{Y}) = exp\left[\ell\left(\hat{\mathbf{b}}_H, \frac{1}{N}(\mathbf{F} + \mathbf{G})\right) - \ell\left(\hat{\mathbf{b}}, \frac{1}{N}\mathbf{F}\right)\right]$$

or the more familiar form

$$\phi(\mathbf{Y})^{2/N} = \frac{|\mathbf{F}|}{|\mathbf{F} + \mathbf{G}|} = \Lambda^*, \tag{9.21}$$

known as *Wilks' lambda*. Note that the distribution of Λ^* does not depend on the covariance matrix $\boldsymbol{\Sigma}$.

Definition 9.4 *Let* $\mathbf{A} \sim Wishart_q(r, \mathbf{I}_q)$, $\mathbf{B} \sim Wishart_q(s, \mathbf{I}_q)$, *with* $\mathbf{A}$ *and* $\mathbf{B}$ *independent, and* $r \geq q$*; then* $U = |\mathbf{A}|/|\mathbf{A} + \mathbf{B}|$ *has the Wilks' lambda distribution with parameters* q, r, *and* s, *denoted* $U(q, r, s)$.

The likelihood ratio test criterion leads to the test that rejects for small values of $\Lambda^* = \frac{|\mathbf{F}|}{|\mathbf{F}+\mathbf{G}|}$ (corresponding in a general sense to large values of $\mathbf{G}$) where the distribution of $\Lambda^* \sim U(q, N - r, s)$ can be found exactly in some special cases.

Result 9.15 *The distributions $U(q, r, s)$ and $U(s, r+s-q, q)$ are the same.*

Proof: Use Result 9.14. ▯

Result 9.16 *The exact distribution of Wilks' lambda can be found in some special cases, where $q = 1$ or $q = 2$, or where $s = 1$ or $s = 2$:*

(a) If $U \sim U(1, N-r, s)$, then $\frac{N-r}{s} \times \frac{1-U}{U} \sim F(s, N-r)$.

(b) If $U \sim U(2, N-r, s)$, then $\frac{N-r-1}{s} \times \frac{1-\sqrt{U}}{\sqrt{U}} \sim F(2s, 2(N-r-1))$.

(c) If $U \sim U(q, N-r, 1)$, then $\frac{N-r-q+1}{q} \times \frac{1-U}{U} \sim F(q, N-r-q+1)$.

(d) If $U \sim U(q, N-r, 2)$, then $\frac{N-r-q+1}{q} \frac{1-\sqrt{U}}{\sqrt{U}} \sim F(2q, 2(N-r-q+1))$.

Proof: For (a) and (c), use the distribution of Hotelling's T^2 from Result 9.13 and Result 9.14. The proofs for (b) and (d) are beyond the realm of this book; see, for example, Anderson [2], Section 8.4. ▯

9.4.3 Other Test Statistics

As outlined previously, the various approaches for testing in the multivariate case are variations on measuring the size of matrix $\mathbf{G}$ compared to $\mathbf{F}$. We could just as well measure the size of $\mathbf{F}^{-1}\mathbf{G}$ or its symmetrized form $\mathbf{F}^{-1/2}\mathbf{G}\mathbf{F}^{-1/2}$ using either the Cholesky or symmetric square root. Again notice that the covariance matrix $\boldsymbol{\Sigma}$ will drop out of the distribution. If we denote the eigenvalues of $\mathbf{F}^{-1}\mathbf{G}$ by $\lambda_1 \geq \lambda_2 \geq \ldots \geq 0$, all of these methods are functions of these eigenvalues. For Wilks' lambda that we just examined, we have the relationship

$$\Lambda^* = |\mathbf{F}|/|\mathbf{F}+\mathbf{G}| = 1/|\mathbf{I}+\mathbf{F}^{-1}\mathbf{G}| = \prod_j (1+\lambda_j).$$

A second approach for testing takes a linear combination across responses, looking for that linear combination that most strongly suggests rejecting the hypothesis. Known as *Roy's union-intersection principle,* this approach views the hypothesis of interest as the infinite intersection of univariate hypotheses arising from each linear combination. If we take the linear combination $\mathbf{d}$ across responses, then we are looking at a univariate response vector $\mathbf{Yd}$ following the usual linear model with mean vector $\mathbf{XBd}$ and variance $\mathbf{d}^T\boldsymbol{\Sigma}\mathbf{d}$. The hypothesis of interest $H : \mathbf{K}^T\mathbf{B}-\mathbf{M} = \mathbf{0}$ can be viewed as the intersection over all possible linear combinations $\mathbf{d}$ of the hypotheses $H_{\mathbf{d}} : \mathbf{K}^T\mathbf{Bd} - \mathbf{Md} = \mathbf{0}$. Under this union-intersection principle, the rejection region will be the union of the rejection regions for each linear combination $\mathbf{d}$. Following the first principles approach, we reject large values of the statistic $\mathbf{d}^T\mathbf{Gd}/\mathbf{d}^T\mathbf{Fd}$. The union of the rejection regions would lead to choosing $\mathbf{d}$ that maximizes this ratio; from Exercise 2.2, the vector $\mathbf{d}$ that maximizes the ratio will produce the largest eigenvalue of $\mathbf{F}^{-1}\mathbf{G}$. In terms of the eigenvalues of $\mathbf{F}^{-1}\mathbf{G}$, obviously *Roy's largest root statistic* corresponds to λ_1. In contrast to Wilks' lambda, there are no easy ways to get the exact distribution of λ_1 except in the trivial case where $q = 1$.

Another simple measure of the size of $\mathbf{F}^{-1}\mathbf{G}$ is its trace, which is attributed to Hotelling and Lawley. In terms of eigenvalues, we can write

$$tr(\mathbf{F}^{-1}\mathbf{G}) = \sum \lambda_j.$$

The last criterion included here is due to (independently) Bartlett, Nanda, and Pillai, but is most commonly known as *Pillai's trace*, $tr(\mathbf{F}+\mathbf{G})^{-1}\mathbf{G} = \sum \frac{\lambda_j}{1+\lambda_j}$. Note that for these three test procedures, we reject large values of the statistic, and no exact small-sample distributions are easily available.

Recall that in the univariate case, when we looked at two ways of writing a hypothesis, as either $H : \mathbf{K}^T\mathbf{b} = \mathbf{m}$ or $H_* : \mathbf{K}_*^T\mathbf{b} = \mathbf{m}_*$, we learned that equivalence between two statements of the hypothesis meant the existence of a nonsingular matrix $\mathbf{Q}$ that related the two in the form $\mathbf{K}_*^T = \mathbf{Q}\mathbf{K}^T$ and $\mathbf{m}_* = \mathbf{Q}\mathbf{m}$. What we showed in Chapter 6 was that the first principles F-test was invariant to nonsingular transformations. In the multivariate case, we also want test procedures that are invariant to the specification of the hypothesis. In addition, we want the test procedure to be invariant to nonsingular transformations of the data, say, $\mathbf{Y}^* = \mathbf{Y}\mathbf{P}$, $\mathbf{B}^* = \mathbf{B}\mathbf{P}$ and $\mathbf{\Sigma}^* = \mathbf{P}^T\mathbf{\Sigma}\mathbf{P}$, where $\mathbf{P}$ is also nonsingular. All four of these test statistics are similarly invariant; following algebra similar to the univariate case, we can show that $\mathbf{G}$ is unchanged by an equivalent statement of the hypothesis

$$\begin{aligned}(\mathbf{K}^T\hat{\mathbf{B}} - \mathbf{M})^T(\mathbf{K}^T(\mathbf{X}^T\mathbf{X})^g\mathbf{K})^{-1}(\mathbf{K}^T\hat{\mathbf{B}} - \mathbf{M}) &= \mathbf{G} \\ &= \left(\mathbf{K}_*^T\hat{\mathbf{B}} - \mathbf{M}_*\right)^T\left(\mathbf{K}_*^T(\mathbf{X}^T\mathbf{X})^g\mathbf{K}_*\right)^{-1}\left(\mathbf{K}_*^T\hat{\mathbf{B}} - \mathbf{M}_*\right)\end{aligned}$$

where $\mathbf{M}_* = \mathbf{Q}\mathbf{M}$. Similar algebra (see Exercise 9.13) shows that transformation by $\mathbf{P}$ leads to $\mathbf{F}_* = \mathbf{P}^T\mathbf{F}\mathbf{P}$ and $\mathbf{G}_* = \mathbf{P}^T\mathbf{G}\mathbf{P}$, so that the eigenvalues of $\mathbf{F}^{-1}\mathbf{G}$ are unaffected, as well as the test statistics.

9.4.4 Power of Tests

In order to assess the power of these potential test statistics, the most difficult thing to grasp is the distribution of matrix $\mathbf{G}$ under the alternative. Under the hypothesis, $\mathbf{G} \sim Wishart_q(s, \mathbf{\Sigma})$, and we can assess the distribution of $\mathbf{G}$ in terms of the inner products of standard normal vectors that would characterize a matrix with a Wishart distribution.

If we construct the Cholesky factor of $\mathbf{W}^T\mathbf{W} = (\mathbf{K}^T(\mathbf{X}^T\mathbf{X})^g\mathbf{K}) = \mathbf{R}\mathbf{R}^T$, then the columns of the $(q \times s)$ matrix $\mathbf{U}$ such that

$$\mathbf{U}^T = \mathbf{R}^{-1}(\mathbf{K}^T\hat{\mathbf{B}} - \mathbf{M}) \tag{9.22}$$

are independent multivariate normal, with mean vectors corresponding to the columns of $E(\mathbf{U}) = \mathbf{C} = (\mathbf{K}^T\mathbf{B} - \mathbf{M})^T\mathbf{R}^{-T}$, and covariance matrix $\mathbf{\Sigma}$. If the hypothesis is true, then $\mathbf{C} = \mathbf{0}$, all of these mean vectors are zero, and since $\mathbf{G} = \mathbf{U}\mathbf{U}^T$, $\mathbf{G}$ has the distribution $Wishart_q(s, \mathbf{\Sigma})$. Under the alternative, $\mathbf{C}$ is not a $(s \times q)$ zero matrix, and

$$E(\mathbf{G}) = E(\mathbf{U}\mathbf{U}^T) = s\mathbf{\Sigma} + \mathbf{C}\mathbf{C}^T. \tag{9.23}$$

As a result, the elements of $\mathbf{C}$ characterize the departure from the hypothesis, and how the distribution of $\mathbf{G} = \mathbf{U}\mathbf{U}^T$ differs from $Wishart_q(s, \boldsymbol{\Sigma})$.

Example 9.2: Iris/One-Way MANOVA continued
Suppose we have a one-way MANOVA where one group is different, say, last group differs by $\mathbf{v}$. For the case $q = 4, a = 3$, and using $\mathbf{K}$ that takes successive differences, say, 1–2, 2–3, we have

$$\mathbf{K}^T\mathbf{B} - \mathbf{M} = \begin{bmatrix} 1 & -1 & 0 \\ 0 & 1 & -1 \end{bmatrix} \begin{bmatrix} \mu_1 & \mu_2 & \mu_3 & \mu_4 \\ \mu_1 & \mu_2 & \mu_3 & \mu_4 \\ \mu_1 + v_1 & \mu_2 + v_2 & \mu_3 + v_3 & \mu_4 + v_4 \end{bmatrix}$$

$$= \begin{bmatrix} 0 & 0 & 0 & 0 \\ v_1 & v_2 & v_3 & v_4 \end{bmatrix} = \mathbf{e}^{(a-1)}\mathbf{v}^T,$$

and we can write $\mathbf{C}$ as a rank 1 matrix, say, $\mathbf{C} = \mathbf{c}\mathbf{d}^T$.

In the univariate case, we were able to capture this departure from the hypothesis in terms of a single number, the noncentrality parameter of the noncentral χ^2 or F random variable. With Results 5.11 and 5.13, we can show that these distributions are stochastically larger than the central distributions they take when the hypothesis is true. From that we can conclude that tests that reject for large values of the statistic are unbiased—the probability of rejection is smallest when the hypothesis is true. Moreover, we could also show that this probability increased monotonically as we went further into the alternative. That is, the probability of a large value—the probability of rejection—increased as the noncentrality parameter increased, so that the noncentrality parameter measured a distance from the hypothesis.

The situation is more difficult in the multivariate case. Not only do we have four test statistics to consider, but we will not be able to capture the distance from the hypothesis in terms of a single number. Nevertheless, we want to show at least that these tests are unbiased, and also that the power increases in some sense as we move further from the hypothesis. Our ability to show this depends on describing the distribution under the alternative, and understanding the properties of the rejection regions or, alternatively, the acceptance regions of each test. The key is a theorem due to Anderson [1] whose proof will not be included.

Theorem 9.2 *Let E be a convex set in the n-dimensional Euclidean space, symmetric about the origin. For $\mathbf{x}$ $(n \times 1)$, let $f(\mathbf{x})$ be a function such that*

1. $f(\mathbf{x}) = f(-\mathbf{x})$

2. $\{\mathbf{x} | f(\mathbf{x}) \geq u\} = K_u$ *is convex for every* u $(0 < u < \infty)$, *and*

3. $\int_E f(\mathbf{x}) d\mathbf{x} < \infty$.

Then

$$\int_E f(\mathbf{x}+k\mathbf{y})\,d\mathbf{x} \geq \int_E f(\mathbf{x}+\mathbf{y})\,d\mathbf{x} \tag{9.24}$$

for every vector $\mathbf{y}$ *and for* $0 \leq k \leq 1$.

Theorem 9.3 *Let* $\mathbf{F} \sim Wishart_q(t, \boldsymbol{\Sigma})$ *on* $\mathcal{V}_q$, *and let the columns of* $\mathbf{U}$ *be independent (and independent of* $\mathbf{F}$*). Let* $\beta(\mathbf{C})$ *be the power of a test when* $E(\mathbf{U}) = \mathbf{C}$, *or when* $\mathbf{U}_{.j} \sim N_q(\mathbf{C}_{.j}, \boldsymbol{\Sigma})$. *Let* $\mathcal{E}_j(\mathbf{F}, \mathbf{U}_{*j})$ *be the acceptance region of the test in* $\mathcal{R}^q$ *of values of* $\mathbf{U}_{.j}$ *for fixed values of* $\mathbf{F}$ *and* $\mathbf{U}_{*j}$, *the remaining elements of* $\mathbf{U}$. *If* $\mathcal{E}_j$ *is symmetric about the origin and convex in* $\mathbf{U}_{.j}$, *then*

1. $\beta(\mathbf{0}) \leq \beta(C_{.j}\mathbf{e}^{(j)T})$ *and*
2. $\beta(\mathbf{C} + (\alpha - 1)\mathbf{C}_{.j}\mathbf{e}^{(j)T})$ *is increasing in* α.

Proof: First note that the expression $\mathbf{C} + (\alpha - 1)\mathbf{C}_{.j}\mathbf{e}^{(j)T}$ merely gives the matrix $\mathbf{C}$ but with the jth column scaled by α. Let $q(\mathbf{U}_{*j}|\mathbf{C}_{*j})$ denote the density of the remaining elements of $\mathbf{U}$, and correspondingly $\mathbf{C}_{*j}$ denote the remaining columns of $\mathbf{C}$. Denote the probability measure of $\mathbf{F}$ by P and denote by f the multivariate normal density with zero mean vector and covariance matrix $\boldsymbol{\Sigma}$. Note that f satisfies the properties in Theorem 9.2. Since the rejection region is the complement of the acceptance region, we have

$$1 - \beta(\mathbf{0}) = \int_{\mathcal{V}_q}\int_{\mathcal{R}^{qs-q}}\int_{\mathcal{E}_j} f(\mathbf{U}_{.j})q(\mathbf{U}_{*j}|\mathbf{0})d\mathbf{U}_{.j}d\mathbf{U}_{*j}dP(\mathbf{F}) \geq$$

$$\int_{\mathcal{V}_q}\int_{\mathcal{R}^{qs-q}}\int_{\mathcal{E}_j} f(\mathbf{U}_{.j} - \mathbf{C}_{.j})q(\mathbf{U}_{*j}|\mathbf{0})d\mathbf{U}_{.j}d\mathbf{U}_{*j}dP(\mathbf{F}) = 1 - \beta(\mathbf{C}_{.j}\mathbf{e}^{(j)T})$$

using Theorem 9.2 with $\mathbf{y} = -\mathbf{C}_{.j}$ and $k = 0$. Similarly, apply the same theorem with $k = \alpha_1/\alpha_2$ and $\mathbf{y} = -\alpha_2\mathbf{C}_{.j}$ to show

$$\begin{aligned}
&1 - \beta\big(\mathbf{C} + (\alpha_1 - 1)\mathbf{C}_{.j}\mathbf{e}^{(j)T}\big)\\
&= \int_{\mathcal{V}_q}\int_{\mathcal{R}^{qs-q}}\int_{\mathcal{E}_j} f(\mathbf{U}_{.j} - \alpha_1\mathbf{C}_{.j})q(\mathbf{U}_{*j}|\mathbf{C}_{*j})d\mathbf{U}_{.j}d\mathbf{U}_{*j}dP(\mathbf{F})\\
&\geq \int_{\mathcal{V}_q}\int_{\mathcal{R}^{qs-q}}\int_{\mathcal{E}_j} f(\mathbf{U}_{.j} - \alpha_2\mathbf{C}_{.j})q(\mathbf{U}_{*j}|\mathbf{C}_{*j})d\mathbf{U}_{.j}d\mathbf{U}_{*j}dP(\mathbf{F})\\
&= 1 - \beta\big(\mathbf{C} + (\alpha_2 - 1)\mathbf{C}_{.j}\mathbf{e}^{(j)T}\big)
\end{aligned}$$

for $0 \leq \alpha_1 \leq \alpha_2$. ∎

So if we can show the acceptance set $\mathcal{E}_j(\mathbf{F}, \mathbf{U}_{*j})$ is convex, then the probability of acceptance increases as we take k smaller, so that the probability of rejection is smallest at $k = 0$ (hence an unbiased test) and increases as k increases to 1. Now the remaining task is to show that the acceptance region of a particular test for given values of $\mathbf{F}$ and other columns of $\mathbf{U}$, say $\mathbf{U}_*$, is convex in $\mathbf{U}_{.j}$ to be denoted by $\mathbf{u}$. Notice that this slight change in notation, along with $\mathbf{G}_* = \mathbf{U}_*\mathbf{U}_*^T$, allows us to write

$$\mathbf{G} = \mathbf{U}\mathbf{U}^T = \sum_k \mathbf{U}_{.k}\mathbf{U}_{.k}^T = \sum_{k \neq j} \mathbf{U}_{.k}\mathbf{U}_{.k}^T + \mathbf{U}_{.j}\mathbf{U}_{.j}^T = \mathbf{U}_*\mathbf{U}_*^T + \mathbf{u}\mathbf{u}^T = \mathbf{G}_* + \mathbf{u}\mathbf{u}^T.$$

In the case of Wilks' lambda, after some analysis the test statistic can be rewritten as

$$\Lambda_* = \frac{|\mathbf{F}|}{|\mathbf{F}+\mathbf{G}|} = \frac{|\mathbf{F}|}{|\mathbf{F}+\mathbf{G}_*+\mathbf{u}\mathbf{u}^T|} = \frac{|\mathbf{F}|}{|\mathbf{F}+\mathbf{G}_*||\mathbf{I}+(\mathbf{F}+\mathbf{G}_*)^{-1}\mathbf{u}\mathbf{u}^T|}$$
$$= \frac{|\mathbf{F}|}{|\mathbf{F}+\mathbf{G}_*|(1+\mathbf{u}^T(\mathbf{F}+\mathbf{G}_*)^{-1}\mathbf{u})},$$

and the acceptance region

$$\frac{|\mathbf{F}|}{|\mathbf{F}+\mathbf{G}|} \geq z$$

for fixed $\mathbf{F}$ and $\mathbf{G}_*$ is equivalent to $\mathbf{u}^T(\mathbf{F}+\mathbf{G}_*)^{-1}\mathbf{u} \leq w$. Since $\mathbf{F}+\mathbf{G}_*$ will be positive definite with probability 1, so will its inverse, and the acceptance region will be an ellipsoid in q dimensions. The acceptance region is symmetric about the origin and clearly is a convex set. Consequently, using Wilks' lambda leads to an unbiased test, and the power of the test increases as $\mathbf{C}$ departs from $\mathbf{0}$.

The Hotelling–Lawley trace is the easiest case, following

$$trace\,[\mathbf{F}^{-1}(\mathbf{G}_* + \mathbf{u}\mathbf{u}^T)] = trace\,[\mathbf{F}^{-1}\mathbf{G}_*] + \mathbf{u}^T\mathbf{F}^{-1}\mathbf{u}$$

so that the acceptance region

$$\{\mathbf{u} : trace\,[\mathbf{F}^{-1}(\mathbf{G}_* + \mathbf{u}\mathbf{u}^T)] \leq z\}$$

also takes the form of an ellipsoid centered at the origin. Pillai's trace criterion is more problematic and the result does not hold; see Exercise 9.18.

For Roy's largest root, the acceptance region for $\mathbf{u}$ is

$$\mathcal{E} = \{\mathbf{u} : \text{largest eigenvalue of } \mathbf{F}^{-1}(\mathbf{G}_* + \mathbf{u}\mathbf{u}^T) \leq z\}.$$

Since we are only interested in the geometry of the region, rotation and rescaling $\mathbf{u}$ will have no effect on whether $\mathcal{E}$ is convex or not, so that we could replace $\mathbf{F}$ with any other nonsingular matrix, even $\mathbf{I}_q$, without changing the geometry of the set. Moreover, we can do the same thing with $\mathbf{G}_*$ except to take care of its rank, since $s-1$ (remember, we're dropping a column) may be less than q. If $\mathbf{G}_*$ is full rank, the same rotation and rescaling argument allows us to convert the matrix to $\mathbf{I}+\mathbf{u}\mathbf{u}^T$, whose largest eigenvalue is $1+\mathbf{u}^T\mathbf{u}$. At this point, the set $\mathcal{E}$ is a sphere centered at the origin. If $\mathbf{G}_*$ has rank $r < p$, then the acceptance region is equivalent to the set of $\mathbf{u}$ such that the largest eigenvalue of

$$\begin{bmatrix} \mathbf{I}_r & \mathbf{0} \\ \mathbf{0} & \mathbf{0} \end{bmatrix} + \mathbf{u}\mathbf{u}^T$$

is less than z. Partitioning a vector $\mathbf{d}^T = (\mathbf{d}_1, \mathbf{d}_2)^T$ in the same fashion and using Exercise 2.2, we have for all $\mathbf{u}$ in the acceptance region

$$\mathbf{d}_1^T\mathbf{d}_1 + (\mathbf{d}^T\mathbf{u})^2 \leq \mathbf{d}^T\mathbf{d}z \tag{9.25}$$

for all $\mathbf{d}$. Taking two points $\mathbf{u}_1$ and $\mathbf{u}_2$ in the acceptance region, to show that this set is convex, we need to show that $p\mathbf{u}_1 + q\mathbf{u}_2$ is also in that set, where $0 \leq p \leq 1$ and $p + q = 1$. Multiplying (9.25) by p and applying to $\mathbf{u}_1$, and adding to a similar expression with q and $\mathbf{u}_2$, we have

$$\mathbf{d}_1^T\mathbf{d}_1 + p(\mathbf{d}^T\mathbf{u}_1)^2 + q(\mathbf{d}^T\mathbf{u}_2)^2 \leq \mathbf{d}^T\mathbf{d}z.$$

Since we can show (see Exercise 9.9) $(ps + qt)^2 \leq ps^2 + qt^2$, for $s = (\mathbf{d}^T\mathbf{u}_1)$ and $t = (\mathbf{d}^T\mathbf{u}_2)$, we have

$$\mathbf{d}_1^T\mathbf{d}_1 + (p\mathbf{d}^T\mathbf{u}_1 + q\mathbf{d}^T\mathbf{u}_2)^2 \leq \mathbf{d}^T\mathbf{d}z,$$

and thus the acceptance region is convex.

The rotation and rescaling operations outlined earlier arise from invariance arguments and are central to the original derivations due to Roy and Anderson and their collaborators. These invariance arguments limit the potential test statistics to functions of the eigenvalues of $\mathbf{F}^{-1}\mathbf{G}$ (recall that all four satisfy that restriction) and lead to the constructing of a canonical form for analysis under the alternative, which depends only on the eigenvalues of $\mathbf{CC}^T$, which serve as the multivariate version of the noncentrality parameter. The results given above are more restrictive; see Anderson [2], Section 8.10 for more general results.

From a practical point of view, the user would like to know which test statistic to use. Following this canonical form, some analysis of the power can be pursued. Nevertheless, simulation studies have reached a consensus. For alternatives that are not diffuse—usually meaning $\mathbf{C}$ has low rank or rank 1—corresponding to MANOVA where the mean vectors lie along a line (see Exercise 9.12), the ordering for power is Roy > Hotelling–Lawley > Wilks > Pillai (see, e.g., Rencher [40]). Recall that Roy's largest root is designed for this scenario. For alternatives that are diffuse—no simple geometry of $\mathbf{C}$ and a more commonly expected alternative—the ordering reverses. Wilks' lambda remains the most popular, partly due to some exact small-sample critical values and partly due to good overall performance.

9.5 Repeated Measures

In general, the q different responses at each design point may correspond to q entirely different types of measurements, for example, height, weight, IQ, and so on, where taking linear combinations across variables must take into account different units of the responses. In certain designed experiments, when the q measurements on an experimental unit or individual may be the same type of measurement, some interesting analysis may be possible, even in cases as simple as a one-sample problem.

Example 9.5: Lead Concentrations
Let Y_{ij} represent the concentration of lead in $j = 1, \cdots, q$ different parts of the body of an individual, say, the blood, brain, muscle, fat tissue, liver, and so on. Since the concentration of the blood is easy to measure from a blood sample, if the concentration is the same throughout the body, then the effect of lead on the brain or other tissues can be more easily assessed *in vivo*.

Example 9.6: Ramus Heights
Elston and Grizzle [9] analyzed the heights of the ramus (jaw) bone, measured in boys every 6 months starting at 8 years of age. Here, Y_{ij} represents heights at four time periods: $j = 1, \cdots, q = 4$ for individual i.

In both of these cases, we have *repeated measures* of the same quantity on the same individual. These measurements are likely to be independent across individuals, but dependent within an individual, and our multivariate linear model (9.1) appears appropriate. In repeated measures problems, we may be interested in testing rather simple hypotheses across these measurements. In the simplest case of a one-sample problem, we have the usual model parameterized as

$$E(\mathbf{Y}) = \mathbf{XB} = \mathbf{1}_N \mu^T,$$

as simple as $\mathbf{X} = \mathbf{1}_N$ and $\mathbf{B} = \mu^T$. Note that this form corresponds to $\mathbf{Y}_{i.}^T iid N_q(\mu, \Sigma)$. We may be interested in testing hypotheses about μ. In the case of lead, we may be interested in testing whether the concentration is the same for all measurements $H : \mu_1 = \mu_2 = \cdots = \mu_q$. In the ramus height example, we may be interested in whether the components of the mean follow a linear trend, such as $H : \mu_j = \beta_0 + \beta_1 j$.

The key to repeated measures analysis is the construction of a $(q-1) \times q$ contrast matrix $\mathbf{C}$, so that $\mathbf{C1}_q = \mathbf{0}$. With a judicious choice of $\mathbf{C}$, we expect to discover a simple model for the mean vector μ, by choosing $\mathbf{C}$ such that $\mathbf{C}\mu$ has many zeros. For the lead example (Example 9.5), a natural choice of the matrix $\mathbf{C}$ may be

$$\mathbf{C}_1 = \begin{bmatrix} 1 & -1 & 0 & 0 & \cdots & 0 \\ 1 & 0 & -1 & 0 & \cdots & 0 \\ 1 & 0 & 0 & -1 & \cdots & 0 \\ & \cdots & & & & \cdots \\ 1 & 0 & 0 & 0 & \cdots & -1 \end{bmatrix},$$

so that if the hypothesis of equal concentrations is true, then $\mathbf{C}\mu = \mathbf{0}$. For the ramus heights case (Example 9.5), the use of quadratic and cubic contrasts is natural, leading to

$$\mathbf{C}_2 = \begin{bmatrix} -3 & -1 & 1 & 3 \\ 1 & -1 & -1 & 1 \\ -1 & -3 & 3 & 1 \end{bmatrix}.$$

In this case, if the components of the mean vector follow a linear trend, then most of the elements of $\mathbf{C}\mu$ will be zero.

Repeated measures analysis, in its simplest case, begins with the basic model (9.1) in a special form, $\mathbf{Y} = \mathbf{1}\mu^T + \mathbf{E}$. Postmultiplying by $\mathbf{C}^T$ leads to

$$\mathbf{Z} = \mathbf{Y}\mathbf{C}^T = \mathbf{1}\mu^T\mathbf{C}^T + \mathbf{E}\mathbf{C}^T, \tag{9.26}$$

which now takes the form of a model for $\mathbf{Z}_{i.}^T$ as iid with constant mean vector $\mathbf{C}\mu$ and covariance matrix $\mathbf{C}\boldsymbol{\Sigma}\mathbf{C}^T$. At this point, we merely test whether subsets of the mean vector $\mathbf{C}\mu$ are zero, as in Example 9.3. From this point, the repeated measures approach can be extended to the usual experimental designs, and the hypotheses of interest are whether subsets of $\mathbf{K}^T\mathbf{B}\mathbf{C}$ are zero.

Example 9.7: Two-Sample Repeated Measures

Suppose we extend the lead example to include two groups of individuals, say, $i = 1, \cdots, n_1$ in the control group and $i = n_1 + 1, \cdots, n_1 + n_2 = N$ in the treatment group. A cell means model would then take the form

$$\mathbf{X}\mathbf{B} = \begin{bmatrix} \mathbf{1}_{n_1} & \mathbf{0} \\ \mathbf{0} & \mathbf{1}_{n_2} \end{bmatrix} \begin{bmatrix} \mu_1^c & \mu_2^c & \mu_3^c & \cdots & \mu_q^c \\ \mu_1^t & \mu_2^t & \mu_3^t & \cdots & \mu_q^t \end{bmatrix}. \tag{9.27}$$

The natural first step would be testing for interaction of the two factors, that is, whether the pattern of lead concentrations was the same in the two groups, and would lead to testing $H : \mathbf{K}_1^T\mathbf{B}\mathbf{C}_1^T = 0$ where $\mathbf{K}_1 = [1 \ -1]$. If no interaction were present, then the main effects would be next, say, testing whether the concentrations were the same, which would be similar to the original problem, testing whether $\mathbf{K}_2^T\mathbf{B}\mathbf{C}_1^T$ were zero where $\mathbf{K}_2 = [1/2 \ 1/2]$. The other main effect is the treatment effect, which could be tested with $\mathbf{K}_2^T\mathbf{B}\frac{1}{q}\mathbf{1}$ looking at the difference in average concentration between the two groups.

As the design matrix $\mathbf{X}$ gets complicated, repeated measures analysis also becomes complicated. Moreover, since the q responses are all the same type of measurement, the question arises how this would be different from a univariate analysis. In the lead example above, a univariate analysis would take the form

$$Y_{ij} = \beta_0 + \beta_1 x_i + \alpha_j + E_{ij}$$

where x_i would be a dummy variable for the two groups. Obviously, the dependence among the observations from the same individual would be lost in moving to a univariate analysis. If the dependence took the form of an additive random individual effect, leading to the model

$$Y_{ij} = \beta_0 + \beta_1 x_i + \alpha_j + \delta_i + E_{ij}, \tag{9.28}$$

then the covariance matrix takes a very structured form:

$$\boldsymbol{\Sigma} = \sigma^2\mathbf{I}_q + \sigma_\delta^2\mathbf{1}\mathbf{1}^T.$$

This covariance matrix should look familiar, as it is the same model as the split plot in Section 8.5. If we insist on using orthogonal contrasts, and include the missing constant row in the contrast matrix, we form

$$\mathbf{C}_* = \begin{bmatrix} \frac{1}{\sqrt{q}}\mathbf{1}^T \\ \mathbf{C} \end{bmatrix} \tag{9.29}$$

as an orthogonal $q \times q$ matrix. Then applying the same postmultiplication step $\mathbf{YC}_*^T = \mathbf{XBC}_*^T + \mathbf{EC}_*^T$ leads to a new $N \times q$ response matrix, but with a covariance matrix of the form

$$\mathbf{C}_*\mathbf{\Sigma}\mathbf{C}_*^T = \mathbf{C}_*\left(\sigma^2\mathbf{I}_q + \sigma_\delta^2\mathbf{1}\mathbf{1}^T\right)\mathbf{C}_*^T = \sigma^2\mathbf{I}_q + q\sigma_\delta^2\mathbf{e}^{(1)}\mathbf{e}^{(1)T}.$$

Recall that the whole plot analysis—corresponding to mean response on an individual subject—and the first of the q responses—uses the variance $\sigma^2 + q\sigma_\delta^2$, while the split-plot analysis uses σ^2 for the within-plot main effects and interactions. Because the covariance matrix $\mathbf{C}_*\mathbf{\Sigma}\mathbf{C}_*^T$ is diagonal, the whole plot response (first component) is independent of the within-plot responses, which are also mutually independent. The split plot is a special case of repeated measures where the dependence arises from an additive individual effect.

9.6 Confidence Intervals

In Chapter 6, the construction of simultaneous confidence intervals for coefficients or estimable linear combination was also discussed in terms of multiple comparisons. The Bonferroni approach is often considered the simplest to understand and apply, but its performance degrades as the number of intervals/comparisons increases. The Scheffé approach has its advantages, but was designed to be pessimistic. Tukey's method is designed for the important case of one-way ANOVA. Much of this remains the same in the multivariate case, although there are situations where these methods can be combined.

Here we want to construct intervals for scalar quantities of the form $\mathbf{x}^T\mathbf{Bd}$, perhaps many at a time. Here $\mathbf{x}$ ($p \times 1$) may be design points or contrast coefficients—$\mathbf{x}^T\mathbf{Bd}$ must be estimable; $\mathbf{d}$ ($q \times 1$) forms linear combinations across response components. We may have many linear combinations or design points $\mathbf{x}$ or many linear combinations across response components $\mathbf{d}$. All of these confidence intervals take the form

$$\mathbf{x}^T\hat{\mathbf{B}}\mathbf{d} \pm c\sqrt{\mathbf{d}^T\hat{\mathbf{\Sigma}}\mathbf{d}\mathbf{x}^T(\mathbf{X}^T\mathbf{X})^g\mathbf{x}}$$

and leading to probability statements of the form

$$Pr\left[\frac{(\mathbf{x}^T(\mathbf{B}-\hat{\mathbf{B}})\mathbf{d})^2}{\mathbf{d}^T\hat{\mathbf{\Sigma}}\mathbf{d}\mathbf{x}^T(\mathbf{X}^T\mathbf{X})^g\mathbf{x}} \le c^2\right] \ge 1-\alpha$$

for various combinations of $\mathbf{x}$ and $\mathbf{d}$. In these two expressions $\hat{\mathbf{\Sigma}}$ is the usual unbiased estimator of the covariance matrix $(N-r)^{-1}\mathbf{F}$. The value of c may change according to the combinations of $\mathbf{x}$ and $\mathbf{d}$. Before proceeding further, consider the simpler case of one-way ANOVA.

Example 9.2: One-Way MANOVA continued
Consider the cell means parameterization of the one-way MANOVA, where $(\mathbf{X}^T\mathbf{X}) = diag(n_i), i = 1, \ldots, a$. Taking $\mathbf{x}$ to be an elementary vector $\mathbf{e}^{(i)}$ leads to $\mathbf{x}^T(\mathbf{X}^T\mathbf{X})^g\mathbf{x} = 1/n_i$, and $\mathbf{x}\hat{\mathbf{B}}$ gives the mean vector for group i as $\hat{\mu}^{(i)} = \overline{Y}^{(i)}$. Taking $\mathbf{x}$ to be a contrast vector leads to $\mathbf{x}\hat{\mathbf{B}}$ as the $(q \times 1)$ vector of contrasts for each measurement, so that $\mathbf{x}^T(\mathbf{X}^T\mathbf{X})^g\mathbf{x} = \sum_i x_i^2/n_i$. Varying $\mathbf{d}$ gives different linear combinations across measurements; taking $\mathbf{d}$ to be the elementary vector $\mathbf{e}^{(j)}$ leads to $\hat{\mathbf{B}}\mathbf{d}$ as the vector of means (for each group) for measurement j. So a confidence interval for the quantity $\sum_{i=1}^a \sum_{j=1}^q x_i d_j \mu_j^{(i)}$ can be constructed as

$$\sum_{i=1}^{a}\sum_{j=1}^{q} x_i d_j \overline{Y}_j^{(i)} \pm c\sqrt{\left(\sum_{j=1}^{q} d_j d_k \hat{\Sigma}_{jk}\right)\left(\sum_{i=1}^{a} x_i^2/n_i\right)}.$$

Taking both $\mathbf{x}$ and $\mathbf{d}$ to be elementary vectors gives $\mathbf{x}^T\mathbf{B}\mathbf{d} = B_{ij} = \mu_j^{(i)}$, the mean of group i $(i = 1, \ldots, a)$ on measurement j $(j = 1, \ldots, q)$, which leads to the simpler form

$$\overline{Y}_j^{(i)} \pm c\sqrt{\hat{\Sigma}_{jj}/n_i}.$$

The Scheffé approach outlined in Chapter 6 was somewhat pessimistic, looking at the worst case. The benefit of such an approach, however, was that the contrasts could be decided *post hoc*. This same attitude can be extended to the multivariate case. Consider confidence intervals $C(\mathbf{x}, \mathbf{d}, c)$ for any linear combination of the form $\mathbf{x}^T\mathbf{B}\mathbf{d}$. To keep things simple, let's assume $\mathbf{X}$ is full-column rank, so that all of these linear combinations are estimable and $\mathbf{X}^T\mathbf{X}$ is nonsingular. These intervals take the form

$$C(\mathbf{x}, \mathbf{d}, c) = \left[\mathbf{x}^T\hat{\mathbf{B}}\mathbf{d} - c\sqrt{\mathbf{d}^T\hat{\boldsymbol{\Sigma}}\mathbf{d}\mathbf{x}^T(\mathbf{X}^T\mathbf{X})^g\mathbf{x}}, \mathbf{x}^T\hat{\mathbf{B}}\mathbf{d} + c\sqrt{\mathbf{d}^T\hat{\boldsymbol{\Sigma}}\mathbf{d}\mathbf{x}^T(\mathbf{X}^T\mathbf{X})^g\mathbf{x}}\right]. \tag{9.30}$$

As before, the issue is how to choose c. The mathematics for extending the Scheffé method follows the same logic as in the univariate case. Since we want to consider any design point/contrast $\mathbf{x}$ and any linear combination $\mathbf{d}$, we need to choose c such that

$$Pr(\mathbf{x}^T\mathbf{B}\mathbf{d} \in C(\mathbf{x}, \mathbf{d}, c) \text{ for all } \mathbf{x} \text{ and all } \mathbf{d}) = 1 - \alpha.$$

$$\begin{aligned}
&Pr(\mathbf{x}^T\mathbf{B}\mathbf{d} \in C(\mathbf{x}, \mathbf{d}, c) \text{ for all } \mathbf{x} \text{ and all } \mathbf{d}) \\
&= Pr\left(max_{\mathbf{x},\mathbf{d}}\frac{(\mathbf{x}^T(\mathbf{B} - \hat{\mathbf{B}})\mathbf{d})^2}{\mathbf{d}^T\hat{\boldsymbol{\Sigma}}\mathbf{d}\mathbf{x}^T(\mathbf{X}^T\mathbf{X})^g\mathbf{x}} \leq c^2\right) \\
&= Pr\left(max_{\mathbf{d}}\frac{\mathbf{d}^T(\mathbf{B} - \hat{\mathbf{B}})^T\mathbf{X}^T\mathbf{X}(\mathbf{B} - \hat{\mathbf{B}})\mathbf{d}}{\mathbf{d}^T\hat{\boldsymbol{\Sigma}}\mathbf{d}} \leq c^2\right) \\
&= Pr(\text{largest eigenvalue of } \hat{\boldsymbol{\Sigma}}^{-1}(\mathbf{B} - \hat{\mathbf{B}})^T\mathbf{X}^T\mathbf{X}(\mathbf{B} - \hat{\mathbf{B}}) \leq c^2)
\end{aligned}$$

At this point, the relationship with Roy's largest root statistic should be clear. Let $\eta_{q,N-r,p}(\alpha)$ be the upper α critical value of the distribution of the largest eigenvalue of $\mathbf{F}^{-1}\mathbf{G}$ for $\mathbf{F} \sim Wishart_q(N-r, \mathbf{I})$ and $\mathbf{G} \sim Wishart_q(p, \mathbf{I})$. Then choosing $c^2 = (N-r)\eta_{q,N-r,p}(\alpha)$ will give the correct coverage.

One serious drawback for the Scheffé generalization just outlined is that the tables of the distribution of Roy's largest root are not widely available. More commonly, we don't have both **x** and **d** varying. If the linear combination **d** were fixed, then we could employ the Scheffé approach from the univariate case to vary across all design points **x** and use $c^2 = rF_{r,n-r}(\alpha)$. If the design point **x** were fixed, then we could allow **d** to vary over R^q and use $c^2 = \frac{q(N-r)}{N-r-q+1}F_{q,N-r-q+1}(\alpha)$, whose mathematics follows Hotelling's T^2. See Exercise 9.19. Finally, the more imaginative approach for where both **x** and **d** vary would be to apply Bonferroni to either of these situations. For this last case, that would mean m design points **x**, and still varying **d**; the only adjustment would be the usual one for Bonferroni, using α/m.

9.7 Summary

1. The linear model is generalized to the case of multiple responses for each individual to the multivariate linear model.

2. Using moment assumptions similar to the Gauss–Markov assumptions, standard estimation results generalize easily to the multivariate case.

3. The chi-square distribution is generalized to the Wishart distribution, and its properties are derived.

4. Hypothesis tests in the multivariate case become much more difficult and complicated, with four commonly used test statistics used for the multivariate version of the general linear hypothesis.

5. Repeated measures analysis tools are appropriate when the multiple responses from each individual are the same type of measurement. The split plot can be viewed as a simplified version of repeated measures.

6. Standard tools for constructing confidence intervals are extended for the multivariate case.

9.8 Notes

- The reader should note that we were able to do quite a bit with the Wishart distribution without ever writing the density, which doesn't exist if the degrees of freedom parameter is less than the dimension.

- A single chapter can hardly summarize the literature on multivariate models. Anderson [2] and Mardia, Kent, and Bibby [29] are both excellent books on the subject, and complementary in style and coverage. The former takes a more theoretical, detailed, and technical approach; the latter, while still covering the theory, is more terse with its mathematics. For more applications but with attention to the theory, see Rencher [40].

9.9 Exercises

9.1. * Fill in the missing step in the proof of Result 9.1 by rewriting $tr[\mathbf{\Sigma}^{-1}(\hat{\mathbf{B}}-\mathbf{B})^T \mathbf{X}^T\mathbf{X}(\hat{\mathbf{B}}-\mathbf{B})]$ as the trace of a nonnegative definite matrix. Hint: Factor the positive definite matrix $\mathbf{\Sigma}^{-1}$.

9.2. Show the missing steps in the proof of Result 9.2 by showing $E\left[\mathbf{Y}_{.k}\mathbf{Y}_{.j}^T\right] = (\mathbf{X}\mathbf{B}_{.k}\mathbf{B}_{.j}^T\mathbf{X}^T + \mathbf{I}_N \Sigma_{jk})$.

9.3. Prove Result 9.3.

9.4. Generalize the results in Equations (9.7) and (9.8) in two ways.

   a. Generalize from a vector λ to a matrix $\mathbf{\Lambda}$ to establish similar results for an estimable vector $\mathbf{\Lambda}^T\mathbf{Bm}$.
   b. Generalize from a vector $\mathbf{m}$ to a matrix $\mathbf{M}$ for $\lambda^T\mathbf{BM}$.

9.5. Prove Result 9.5.

9.6. Prove that if $\mathbf{A}$ and $\mathbf{B}$ are positive definite, the eigenvalues of $\mathbf{A}^{-1}\mathbf{B}$ are positive.

9.7. Finish the proof of Result 9.7 by finding the complete sufficient statistic for $(\mathbf{B}, \mathbf{\Sigma})$.

9.8. Extend the Elston and Grizzle ramus height example with an expected linear trend to a two-sample problem, suppose $n_1 = n_2 = 20$, and give the tests and their degrees of freedom for interaction, and main effects of time and treatment.

9.9. Prove that $(ps+qt)^2 \le ps^2 + qt^2$ where $p+q=1$.

9.10. Can you extend Bartlett decomposition (Result 9.11) for the case where $m \le p$ where $\mathbf{F}$ is singular (nonnegative definite)?

9.11. Consider the one-sample problem with $\mathbf{XB} = \mathbf{1}\mu^T$. Test $H : \mu = \mathbf{0}$ using the one-sample Hotelling's T^2-statistic. Hint: Use $\mathbf{k} = 1$ (scalar).

9.12. Suppose we have MANOVA where $\mu^i = \mathbf{a} + i\mathbf{b}$; show that $\mathbf{C} = E(\mathbf{U})$ from Equation (9.22) can be written as the rank 1 matrix $\mathbf{cd}^T$.

9.13. Consider the multivariate linear model $\mathbf{Y} = \mathbf{XB} + \mathbf{E}$ under normality. Show that if we apply a nonsingular transformation $\mathbf{P}$ to the data $\mathbf{Y}_* = \mathbf{YP}$ and parameters $\mathbf{B}_* = \mathbf{BP}$, the four tests for the hypothesis $H : \mathbf{K}^T\mathbf{B} = \mathbf{M}$ considered in this chapter are unaffected.

9.14. Consider the one-sample repeated measures situation, where $E(\mathbf{Y}) = \mathbf{XB} = \mathbf{1}\mu^T$ where the mean vector takes the form $\mu = \mathbf{Ac}$ with $\mathbf{A}$ a known design matrix across responses. Find the BLUE for $\mathbf{c}$.

9.15. Confirm the statement that all four test statistics coincide with the usual F-test in the case where $q = 1$.

9.16. Do all four test statistics (Wilks' lambda, Roy's largest root, Hotelling-Lawley trace, Pillai's trace) coincide with Hotelling's T^2 when $s = 1$?

9.17. Prove Equation (9.22)—the multivariate version of Theorem 6.1.

9.18. The acceptance region for Pillai's trace criterion takes the form

$$\mathit{trace}\,(\mathbf{F} + \mathbf{G}_* + \mathbf{u}\mathbf{u}^T)^{-1}(\mathbf{G}_* + \mathbf{u}\mathbf{u}^T) \le z.$$

a. Using Exercise A.75 find an expression for $(\mathbf{F} + \mathbf{G}_* + \mathbf{u}\mathbf{u}^T)^{-1}$ in terms of $(\mathbf{F} + \mathbf{G}_*)^{-1}$.
b. Show that

$$\begin{aligned}&\mathit{trace}\,(\mathbf{F} + \mathbf{G}_* + \mathbf{u}\mathbf{u}^T)^{-1}(\mathbf{G}_* + \mathbf{u}\mathbf{u}^T)\\ &= \mathit{trace}\,(\mathbf{F} + \mathbf{G}_*)^{-1}\mathbf{G}_* + \frac{\mathbf{u}^T(\mathbf{F} + \mathbf{G}_*)^{-1}\mathbf{F}(\mathbf{F} + \mathbf{G}_*)^{-1}\mathbf{u}}{1 + \mathbf{u}'(\mathbf{F} + \mathbf{G}_*)^{-1}\mathbf{u}}.\end{aligned}$$

c. Are there values of z for which the acceptance region is an ellipsoid?

9.19. Following the Scheffé approach for constructing confidence intervals, but a fixed design point $\mathbf{x}$, find the distribution of the maximum of the square of a t-statistic

$$max_{\mathbf{d}} \frac{(\mathbf{x}^T(\mathbf{B} - \hat{\mathbf{B}})\mathbf{d})^2}{\mathbf{d}^T\hat{\Sigma}\mathbf{d}\mathbf{x}^T(\mathbf{X}^T\mathbf{X})^g\mathbf{x}}$$

and verify that choosing $c^2 = \frac{q(N-r)}{N-r-q+1}F_{q,N-r-q+1}(\alpha)$ gives the appropriate coverage.

Appendix A

Review of Linear Algebra

A.1 Notation and Fundamentals

Vectors are usually represented by boldface lowercase roman letters, for example, $\mathbf{a}$, and individual components are subscripted, as a_j. Occasionally, parentheses will be used for emphasis, as $(\mathbf{Ax})_j$ to denote the jth component of the vector $\mathbf{Ax}$. Parenthesized superscripts will occasionally be used to enumerate a list of vectors, as $\mathbf{x}^{(1)}$ and $\mathbf{x}^{(2)}$. Beginning in Chapter 5, random vectors may be written in uppercase, but still boldface. Unless otherwise noted, all vectors are column vectors. Matrices are always represented by boldface uppercase roman letters, as $\mathbf{A}$. A column of such a matrix will be denoted by subscripts, as $\mathbf{A}_{.j}$ denotes the jth column. For both vectors and matrices, the dimensions may be emphasized parenthetically as (m by n) or ($m \times n$). Transposes are denoted by a superscript T, as $\mathbf{x}^T$ for vectors, and $\mathbf{A}^T$ for matrices. A vector with all components equal to 1 will be denoted by $\mathbf{1}$, and if its dimension needs emphasis, a subscript will be employed, $\mathbf{1}_n$. Similarly, the boldface $\mathbf{0}$ represents a vector with all components zero. An identity matrix of order n will be denoted by $\mathbf{I}_n$, with the subscript emphasizing the shape.

A set of vectors $\mathbf{x}^{(1)}, \mathbf{x}^{(2)}, \ldots, \mathbf{x}^{(n)}$ is *linearly dependent* if some nontrivial linear combination of them yields the zero vector; that is, there exist coefficients c_j, $j = 1, \ldots, n$, not all zero, such that

$$\sum_{j=1}^{n} c_j \mathbf{x}^{(j)} = \mathbf{0}. \tag{A.1}$$

This set of vectors is *linearly independent* if (A.1) implies $c_j \equiv 0$. One view of linear dependence is that at least one of the vectors can be written as a linear combination of the others in the set; that is, if $c_j \neq 0$, then

$$\mathbf{x}^{(j)} = -(1/c_j) \sum_{k \neq j} c_k \mathbf{x}^{(k)}. \tag{A.2}$$

For vectors in R^d, the space of all vectors of dimension d, no set with more than d vectors can be linearly independent.

Two vectors are *orthogonal* to each other, written $\mathbf{x} \perp \mathbf{y}$, if their inner product is zero, that is,

$$\mathbf{x}^T \mathbf{y} = \mathbf{y}^T \mathbf{x} = \sum_j x_j y_j = 0.$$

The *length* of a vector, its Euclidean norm, will be denoted by

$$\|\mathbf{x}\| = (\mathbf{x}^T\mathbf{x})^{1/2} = \left(\sum_j x_j^2\right)^{1/2}.$$

A set of vectors $\mathbf{x}^{(1)}, \mathbf{x}^{(2)}, \ldots, \mathbf{x}^{(n)}$ are mutually orthogonal iff $\mathbf{x}^{(i)T}\mathbf{x}^{(j)} = 0$ for all $i \neq j$. It can also be shown that a set of mutually orthogonal nonzero vectors are also linearly independent (see Exercise A.1). The most common set of vectors that are mutually orthogonal are the *elementary* vectors $\mathbf{e}^{(1)}, \mathbf{e}^{(2)}, \ldots, \mathbf{e}^{(n)}$, which are all zero, except for one element equal to 1, so that $(\mathbf{e}^{(i)})_i = 1$ and $(\mathbf{e}^{(i)})_j = 0$ for $j \neq i$. A matrix $\mathbf{Q}$ composed of columns that are mutually orthogonal and normalized to have unit length is said to have *orthonormal columns*. Algebraically, mutual orthogonality of the columns means that $\mathbf{Q}^T\mathbf{Q}$ is diagonal; adding normalization means that $\mathbf{Q}^T\mathbf{Q} = \mathbf{I}$. If the matrix $\mathbf{Q}$ is square, then the columns are also linearly independent; $\mathbf{Q}$ is nonsingular and called an *orthogonal matrix*. Nonsingularity also means that $\mathbf{Q}^T\mathbf{Q} = \mathbf{I} = \mathbf{Q}\mathbf{Q}^T$, since the left inverse $\mathbf{Q}^T$ is also a right inverse (besides being its transpose). Multiplying a vector by an orthogonal matrix can be viewed as a rotation of the vector, since the norm is preserved: $\|\mathbf{Q}\mathbf{x}\| = \|\mathbf{x}\|$.

A *vector space* $\mathcal{S}$ is a set of vectors that are closed under addition and scalar multiplication, that is, if $\mathbf{x}^{(1)}$ and $\mathbf{x}^{(2)}$ are in $\mathcal{S}$, then $c_1\mathbf{x}^{(1)} + c_2\mathbf{x}^{(2)}$ is in $\mathcal{S}$. A vector space $\mathcal{S}$ is said to be *generated* by a set of vectors $\mathbf{x}^{(1)}, \mathbf{x}^{(2)}, \ldots, \mathbf{x}^{(n)}$ if for every $\mathbf{x} \in \mathcal{S}$, there exist coefficients c_j so that we can write

$$\mathbf{x} = \sum_j c_j \mathbf{x}^{(j)}.$$

If a vector space $\mathcal{S}$ is generated by a set of linearly independent vectors $\mathbf{x}^{(1)}, \mathbf{x}^{(2)}, \ldots, \mathbf{x}^{(n)}$, then this set of vectors form a *basis* for the space $\mathcal{S}$. Viewed another way, the set of all linear combinations of a set of vectors is called the *span* of that set:

$$\mathbf{x} \in \mathit{span}\{\mathbf{x}^{(1)}, \mathbf{x}^{(2)}, \ldots, \mathbf{x}^{(n)}\}$$

if and only if there exists constants c_j such that $\mathbf{x} = \sum_j c_j \mathbf{x}^{(j)}$. If the vectors $\mathbf{x}^{(j)}$ have dimension d, then the span is a subspace of R^d. The number of vectors in the basis for a vector space $\mathcal{S}$ is the dimension of the space $\mathcal{S}$, written *dim* $(\mathcal{S})$. The elementary vectors $\mathbf{e}^{(j)}$, $j = 1, \ldots, d$, which are zero except for one in component j, form a basis for R^d, the space of d-dimensional vectors.

Example A.1: Linear Dependence

Let

$$\mathbf{x}^{(1)} = \begin{bmatrix} 1 \\ 1 \\ 1 \\ 1 \end{bmatrix}, \quad \mathbf{x}^{(2)} = \begin{bmatrix} 1 \\ 2 \\ 3 \\ 4 \end{bmatrix}, \quad \text{and} \quad \mathbf{x}^{(3)} = \begin{bmatrix} -3 \\ -1 \\ 1 \\ 3 \end{bmatrix}.$$

Then $\mathbf{x}^{(1)}$ and $\mathbf{x}^{(2)}$ are linearly independent, but $\mathbf{x}^{(1)}$, $\mathbf{x}^{(2)}$, and $\mathbf{x}^{(3)}$ are linearly dependent since $5\mathbf{x}^{(1)} - 2\mathbf{x}^{(2)} + x^{(3)} = \mathbf{0}$, or $\mathbf{x}^{(3)} = 2\mathbf{x}^{(2)} - 5\mathbf{x}^{(1)}$. Also, $\mathbf{x}^{(3)} \in$ span $\{\mathbf{x}^{(1)}, \mathbf{x}^{(2)}\}$, this set has dimension 2, and $\mathbf{x}^{(1)} \perp \mathbf{x}^{(3)}$.

A.2 Rank, Column Space, and Nullspace

Some matrix concepts arise from viewing columns or rows of the matrix as vectors.

Definition A.1 *The* rank *of a matrix* $\mathbf{A}$ *is the number of linear independent rows or columns, and denoted by* rank $(\mathbf{A})$ *or* $r(\mathbf{A})$.

If a matrix $\mathbf{A}$ with shape m by n has rank m, then it is said to have full-row rank; similarly, if $r(\mathbf{A}) = n$, then $\mathbf{A}$ has full-column rank. A square matrix $\mathbf{A}$ with shape n by n with full-row or -column rank is *nonsingular* and an inverse, denoted by $\mathbf{A}^{-1}$, exists, such that $\mathbf{A} \times \mathbf{A}^{-1} = \mathbf{I}_n = \mathbf{A}^{-1} \times \mathbf{A}$. A square matrix $\mathbf{A}$ of dimension n with rank less than n is called *singular*.

Definition A.2 *The* column space *of a matrix, denoted by* $\mathcal{C}(\mathbf{A})$, *is the vector space spanned by the columns of the matrix, that is,*

$$\mathcal{C}(\mathbf{A}) = \{\mathbf{x} : \textit{ there exists a vector } \mathbf{c} \textit{ such that } \mathbf{x} = \mathbf{A}\mathbf{c}\}.$$

Denote the columns of $\mathbf{A}$ as follows:

$$\mathbf{a}^{(j)} = \mathbf{A}_{.j}$$

for $j = 1, \ldots, n$, so that (component i of the vector $\mathbf{a}^{(j)}$) $(\mathbf{a}^{(j)})_i = \mathbf{A}_{ij}$. Then if $\mathbf{x} \in \mathcal{C}(\mathbf{A})$, we can find coefficients c_j such that

$$\mathbf{x} = \sum_j c_j \mathbf{a}^{(j)}$$

so that $\mathbf{x}$ is a linear combination of columns of $\mathbf{A}$, or

$$x_i = \sum_j A_{ij} c_j$$

by the rules of matrix-vector multiplication. The column space of a matrix consists of all vectors formed by multiplying that matrix by any vector. Notice that the vectors in $\mathcal{C}(\mathbf{A})$ have dimension m. The number of basis vectors for $\mathcal{C}(\mathbf{A})$ is then the number of linearly independent columns of the matrix $\mathbf{A}$, and so

$$dim(\mathcal{C}(\mathbf{A})) = rank(\mathbf{A}).$$

Be careful of the use of the term *dimension*: vectors in $\mathcal{C}(\mathbf{A})$ have dimension m, while the dimension of the subspace $\mathcal{C}(\mathbf{A})$ is rank($\mathbf{A}$).

Example A.2: Matrix-Vector Multiplication
The usual view of matrix-vector multiplication is stacked row-by-column inner products:

$$\begin{bmatrix} 1 & 1 & -3 \\ 1 & 2 & -1 \\ 1 & 3 & 1 \\ 1 & 4 & 3 \end{bmatrix} \begin{bmatrix} 5 \\ 4 \\ 3 \end{bmatrix} = \begin{bmatrix} 1\times 5 + 1 \times 4 + (-3) \times 3 \\ 1 \times 5 + 2 \times 4 + (-1) \times 3 \\ 1 \times 5 + 3 \times 4 + 1 \times 3 \\ 1 \times 5 + 4 \times 4 + 3 \times 3 \end{bmatrix} = \begin{bmatrix} 0 \\ 10 \\ 20 \\ 30 \end{bmatrix}.$$

However, viewing $\mathbf{Ac}$ as a linear combination of columns presents the calculations as

$$\mathbf{a}^{(1)} = \begin{bmatrix} 1 \\ 1 \\ 1 \\ 1 \end{bmatrix}, \quad \mathbf{a}^{(2)} = \begin{bmatrix} 1 \\ 2 \\ 3 \\ 4 \end{bmatrix}, \quad \mathbf{a}^{(3)} = \begin{bmatrix} -3 \\ -1 \\ 1 \\ 3 \end{bmatrix}$$

so that

$$\mathbf{Ac} = \begin{bmatrix} 1 \\ 1 \\ 1 \\ 1 \end{bmatrix} (5) + \begin{bmatrix} 1 \\ 2 \\ 3 \\ 4 \end{bmatrix} (4) + \begin{bmatrix} -3 \\ -1 \\ 1 \\ 3 \end{bmatrix} (3) = \begin{bmatrix} 5 \\ 5 \\ 5 \\ 5 \end{bmatrix} + \begin{bmatrix} 4 \\ 8 \\ 12 \\ 16 \end{bmatrix} + \begin{bmatrix} -9 \\ -3 \\ 3 \\ 9 \end{bmatrix} = \begin{bmatrix} 0 \\ 10 \\ 20 \\ 30 \end{bmatrix}$$

Result A.1 $\quad$ *rank*$(\mathbf{AB}) \leq$ *min*(*rank*$(\mathbf{A})$, *rank*$(\mathbf{B})$).

Proof: Each column of the product $\mathbf{AB}$ is a linear combination of columns of $\mathbf{A}$, so the number of linearly independent columns of $\mathbf{AB}$ cannot be greater than that of $\mathbf{A}$. Similarly, since *rank*$(\mathbf{AB})$ = *rank*$(\mathbf{B}^T\mathbf{A}^T)$, the same argument gives *rank*$(\mathbf{B}^T)$ as an upper bound. □

Example A.3: Rank of Product
The product can have smaller rank, as can be seen by taking

$$\mathbf{A} = \begin{bmatrix} 1 & 1 & 1 \\ 0 & 1 & 1 \end{bmatrix} \quad \text{and} \quad \mathbf{B} = \begin{bmatrix} 1 & 1 \\ 0 & 1 \\ 1 & 0 \end{bmatrix},$$

each with rank 2, while the product

$$\mathbf{AB} = \begin{bmatrix} 2 & 2 \\ 1 & 1 \end{bmatrix}$$

has only rank 1.

Result A.2   *(a) If $\mathbf{A} = \mathbf{BC}$, then $\mathcal{C}(\mathbf{A}) \subseteq \mathcal{C}(\mathbf{B})$.*
(b) If $\mathcal{C}(\mathbf{A}) \subseteq \mathcal{C}(\mathbf{B})$, then there exists a matrix $\mathbf{C}$ such that $\mathbf{A} = \mathbf{BC}$.

Proof: For (a), any vector $\mathbf{x}$ in $\mathcal{C}(\mathbf{A})$ can be written as $\mathbf{Ad} = \mathbf{B(Cd)}$, which is clearly in $\mathcal{C}(\mathbf{B})$. For (b), let $\mathbf{A}_{.j}$ be the jth column of the matrix $\mathbf{A}$. Clearly $\mathbf{A}_{.j} \in \mathcal{C}(\mathbf{A})$, and hence in $\mathcal{C}(\mathbf{B})$ by hypothesis. Therefore $\mathbf{A}_{.j}$ can be written as a linear combination of columns of $\mathbf{B}$, that is, there exists a vector $\mathbf{c}^{(j)}$ such that $\mathbf{A}_{.j} = \mathbf{Bc}^{(j)}$. Each column of $\mathbf{A}$ determines a vector $\mathbf{c}^{(j)}$, which forms a column of the matrix $\mathbf{C}$, that is, $\mathbf{C}_{.j} = \mathbf{c}^{(j)}$.
$\square$

Definition A.3   *The* null space *of a matrix, denoted by $\mathcal{N}(\mathbf{A})$, $\mathcal{N}(\mathbf{A}) = \{\mathbf{y}: \mathbf{Ay} = \mathbf{0}\}$.*

Notice that if the matrix $\mathbf{A}$ is $m \times n$, then vectors in $\mathcal{N}(\mathbf{A})$ have dimension n while vectors in $\mathcal{C}(\mathbf{A})$ have dimension m. Mathematically, this may be expressed as $\mathcal{C}(\mathbf{A}) \subseteq R^m$ and $\mathcal{N}(\mathbf{A}) \subseteq R^n$.

Result A.3   *If $\mathbf{A}$ has full-column rank, then $\mathcal{N}(\mathbf{A}) = \{\mathbf{0}\}$.*

Proof: If A has full-column rank, then the columns of $\mathbf{A}$, say, $\mathbf{a}^{(j)}$, are linearly independent, and if $\mathbf{c} \in \mathcal{N}(\mathbf{A})$, then $\mathbf{Ac} = \sum_j c_j \mathbf{a}^{(j)} = \mathbf{0}$ means $\mathbf{c} = \mathbf{0}$.
$\square$

Notice also that determining the dimension and basis of the nullspace of a matrix is the same work as finding its rank. In determining the rank, we discover the linear dependencies of columns. Finding a column that can be written as a linear combination of other columns produces a vector in the nullspace of the matrix. Each new dependency that is discovered reduces the rank by one and produces another basis vector for the nullspace.

Example A.4: Nullspace
Construct a matrix from the vectors in Example A.1, that is, let

$$\mathbf{A} = \begin{bmatrix} 1 & 1 & -3 \\ 1 & 2 & -1 \\ 1 & 3 & 1 \\ 1 & 4 & 3 \end{bmatrix} \quad \text{and} \quad \mathbf{y} = \begin{bmatrix} 5 \\ -2 \\ 1 \end{bmatrix}$$

then $\mathbf{A}$ has two linearly independent columns, so that $rank\,(\mathbf{A}) = 2$. Vectors in $\mathcal{N}(\mathbf{A})$ have dimension 3, that is, $\mathcal{N}(\mathbf{A}) \subset R^3$, and $\mathbf{y}$ is a basis vector for $\mathcal{N}(\mathbf{A})$.

The following theorem is the cornerstone of what we use in linear algebra, and its spirit will be seen in Section A.3 in the construction of solutions to linear equations.

Theorem A.1   *If the matrix $\mathbf{A}$ is m by n, then $dim\,(\mathcal{N}(\mathbf{A})) = n - r$ where $r = rank\,(\mathbf{A})$, or, more elegantly,*

$$dim\,(\mathcal{N}(\mathbf{A})) + dim\,(\mathcal{C}(\mathbf{A})) = n.$$

Proof: Denote $dim(\mathcal{N}(\mathbf{A}))$ by k, to be determined, and construct a set of basis vectors for $\mathcal{N}(\mathbf{A})$: $\{\mathbf{u}^{(1)}, \mathbf{u}^{(2)}, \ldots, \mathbf{u}^{(k)}\}$, so that $\mathbf{A}\mathbf{u}^{(i)} = \mathbf{0}$, for $i = 1, \ldots, k$. Now, construct a basis for R^n by adding the vectors $\{\mathbf{u}^{(k+1)}, \ldots, \mathbf{u}^{(n)}\}$, which are not in $\mathcal{N}(\mathbf{A})$. Clearly, $\mathbf{A}\mathbf{u}^{(i)} \in \mathcal{C}(\mathbf{A})$ for $i = k+1, \ldots, n$, and so the span of these vectors form a subspace of $\mathcal{C}(\mathbf{A})$. These vectors $\{\mathbf{A}\mathbf{u}^{(i)}, i = k+1, \ldots, n\}$ are also linearly independent from the following argument: suppose $\sum_{i=k+1}^{n} c_i \mathbf{A}\mathbf{u}^{(i)} = \mathbf{0}$; then $\sum_{i=k+1}^{n} c_i \mathbf{A}\mathbf{u}^{(i)} = \mathbf{A}[\sum_{i=k+1}^{n} c_i \mathbf{u}^{(i)}] = \mathbf{0}$, and hence $\sum_{i=k+1}^{n} c_i \mathbf{u}^{(i)}$ is a vector in $\mathcal{N}(\mathbf{A})$. Therefore there exist b_i such that $\sum_{i=k+1}^{n} c_i \mathbf{u}^{(i)} = \sum_{i=1}^{k} b_i \mathbf{u}^{(i)}$, or $\sum_{i=1}^{k} b_i \mathbf{u}^{(i)} - \sum_{i=k+1}^{n} c_i \mathbf{u}^{(i)} = \mathbf{0}$. Since $\{\mathbf{u}^{(i)}\}$ form a basis for R^n, c_i must all be zero. Therefore $\mathbf{A}\mathbf{u}^{(i)}$, $i = k+1, \ldots, n$ are linearly independent. At this point, since $span\{\mathbf{A}\mathbf{u}^{(k+1)}, \ldots, \mathbf{A}\mathbf{u}^{(n)}\} \subseteq \mathcal{C}(\mathbf{A})$, $dim(\mathcal{C}(\mathbf{A}))$ is at least $n-k$. Suppose there is a vector $\mathbf{y}$ that is in $\mathcal{C}(\mathbf{A})$, but not in the span; then there exists $\mathbf{u}^{(n+1)}$ so that $\mathbf{y} = \mathbf{A}\mathbf{u}^{(n+1)}$ and $\mathbf{u}^{(n+1)}$ is linearly independent of $\{\mathbf{u}^{(k+1)}, \ldots, \mathbf{u}^{(n)}\}$ (and clearly not in $\mathcal{N}(\mathbf{A})$), making $n+1$ linearly independent vectors in R^n. Since that is not possible, the span is equal to $\mathcal{C}(\mathbf{A})$ and $dim(\mathcal{C}(\mathbf{A})) = n - k = r = rank(\mathbf{A})$, so that $k = dim(\mathcal{N}(\mathbf{A})) = n - r$.□

Another view of this relationship is that there exists a nonsingular matrix $\mathbf{C}$ from elementary column operations, so that $\mathbf{AC} = \mathbf{B}$ where the first r columns of $\mathbf{B}$ are linearly independent and the last $n-r$ columns of $\mathbf{B}$ are all zero. Then the elementary vectors $\mathbf{e}^{(j)}$, $j = r+1, \ldots, n$ form a basis for $\mathcal{N}(\mathbf{B})$, which then has dimension $n-r$, since these vectors span the space of all vectors whose first r components are all zero. Then the vectors $\mathbf{C}\mathbf{e}^{(j)}$, $j = r+1, \ldots, n$ form a basis for $\mathcal{N}(\mathbf{A})$.

Example A.4: continued
Using the same matrix as above, that is, let

$$\mathbf{A} = \begin{bmatrix} 1 & 1 & -3 \\ 1 & 2 & -1 \\ 1 & 3 & 1 \\ 1 & 4 & 3 \end{bmatrix},$$

then

$$\mathbf{y} = \begin{bmatrix} 5 \\ -2 \\ 1 \end{bmatrix}$$

is a basis vector for $\mathcal{N}(\mathbf{A})$. Taking

$$\mathbf{C} = \begin{bmatrix} 1 & 0 & 5 \\ 0 & 1 & -2 \\ 0 & 0 & 1 \end{bmatrix},$$

then

$$\mathbf{B} = \mathbf{AC} = \begin{bmatrix} 1 & 1 & 0 \\ 1 & 2 & 0 \\ 1 & 3 & 0 \\ 1 & 4 & 0 \end{bmatrix}$$

where the first $r = 2$ columns are linearly independent and whose last $n - r = 1$ column is zero. If we were to do elementary row (instead of column) operations, then we would produce

$$\mathbf{E} = \mathbf{FA} = \begin{bmatrix} 1 & 0 & 0 & 0 \\ -1 & 1 & 0 & 0 \\ 1 & -2 & 1 & 0 \\ 2 & -3 & 0 & 1 \end{bmatrix} \begin{bmatrix} 1 & 1 & -3 \\ 1 & 2 & -1 \\ 1 & 3 & 1 \\ 1 & 4 & 3 \end{bmatrix} = \begin{bmatrix} 1 & 1 & -3 \\ 0 & 1 & 2 \\ 0 & 0 & 0 \\ 0 & 0 & 0 \end{bmatrix}$$

which shows little promise. However, consider now solving $\mathbf{Ez} = \mathbf{0}$:

$$z_1 + z_2 - 3z_3 = 0$$

and

$$z_2 + 2z_3 = 0$$

which is solved with $z_2 = -2z_3$ and, substituting into the previous equation, yields

$$z_1 - 2z_3 - 3z_3 = 0$$

and then $z_1 = 5z_3$, and a solution vector

$$\begin{bmatrix} 5 \\ -2 \\ 1 \end{bmatrix} z_3$$

with z_3 parameterizing $\mathcal{N}(\mathbf{E}) = \mathcal{N}(\mathbf{A})$.

Definition A.4 *Two vector spaces $\mathcal{S}$ and $\mathcal{T}$ form* orthogonal complements *in R^m if and only if $\mathcal{S}, \mathcal{T} \subseteq R^m$, $\mathcal{S} \cap \mathcal{T} = \{\mathbf{0}\}$, $dim(\mathcal{S}) = r$, $dim(\mathcal{T}) = m - r$, and every vector in $\mathcal{S}$ is orthogonal to every vector in $\mathcal{T}$.*

Result A.4 *Let $\mathcal{S}$ and $\mathcal{T}$ be orthogonal complements in R^m, then any vector $\mathbf{x} \in R^m$ can be written as*

$$\mathbf{x} = \mathbf{s} + \mathbf{t}$$

where $\mathbf{s} \in \mathcal{S}$ and $\mathbf{t} \in \mathcal{T}$, and this decomposition is unique.

Proof: First, suppose that a decomposition is not possible, that is, $\mathbf{x}$ is linearly independent of basis vectors of both $\mathcal{S}$ and $\mathcal{T}$, but that would give $m+1$ linearly independent vectors in R^m, which is not possible. For uniqueness, suppose $\mathbf{x} = \mathbf{s}^{(1)} + \mathbf{t}^{(1)} = \mathbf{s}^{(2)} + \mathbf{t}^{(2)}$ with $\mathbf{s}^{(1)}, \mathbf{s}^{(2)} \in \mathcal{S}$ and $\mathbf{t}^{(1)}, \mathbf{t}^{(2)} \in \mathcal{T}$, then $\mathbf{x} - \mathbf{x} = \mathbf{0}$ or

$$\mathbf{s}^{(1)} - \mathbf{s}^{(2)} = \mathbf{t}^{(2)} - \mathbf{t}^{(1)}$$

where the left-hand side is in $\mathcal{S}$ and the right in $\mathcal{T}$; both sides must be zero. □

The Pythagorean Theorem applies to the decomposition in (A.4), since if $\mathbf{x} = \mathbf{s}+\mathbf{t}$ with $\mathbf{s} \in \mathcal{S}$ and $\mathbf{t} \in \mathcal{T}$, then

$$\|\mathbf{x}\|^2 = \mathbf{x}^T\mathbf{x} = (\mathbf{s}+\mathbf{t})^T(\mathbf{s}+\mathbf{t}) = \mathbf{s}^T\mathbf{s} + 2\mathbf{s}^T\mathbf{t} + \mathbf{t}^T\mathbf{t} = \|\mathbf{s}\|^2 + \|\mathbf{t}\|^2$$

owing to the orthogonality of $\mathbf{s}$ and $\mathbf{t}$.

Result A.5 *If $\mathbf{A}$ is an $m \times n$ matrix, then $\mathcal{C}(\mathbf{A})$ and $\mathcal{N}(\mathbf{A}^T)$ are orthogonal complements in R^m.*

Proof: Both $\mathcal{C}(\mathbf{A})$ and $\mathcal{N}(\mathbf{A}^T)$ are vector spaces with vectors in R^m. From Theorem A.1, $dim\,(\mathcal{C}(\mathbf{A})) = r = rank\,(\mathbf{A})$, and $dim\,(\mathcal{N}(\mathbf{A}^T)) = m - r$, since $rank\,(\mathbf{A}) = rank\,(\mathbf{A}^T)$. Suppose $\mathbf{v}$ is in both spaces; then $\mathbf{v} = \mathbf{Ac}$ for some vector $\mathbf{c}$, and $\mathbf{A}^T\mathbf{v} = \mathbf{0} = \mathbf{A}^T\mathbf{Ac}$, so also $\mathbf{c}^T\mathbf{A}^T\mathbf{Ac} = \|\mathbf{Ac}\|^2 = \mathbf{0}$, which implies $\mathbf{Ac} = \mathbf{0} = \mathbf{v}$. Lastly, if $\mathbf{v} \in \mathcal{C}(\mathbf{A})$, then $\mathbf{v} = \mathbf{Ac}$ for some vector $\mathbf{c}$, and for any $\mathbf{w} \in \mathcal{N}(\mathbf{A}^T)$, $\mathbf{v}^T\mathbf{w} = \mathbf{c}^T\mathbf{A}^T\mathbf{w} = \mathbf{0}$. □

Result A.6 *Let $\mathcal{S}_1$ and $\mathcal{T}_1$ be orthogonal complements, as well as $\mathcal{S}_2$ and $\mathcal{T}_2$; then if $\mathcal{S}_1 \subseteq \mathcal{S}_2$, then $\mathcal{T}_2 \subseteq \mathcal{T}_1$.*

Proof: Construct $\mathbf{A}_1$ and $\mathbf{A}_2$ so that $\mathcal{S}_1 = \mathcal{C}(\mathbf{A}_1)$ and $\mathcal{S}_2 = \mathcal{C}(\mathbf{A}_2)$; each matrix can be formed by stacking basis vectors of the spaces as columns. Let $\mathbf{v} \in \mathcal{T}_2$, then $\mathbf{v} \in \mathcal{N}(\mathbf{A}_2^T)$ or $\mathbf{A}_2^T\mathbf{v} = \mathbf{0}$. Now $\mathcal{S}_1 \subseteq \mathcal{S}_2$ means (from Result A.2(b)) there exists a matrix $\mathbf{B}$ such that $\mathbf{A}_1 = \mathbf{A}_2\mathbf{B}$, so that $\mathbf{A}_1^T\mathbf{v} = \mathbf{B}^T\mathbf{A}_2^T\mathbf{v} = \mathbf{0}$, so that $\mathbf{v} \in \mathcal{T}_1$. □

A.3 Some Useful Results

The following results are powerful tools that deserve special attention, although they may, at times, appear obvious.

Result A.7 *Consider two vector spaces $\mathcal{S}$ and $\mathcal{T}$. If $\mathcal{S} \subseteq \mathcal{T}$ and $dim\,(\mathcal{S}) = dim\,(\mathcal{T}) = k$, then $\mathcal{S} = \mathcal{T}$.*

Proof: Suppose there is a vector $\mathbf{t} \in \mathcal{T}$, but not in $\mathcal{S}$, and let $\{\mathbf{s}^{(1)}, \ldots, \mathbf{s}^{(k)}\}$ be a basis for $\mathcal{S}$. Then $\{\mathbf{s}^{(1)}, \ldots, \mathbf{s}^{(k)}, \mathbf{t}\}$ would then be a set of $k+1$ linearly independent vectors in $\mathcal{T}$, which would make $dim\,(\mathcal{T})$ at least $k+1$. This is a contradiction, and hence there cannot be a vector $\mathbf{t}$ that is in $\mathcal{T}$ and not also in $\mathcal{S}$. ∎

Result A.8 *Let* $\mathbf{A}$ *be an* $m \times n$ *matrix and* $\mathbf{b}$ *a fixed vector. If* $\mathbf{Ax} + \mathbf{b} = \mathbf{0}$ *for all* $\mathbf{x} \in R^n$, *then* $\mathbf{A} = \mathbf{0}$ *and* $\mathbf{b} = \mathbf{0}$.

Proof: Taking $\mathbf{x} = \mathbf{0}$ produces $\mathbf{b} = \mathbf{0}$. Then take $\mathbf{x} = \mathbf{e}^{(i)}$, the ith elementary vector, yielding $\mathbf{Ae}^{(i)} = \mathbf{0}$, which says that column i of $\mathbf{A}$ is zero; continuing for $i = 1, \ldots, n$ produces the desired result. ∎

Corollary A.1 *If* $\mathbf{Bx} = \mathbf{Cx}$ *for all* $\mathbf{x}$, *then* $\mathbf{B} = \mathbf{C}$.

Proof: Apply Result A.8 with $\mathbf{A} = \mathbf{B} - \mathbf{C}$ and $\mathbf{b} = \mathbf{0}$. ∎

Corollary A.2 *Let* $\mathbf{A}$ *have full-column rank; then if* $\mathbf{AB} = \mathbf{AC}$, *then* $\mathbf{B} = \mathbf{C}$.

Proof: From Result A.3, $\mathcal{N}(\mathbf{A}) = \mathbf{0}$. Since $\mathbf{A}\,(\mathbf{B} - \mathbf{C}) = \mathbf{0}$, then each column of $(\mathbf{B} - \mathbf{C})$ must be in $\mathcal{N}(\mathbf{A})$ and hence must be zero. ∎

Lemma A.1 *If* $\mathbf{C}^T\mathbf{C} = \mathbf{0}$, *then* $\mathbf{C} = \mathbf{0}$.

Proof: The diagonal elements of $\mathbf{C}^T\mathbf{C}$ are squared lengths of columns of $\mathbf{C}$; if they are all zero, then each column of $\mathbf{C}$ is the zero vector. ∎

A.4 Solving Equations and Generalized Inverses

Consider a system of linear equations

$$\mathbf{Ax} = \mathbf{c} \tag{A.3}$$

where $\mathbf{A}$ is an $m \times n$ matrix, and the right-hand-side $(m \times 1)$ vector $\mathbf{c}$ is a given vector of constants. We wish to find, if possible, a solution vector $\mathbf{x}$ that satisfies (A.3). If the matrix $\mathbf{A}$ is a square matrix $(m = n)$ and nonsingular, then an inverse $\mathbf{A}^{-1}$ exists, and $\mathbf{A}^{-1}\mathbf{c}$ is the unique solution to (A.3). If $\mathbf{A}$ is singular or not square $(m \neq n)$, then (A.3) may not have a solution, or it may have infinitely many solutions.

Definition A.5 *A system of equations* $\mathbf{Ax} = \mathbf{c}$ *is* consistent *iff there exists a solution* $\mathbf{x}^*$ *such that* $\mathbf{Ax}^* = \mathbf{c}$.

Notice that if $\mathbf{Ax}^* = \mathbf{c}$ for some vector $\mathbf{x}^*$, then $\mathbf{c}$ is a linear combination of the columns of $\mathbf{A}$, and hence $\mathbf{c}$ belongs to the column space of $\mathbf{A}$, denoted as $\mathcal{C}(\mathbf{A})$. Conversely, if $\mathbf{c}$ is an element of $\mathcal{C}(\mathbf{A})$, then there exists some vector $\mathbf{x}^*$ such that $\mathbf{Ax}^* = \mathbf{c}$, and hence the system of Equations (A.3) is consistent. We have thus established the following result:

Result A.9 *A system of equations* $\mathbf{Ax} = \mathbf{c}$ *is consistent if and only if* $\mathbf{c} \in \mathcal{C}(\mathbf{A})$.

Example A.5: Inconsistent Equations

The system of equations below is inconsistent since there is no linear combination of the columns of the matrix that can produce the right-hand side.

$$\begin{bmatrix} 1 & 0 \\ 1 & 1 \\ 0 & 1 \end{bmatrix} \begin{bmatrix} x_1 \\ x_2 \end{bmatrix} = \begin{bmatrix} 1 \\ 1 \\ 1 \end{bmatrix}$$

On the other hand, the following equations are consistent, as $\mathbf{x}^* = (1, 1)^T$ is a solution:

$$\begin{bmatrix} 1 & 0 \\ 1 & 1 \\ 0 & 1 \end{bmatrix} \begin{bmatrix} x_1 \\ x_2 \end{bmatrix} = \begin{bmatrix} 1 \\ 2 \\ 1 \end{bmatrix}.$$

Similar to the inverse of a square, nonsingular matrix, we now define a generalized inverse of a matrix. The generalized inverse exists for any matrix, whether or not the matrix is square, whether or not the matrix has full rank. The most common notations employed for a generalized inverse of the matrix $\mathbf{A}$ are $\mathbf{A}^g$ or $\mathbf{A}^-$ or $\mathbf{G}$. If $\mathbf{A}$ is $m \times n$, then a generalized inverse $\mathbf{A}^g$ is $n \times m$—the same shape as its transpose and may have the same or larger rank than $\mathbf{A}$. The purpose of defining such an inverse is to find solutions to a consistent system of equations $\mathbf{Ax} = \mathbf{c}$.

Definition A.6 *A matrix* $\mathbf{G}$ *is a* generalized inverse *of the matrix* $\mathbf{A}$ *if and only if it satisfies* $\mathbf{AGA} = \mathbf{A}$.

Result A.10 *Let* $\mathbf{A}$ *be an* $m \times n$ *matrix with rank* r. *If* $\mathbf{A}$ *can be partitioned as below, with rank* $(\mathbf{A}) =$ *rank* $(\mathbf{C}) = r$,

$$\mathbf{A} = \begin{bmatrix} \mathbf{C} & \mathbf{D} \\ \mathbf{E} & \mathbf{F} \end{bmatrix} \begin{matrix} r \\ m - r \end{matrix}$$

so that $\mathbf{C}$ *is nonsingular, then the matrix*

$$\mathbf{G} = \begin{bmatrix} \mathbf{C}^{-1} & \mathbf{0} \\ \mathbf{0} & \mathbf{0} \end{bmatrix} \begin{matrix} r \\ n - r \end{matrix} \tag{A.4}$$

is a generalized inverse of $\mathbf{A}$.

Proof: Multiplying them out as **AGA**, we obtain

$$\begin{bmatrix} \mathbf{C} & \mathbf{D} \\ \mathbf{E} & \mathbf{F} \end{bmatrix} \begin{bmatrix} \mathbf{C}^{-1} & \mathbf{0} \\ \mathbf{0} & \mathbf{0} \end{bmatrix} \begin{bmatrix} \mathbf{C} & \mathbf{D} \\ \mathbf{E} & \mathbf{F} \end{bmatrix} = \begin{bmatrix} \mathbf{C} & \mathbf{D} \\ \mathbf{E} & \mathbf{E}\mathbf{C}^{-1}\mathbf{D} \end{bmatrix}$$

□

Since the full matrix **A** and the submatrix **C** have the same rank r, the rows [**E** **F**] are linearly dependent on the rows [**C** **D**], so that

$$[\mathbf{E} \quad \mathbf{F}] = \mathbf{K}[\mathbf{C} \quad \mathbf{D}]$$

for some matrix **K**. That is, $\mathbf{E} = \mathbf{KC}$ and $\mathbf{F} = \mathbf{KD}$. Since **C** is nonsingular, we can find $\mathbf{K} = \mathbf{EC}^{-1}$, so that $\mathbf{F} = \mathbf{KD} = (\mathbf{EC}^{-1})\mathbf{D}$, and the generalized inverse definition is satisfied.

Corollary A.3 *Let* **A** *be an* $m \times n$ *matrix with rank* r. *If* **A** *can be partitioned as below, with rank* $(\mathbf{A}) = $ *rank* $(\mathbf{F}) = r$, *and* **F** *nonsingular,*

$$\mathbf{A} = \begin{bmatrix} \mathbf{C} & \mathbf{D} \\ \mathbf{E} & \mathbf{F} \end{bmatrix} \begin{matrix} m-r \\ r \end{matrix},$$

then the matrix

$$\mathbf{G} = \begin{bmatrix} \mathbf{0} & \mathbf{0} \\ \mathbf{0} & \mathbf{F}^{-1} \end{bmatrix} \begin{matrix} n-r \\ r \end{matrix} \tag{A.5}$$

is also a generalized inverse of **A**.

Notice that the shape of **G** $(n \times m)$ dictates the sizes of the blocks of zeros in its definition in (A.4); moreover, if $r = m = n$, then $\mathbf{A} = \mathbf{C}, \mathbf{G} = \mathbf{A}^{-1}$ and the blocks of zeros disappear.

To extend Result A.10 for the construction of a generalized inverse for any matrix, notice that we can find r linearly independent rows and columns and then permute the rows and columns so that the upper-left-hand corner of the transformed matrix is nonsingular and has the same rank as that of the original matrix. In other words, given any matrix **A**, we can find permutation matrices **P** and **Q** so that

$$\mathbf{PAQ} = \begin{bmatrix} \mathbf{C} & \mathbf{D} \\ \mathbf{E} & \mathbf{F} \end{bmatrix} \tag{A.6}$$

where *rank* $(\mathbf{A}) = $ *rank* $(\mathbf{C})$ and **C** is nonsingular. See Exercises A.27 and A.30 for other general approaches for constructing generalized inverses.

Result A.11 *For a given* $(m \times n)$ *matrix* **A** *with rank* r, *let* **P** *and* **Q** *be permutation matrices such that*

$$\mathbf{PAQ} = \begin{bmatrix} \mathbf{C} & \mathbf{D} \\ \mathbf{E} & \mathbf{F} \end{bmatrix}$$

where rank $(\mathbf{A}) = $ *rank* $(\mathbf{C}) = r$ *and* $\mathbf{C}$ *is nonsingular. Then the matrix* $\mathbf{G}$ *below is a generalized inverse of* $\mathbf{A}$*:*

$$\mathbf{G} = \mathbf{Q}\begin{bmatrix} \mathbf{C}^{-1} & \mathbf{0} \\ \mathbf{0} & \mathbf{0} \end{bmatrix}\mathbf{P} \tag{A.7}$$

Proof: Note that

$$\mathbf{AGA} = \mathbf{P}^{-1}\begin{bmatrix} \mathbf{C} & \mathbf{D} \\ \mathbf{E} & \mathbf{F} \end{bmatrix}\mathbf{Q}^{-1}\mathbf{Q}\begin{bmatrix} \mathbf{C}^{-1} & \mathbf{0} \\ \mathbf{0} & \mathbf{0} \end{bmatrix}\mathbf{P}\mathbf{P}^{-1}\begin{bmatrix} \mathbf{C} & \mathbf{D} \\ \mathbf{E} & \mathbf{F} \end{bmatrix}\mathbf{Q}^{-1}$$

$$= \mathbf{P}^{-1}\begin{bmatrix} \mathbf{C} & \mathbf{D} \\ \mathbf{E} & \mathbf{F} \end{bmatrix}\begin{bmatrix} \mathbf{C}^{-1} & \mathbf{0} \\ \mathbf{0} & \mathbf{0} \end{bmatrix}\begin{bmatrix} \mathbf{C} & \mathbf{D} \\ \mathbf{E} & \mathbf{F} \end{bmatrix}\mathbf{Q}^{-1} = \mathbf{P}^{-1}\begin{bmatrix} \mathbf{C} & \mathbf{D} \\ \mathbf{E} & \mathbf{F} \end{bmatrix}\mathbf{Q}^{-1} = \mathbf{A}$$

following the arguments of Result A.10. □

Example A.6: Generalized Inverses

Let us construct some generalized inverses for the matrix $\mathbf{A}$ below:

$$\mathbf{A} = \begin{bmatrix} 1 & 1 & 1 & 1 \\ 0 & 1 & 0 & -1 \\ 1 & 0 & 1 & 2 \end{bmatrix}$$

From Exercise A.14, we find that *rank* $(A) = 2$, and following Result A.10, we find

$$\mathbf{G}_1 = \begin{bmatrix} 1 & -1 & 0 \\ 0 & 1 & 0 \\ 0 & 0 & 0 \\ 0 & 0 & 0 \end{bmatrix}$$

is a generalized inverse of $\mathbf{A}$. Following the extension using permutation matrices, we have

$$\mathbf{PAQ} = \begin{bmatrix} 0 & 1 & 0 \\ 0 & 0 & 1 \\ 1 & 0 & 0 \end{bmatrix}\begin{bmatrix} 1 & 1 & 1 & 1 \\ 0 & 1 & 0 & -1 \\ 1 & 0 & 1 & 2 \end{bmatrix}\begin{bmatrix} 0 & 0 & 1 & 0 \\ 0 & 0 & 0 & 1 \\ 1 & 0 & 0 & 0 \\ 0 & 1 & 0 & 0 \end{bmatrix} = \begin{bmatrix} 0 & -1 & 0 & 1 \\ 1 & 2 & 1 & 0 \\ 1 & 1 & 1 & 1 \end{bmatrix}$$

and obtain a second generalized inverse $\mathbf{G}_2$:

$$\mathbf{G}_2 = \mathbf{Q}\begin{bmatrix} 2 & 1 & 0 \\ -1 & 0 & 0 \\ 0 & 0 & 0 \\ 0 & 0 & 0 \end{bmatrix}\mathbf{P} = \begin{bmatrix} 0 & 0 & 0 \\ 0 & 0 & 0 \\ 0 & 2 & 1 \\ 0 & -1 & 0 \end{bmatrix}$$

Now that we have a definition of a generalized inverse, and have demonstrated that a generalized inverse exists for any matrix, we can begin to construct solutions of consistent systems of linear equations.

Result A.12 *Let* $\mathbf{Ax} = \mathbf{c}$ *be a consistent system of equations and let* $\mathbf{G}$ *be a generalized inverse of* $\mathbf{A}$*, then* $\mathbf{Gc}$ *is a solution to the equations* $\mathbf{Ax} = \mathbf{c}$.

Proof: If the equations $\mathbf{Ax} = \mathbf{c}$ are consistent, then $\mathbf{c} \in \mathcal{C}(\mathbf{A})$, or there exists some vector $\mathbf{z}$ such that $\mathbf{c} = \mathbf{Az}$. Then $\mathbf{A}(\mathbf{Gc}) = \mathbf{AGAz} = \mathbf{Az} = \mathbf{c}$. The first equality follows from $\mathbf{c} \in \mathcal{C}(\mathbf{A})$, the second from the definition of a generalized inverse. □

Result A.13 *Let* $\mathbf{Ax} = \mathbf{c}$ *be a consistent system of equations and let* $\mathbf{G}$ *be a generalized inverse of* $\mathbf{A}$*; then* $\tilde{\mathbf{x}}$ *is a solution to the equations* $\mathbf{Ax} = \mathbf{c}$ *if and only if there exists a vector* $\mathbf{z}$ *such that* $\tilde{\mathbf{x}} = \mathbf{Gc} + (\mathbf{I} - \mathbf{GA})\mathbf{z}$.

Proof: (If) If $\tilde{\mathbf{x}} = \mathbf{Gc} + (\mathbf{I} - \mathbf{GA})\mathbf{z}$, then

$$\mathbf{A}\tilde{\mathbf{x}} = \mathbf{AGc} + \mathbf{A}(\mathbf{I} - \mathbf{GA})\mathbf{z} = \mathbf{AGc} + (\mathbf{A} - \mathbf{AGA})\mathbf{z} = \mathbf{c} + \mathbf{0z} = \mathbf{c}$$

since $\mathbf{AGc} = \mathbf{c}$ from Result A.12, and $\mathbf{AGA} = \mathbf{A}$ from the definition of a generalized inverse.

(Only if) If $\tilde{\mathbf{x}}$ is a solution, then the following algebra,

$$\tilde{\mathbf{x}} = \mathbf{Gc} + [\tilde{\mathbf{x}} - \mathbf{Gc}] = \mathbf{Gc} + [\tilde{\mathbf{x}} - \mathbf{GA}\tilde{\mathbf{x}}] = \mathbf{Gc} + [\mathbf{I}_n - \mathbf{GA}]\,\tilde{\mathbf{x}},$$

shows that any solution can be written in this form, here with $\mathbf{z} = \tilde{\mathbf{x}}$. □

Result A.13 has many ramifications. First, by varying $\mathbf{z}$ over R^n, we can sweep out all possible solutions to a system of equations (see Exercise A.26). Second, the collection of solutions does not depend on the choice of generalized inverse. Furthermore, we can view the family of solutions as the sum of two pieces: here $\mathbf{Gc}$ is a particular solution to the (nonhomogeneous) system of equations $\mathbf{Ax} = \mathbf{c}$, and $(\mathbf{I} - \mathbf{GA})\mathbf{z}$ solves the (homogeneous) system of equations $\mathbf{Ax} = \mathbf{0}$, producing vectors in $\mathcal{N}(\mathbf{A})$.

Example A.6: continued

Use $\mathbf{A}$ as before and take $\mathbf{c} = (3, 2, 1)^T$. Using $\mathbf{G}_1$ as $\mathbf{A}^g$ to form all solutions $\mathbf{A}^g\mathbf{c} + (\mathbf{I} - \mathbf{A}^g\mathbf{A})\mathbf{z}$, we have

$$\begin{bmatrix} 1 & -1 & 0 \\ 0 & 1 & 0 \\ 0 & 0 & 0 \\ 0 & 0 & 0 \end{bmatrix} \begin{bmatrix} 3 \\ 2 \\ 1 \end{bmatrix} + \begin{bmatrix} 0 & 0 & -1 & -2 \\ 0 & 0 & 0 & 1 \\ 0 & 0 & 1 & 0 \\ 0 & 0 & 0 & 1 \end{bmatrix} \begin{bmatrix} z_1 \\ z_2 \\ z_3 \\ z_4 \end{bmatrix} = \begin{bmatrix} 1 \\ 2 \\ 0 \\ 0 \end{bmatrix} + \begin{bmatrix} -1 & -2 \\ 0 & 1 \\ 1 & 0 \\ 0 & 1 \end{bmatrix} \begin{bmatrix} z_3 \\ z_4 \end{bmatrix}.$$

In elementary linear algebra, the usual route for solving a general system of equations is to perform elementary row operations (permuting or rescaling rows, or adding multiples of one row to another) to the matrix augmented with the right-hand side appended as an extra column. These row operations are continued until each row of the matrix has 1 as its leftmost nonzero entry. Since these elementary row operations can be viewed as premultiplying by a matrix, say, $\mathbf{M}$, we can view this approach as constructing $\mathbf{M}\,(\mathbf{A}|\mathbf{c})$. Making the simplifying assumption that the first r columns of $\mathbf{A}$ are linearly independent, we will arrive at

$$\mathbf{M}\,(\mathbf{A}|\mathbf{c}) = \begin{bmatrix} \mathbf{B}_{11} & \mathbf{B}_{12} & \mathbf{b}_1 \\ \mathbf{0} & \mathbf{0} & \mathbf{b}_2 \end{bmatrix} \begin{matrix} r \\ m-r \end{matrix}$$

where the matrix $\mathbf{B}_{11}$ is an $r \times r$ upper triangular matrix with all its diagonal elements equal to one and $r = rank\,(\mathbf{A})$. Now we recognize that if $m = r$ or $\mathbf{A}$ is full-row rank, then the last row above disappears and a solution to $\mathbf{Ax} = \mathbf{c}$ can be found by solving the upper triangular system $\mathbf{B}_{11}\mathbf{x} = \mathbf{b}_1$. If $r < m$, then $\mathbf{b}_2$ must be $\mathbf{0}$; otherwise the equations are not consistent. If the equations are consistent, then these operations lead to the equation

$$\mathbf{B}_{11}\mathbf{x}_1 + \mathbf{B}_{12}\mathbf{x}_2 = \mathbf{b}_1$$

and the general solution takes the form

$$\begin{bmatrix} \mathbf{x}_1 \\ \mathbf{x}_2 \end{bmatrix} = \begin{bmatrix} \mathbf{B}_{11}^{-1}\mathbf{b}_1 - \mathbf{B}_{11}^{-1}\mathbf{B}_{12}\mathbf{x}_2 \\ \mathbf{x}_2 \end{bmatrix} = \begin{bmatrix} \mathbf{B}_{11}^{-1}\mathbf{b}_1 \\ \mathbf{0} \end{bmatrix} + \begin{bmatrix} -\mathbf{B}_{11}^{-1}\mathbf{B}_{12} \\ \mathbf{I}_{m-r} \end{bmatrix} \mathbf{x}_2$$

with $\mathbf{x}_2$ arbitrary and taking the role of the vector $\mathbf{z}$ in Result A.13. (If the simplifying assumption above is incorrect, some column permutations will be needed.)

Example A.6: continued
Returning to solving $\mathbf{Ax} = \mathbf{c}$, as

$$\begin{bmatrix} 1 & 1 & 1 & 1 \\ 0 & 1 & 0 & -1 \\ 1 & 0 & 1 & 2 \end{bmatrix} \begin{bmatrix} x_1 \\ x_2 \\ x_3 \\ x_4 \end{bmatrix} = \begin{bmatrix} 3 \\ 2 \\ 1 \end{bmatrix},$$

applying elementary row operations (subtract row 1 from row 3, subtract 2 from 1, add 2 to 3) to the augmented matrix as described above leads to

$$\mathbf{M} \begin{bmatrix} 1 & 1 & 1 & 1 & 3 \\ 0 & 1 & 0 & -1 & 2 \\ 1 & 0 & 1 & 2 & 1 \end{bmatrix} = \begin{bmatrix} 1 & 0 & 1 & 2 & 1 \\ 0 & 1 & 0 & -1 & 2 \\ 0 & 0 & 0 & 0 & 0 \end{bmatrix}.$$

Now we see that the last row is zero and with $\mathbf{B}_{12}$ nonzero, a family of solutions exists and matches what we found previously, where $\mathbf{B}_{11} = \mathbf{I}_2$ and

$$\begin{bmatrix} \mathbf{B}_{11}^{-1}\mathbf{b}_1 \\ \mathbf{0} \end{bmatrix} + \begin{bmatrix} -\mathbf{B}_{11}^{-1}\mathbf{B}_{12} \\ \mathbf{I}_{m-r} \end{bmatrix} \mathbf{x}_2 = \begin{bmatrix} 1 \\ 2 \\ 0 \\ 0 \end{bmatrix} + \begin{bmatrix} -1 & -2 \\ 0 & 1 \\ 1 & 0 \\ 0 & 1 \end{bmatrix} \begin{bmatrix} x_3 \\ x_4 \end{bmatrix},$$

with $x_3 = z_3$ and $x_4 = z_4$.

A.5 Projections and Idempotent Matrices

Projections help to clarify the geometry of both the linear least squares problem and the solutions of linear equations. We will begin with the general case of projections and proceed to the symmetric case, which is more relevant to the orthogonal decomposition in the least squares problem. We begin with a simple class of matrices.

Definition A.7 *A square matrix* $\mathbf{P}$ *is* idempotent *iff* $\mathbf{P}^2 = \mathbf{P}$.

Definition A.8 *A square matrix* $\mathbf{P}$ *is a* projection *onto the vector space* $\mathcal{S}$ *iff*

(a) $\mathbf{P}$ *is idempotent,*

(b) for any $\mathbf{x}$, $\mathbf{Px} \in \mathcal{S}$, *and*

(c) if $\mathbf{z} \in \mathcal{S}$, $\mathbf{Pz} = \mathbf{z}$.

From the definition, we can see that any idempotent matrix is a projection matrix onto its own column space. That is, if $\mathbf{P}$ is idempotent, then $\mathbf{P}$ projects onto $\mathcal{C}(\mathbf{P})$. For two classes of projection matrices, its column space is interesting: one projects onto column spaces, the other onto null spaces.

Result A.14 $\mathbf{AA}^g$ *is a projection onto* $\mathcal{C}(\mathbf{A})$.

Proof: (a) $(\mathbf{AA}^g)^2 = \mathbf{AA}^g\mathbf{AA}^g = \mathbf{AA}^g$, so it is idempotent; (b) $\mathbf{AA}^g\mathbf{x} = \mathbf{A}\,(\mathbf{A}^g\mathbf{x}) \in \mathcal{C}(\mathbf{A})$; and (c) If $\mathbf{z} \in \mathcal{C}(\mathbf{A})$; then we can write $\mathbf{z} = \mathbf{Ay}$ for some $\mathbf{y}$. Hence $\mathbf{AA}^g\mathbf{z} = \mathbf{AA}^g\mathbf{Ay} = \mathbf{Ay} = \mathbf{z}$. ☐

The projection in Result A.14 may not be unique, since it depends on the choice of the generalized inverse $\mathbf{A}^g$; hence the use of the indefinite article "a." This result can be used to show that a particular vector $\mathbf{v}$ is in a column space $\mathcal{C}(\mathbf{A})$, by multiplying $\mathbf{AA}^g\mathbf{v}$. If the product $\mathbf{AA}^g\mathbf{v}$ equals $\mathbf{v}$, then $\mathbf{v}$ is in $\mathcal{C}(\mathbf{A})$, and vice versa.

In constructing a family of solutions in Section A.4, we constructed a projection onto $\mathcal{N}(\mathbf{A})$. If we have two solutions to the system of equations $\mathbf{Ax} = \mathbf{c}$, constructed from the same generalized inverse $\mathbf{A}^g$, say $\mathbf{x}^{(1)} = \mathbf{A}^g\mathbf{c}$ and $\mathbf{x}^{(2)} = \mathbf{A}^g\mathbf{c} + (\mathbf{I} - \mathbf{A}^g\mathbf{A})\,\mathbf{z}$, then

$$\mathbf{A}\left(\mathbf{x}^{(1)} - \mathbf{x}^{(2)}\right) = \mathbf{c} - \mathbf{c} = \mathbf{0},$$

so that $(\mathbf{x}^{(1)} - \mathbf{x}^{(2)}) \in \mathcal{N}(\mathbf{A})$, or, more interestingly, $(\mathbf{I} - \mathbf{A}^g\mathbf{A})\,\mathbf{z} \in \mathcal{N}(\mathbf{A})$ for any $\mathbf{z}$. This leads to the following complementary result.

Result A.15 $(\mathbf{I} - \mathbf{A}^g\mathbf{A})$ *is a projection onto* $\mathcal{N}(\mathbf{A})$.

Proof: (a) $(\mathbf{I} - \mathbf{A}^g\mathbf{A})^2 = (\mathbf{I} - \mathbf{A}^g\mathbf{A})(\mathbf{I} - \mathbf{A}^g\mathbf{A}) = \mathbf{I} - 2\mathbf{A}^g\mathbf{A} + \mathbf{A}^g\mathbf{A} = (\mathbf{I} - \mathbf{A}^g\mathbf{A})$, so it is idempotent; (b) for any $\mathbf{x}$, $(\mathbf{I} - \mathbf{A}^g\mathbf{A})\mathbf{x} \in \mathcal{N}(\mathbf{A})$ since $\mathbf{A}(\mathbf{I} - \mathbf{A}^g\mathbf{A})\mathbf{x} = \mathbf{0}$; and (c) if $\mathbf{z} \in \mathcal{N}(\mathbf{A})$, then $(\mathbf{I} - \mathbf{A}^g\mathbf{A})\mathbf{z} = \mathbf{z}$. □

Often, just a quick examination is all that is needed to find a basis for the nullspace of a matrix. But in other cases, finding these basis vectors can be quite difficult, and a set of linearly independent columns from a projection matrix of the form $(\mathbf{I} - \mathbf{A}^g\mathbf{A})$ will form a basis for $\mathcal{N}(\mathbf{A})$.

Now both $\mathbf{A}\mathbf{A}^g$ and $(\mathbf{I} - \mathbf{A}^g\mathbf{A})$ are projection matrices, but they are not unique projectors onto their respective subspaces, since they depend on the choice of generalized inverse. Again, they are referred to as "a projector," emphasizing the indefinite article "a."

Example A.7: Some Simple Projections

Let $\mathbf{A} = \begin{bmatrix} 1 \\ 2 \end{bmatrix}$, so that $\mathcal{C}(\mathbf{A})$ is a line with slope 2 through the origin in R^2. To find a generalized inverse for $\mathbf{A}$, let $\mathbf{G} = [u \;\; v]$. The definition $\mathbf{AGA} = \mathbf{A}$ becomes $u + 2v = 1$ or $v = (1-u)/2$, or $\mathbf{G}_u = [u \;\; (1-u)/2]$ is a generalized inverse for $\mathbf{A}$ for any u. Constructing a projection matrix onto $\mathcal{C}(\mathbf{A})$ via Result A.14 leads to

$$\mathbf{AG}_u = \begin{bmatrix} u & (1-u)/2 \\ 2u & 1-u \end{bmatrix},$$

which the reader should recognize as idempotent with rank 1. Taking some special cases with u, we find

$$\mathbf{AG}_0 = \begin{bmatrix} 0 & 1/2 \\ 0 & 1 \end{bmatrix} \quad \text{and} \quad \mathbf{AG}_1 = \begin{bmatrix} 1 & 0 \\ 2 & 0 \end{bmatrix}.$$

In Figure A.1, we see how a typical point in R^2 such as $\mathbf{x} = \begin{bmatrix} 3 \\ 2 \end{bmatrix}$ would be projected differently onto $\mathcal{C}(\mathbf{A})$, depending on the projector:

$$\mathbf{AG}_0\mathbf{x} = \begin{bmatrix} 0 & 1/2 \\ 0 & 1 \end{bmatrix}\begin{bmatrix} 3 \\ 2 \end{bmatrix} = \begin{bmatrix} 1 \\ 2 \end{bmatrix}, \text{ projecting horizontally,}$$

$$\mathbf{AG}_1\mathbf{x} = \begin{bmatrix} 1 & 0 \\ 2 & 0 \end{bmatrix}\begin{bmatrix} 3 \\ 2 \end{bmatrix} = \begin{bmatrix} 3 \\ 6 \end{bmatrix}, \text{ projecting vertically.}$$

For the least squares problem, we seek a symmetric projection matrix that will be referred to as "the projector" because of its uniqueness. In Chapter 2, a symmetric projection matrix will permit orthogonal decompositions and generalizations of the Pythagorean Theorem.

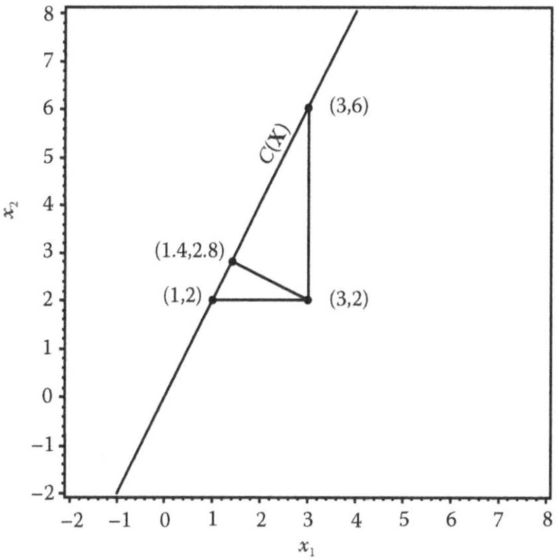

Figure A.1: Three projections onto a column space.

Result A.16 *A symmetric, idempotent matrix* $\mathbf{P}$ *that projects onto the vector space* $\mathcal{S}$ *is unique.*

Proof: Suppose both $\mathbf{P}$ and $\mathbf{Q}$ project onto the same vector space $\mathcal{S}$. Then we wish to show $\mathbf{P} = \mathbf{Q}$, by showing

$$(\mathbf{P}-\mathbf{Q})^T(\mathbf{P}-\mathbf{Q}) = \mathbf{P}^T\mathbf{P} - \mathbf{Q}^T\mathbf{P} - \mathbf{P}^T\mathbf{Q} + \mathbf{Q}^T\mathbf{Q} = \mathbf{0}.$$

For any $\mathbf{z}$, $\mathbf{Qz} \in \mathcal{S}$, and so $\mathbf{P}(\mathbf{Qz}) = \mathbf{PQz} = \mathbf{Qz}$ for all $\mathbf{z}$. Hence by Corollary A.1, $\mathbf{PQ} = \mathbf{Q}$. By reversing roles, we have $\mathbf{Pz} \in \mathcal{S}$ and so $\mathbf{QP} = \mathbf{P}$. Employing these results, as well as idempotency and symmetry, we have

$$(\mathbf{P}-\mathbf{Q})^T(\mathbf{P}-\mathbf{Q}) = \mathbf{P}^2 - \mathbf{QP} - \mathbf{PQ} + \mathbf{Q}^2 = \mathbf{P} - \mathbf{P} - \mathbf{Q} + \mathbf{Q} = \mathbf{0}.$$

The conclusion then follows from Lemma A.1. □

Example A.7: continued

Notice $\mathbf{A}^T\mathbf{A} = 5$, and as a preview to Chapter 2 where a method is presented to construct a symmetric projection matrix, $(\mathbf{A}^T\mathbf{A})^g\mathbf{A}^T = \mathbf{G}_{1/5} = [1/5 \;\; 2/5]$ so that

$$\mathbf{P_A} = \mathbf{AG}_{1/5} = \begin{bmatrix} 1/5 & 2/5 \\ 2/5 & 4/5 \end{bmatrix}$$

is the symmetric, idempotent matrix projecting onto $\mathcal{C}(\mathbf{A})$. In Figure A.1, we see

$$\mathbf{AG}_{1/5}\mathbf{x} = \begin{bmatrix} 1/5 & 2/5 \\ 2/5 & 4/5 \end{bmatrix} \begin{bmatrix} 3 \\ 2 \end{bmatrix} = \begin{bmatrix} 7/5 \\ 14/5 \end{bmatrix},$$

projecting perpendicular to $\mathcal{C}(\mathbf{A})$. Note also

$$(\mathbf{I} - \mathbf{G}_{1/5}\mathbf{A}) = \begin{bmatrix} 4/5 & -2/5 \\ -2/5 & 1/5 \end{bmatrix},$$

is symmetric, idempotent, and projecting onto $\mathcal{N}(\mathbf{A})$.

Corollary A.4 *If a symmetric, idempotent matrix* $\mathbf{P}$ *projects onto* $\mathcal{S}$*, then* $\mathbf{I} - \mathbf{P}$ *projects onto its orthogonal complement.*

Proof: Since $\mathcal{S} = \mathcal{C}(\mathbf{P})$, its orthogonal complement is $\mathcal{N}(\mathbf{P}^T) = \mathcal{N}(\mathbf{P})$. So we need to show that $\mathbf{I} - \mathbf{P}$ projects onto $\mathcal{N}(\mathbf{P})$. First, since $\mathbf{P}$ is idempotent, so is $\mathbf{I} - \mathbf{P}$ (Exercise A.36); also note the symmetry of $\mathbf{I}-\mathbf{P}$. Second, for any vector $\mathbf{x}$, $(\mathbf{I} - \mathbf{P})\,\mathbf{x} \in \mathcal{N}(\mathbf{P})$ since $\mathbf{P}\,(\mathbf{I} - \mathbf{P})\,\mathbf{x} = \mathbf{0}$ for all $\mathbf{x}$. Third, if $\mathbf{z} \in \mathcal{N}(\mathbf{P})$, then $(\mathbf{I} - \mathbf{P})\,\mathbf{z} = \mathbf{z}$ since $\mathbf{P}\mathbf{z} = \mathbf{0}$. □

A.6 Trace, Determinants, and Eigenproblems

The trace and determinant are two functions that are only defined for square matrices, and both can be viewed as a measure of the size of a matrix. The trace is just the sum of the diagonal elements of the matrix,

$$\mathit{trace}\,(\mathbf{A}) = \mathit{tr}\,(\mathbf{A}) = \sum_i A_{ii}.$$

The trace is most often used here to permit some clever manipulations of matrices. The tools for these manipulations are summarized in the following result.

Result A.17

(a) $\mathit{trace}\,(\mathbf{AB}) = \mathit{trace}\,(\mathbf{BA})$

(b) $\mathit{trace}\,(\mathbf{A}^T\mathbf{A}) = \sum_i \sum_j A_{ij}^2$

The determinant of a matrix measures the change in unit volume under a linear transformation, and is only defined for square matrices. The determinant of an identity matrix, of course, is 1, and either ± 1 for orthogonal or permutation matrices. The determinant of a singular matrix is zero. For a diagonal or triangular matrix, the determinant is the product of the diagonal elements, and the determinant is zero if the matrix is singular. Some other rules for determinants are collected in the following result:

Result A.18

(a) $\mathit{det}\left(\begin{bmatrix} a & b \\ c & d \end{bmatrix}\right) = ad - bc$

(b) $|\mathbf{AB}| = |\mathbf{A}||\mathbf{B}|$

(c) $|\mathbf{A}^{-1}| = 1/|\mathbf{A}|$

(d) $|c\mathbf{A}| = c^n|\mathbf{A}|$ *where* $\mathbf{A}$ *is* $n \times n$

(e) $det\left(\begin{bmatrix} \mathbf{A} & \mathbf{B} \\ \mathbf{C} & \mathbf{D} \end{bmatrix}\right) = |\mathbf{A}||\mathbf{D} - \mathbf{C}\mathbf{A}^{-1}\mathbf{B}| = |\mathbf{D}||\mathbf{A} - \mathbf{B}\mathbf{D}^{-1}\mathbf{C}|$

Eigenvalues and eigenvectors are functions of a square matrix. The defining equation for all eigenproblems is

$$\mathbf{A}\mathbf{x} = \lambda\mathbf{x} \tag{A.8}$$

where λ is the *eigenvalue* and $\mathbf{x}$ is its associated *eigenvector*. Taking the difference of the two sides leads to

$$(\mathbf{A} - \lambda\mathbf{I}_n)\,\mathbf{x} = \mathbf{0}, \tag{A.9}$$

which can be viewed as values of λ that make the matrix $(\mathbf{A} - \lambda\mathbf{I}_n)$ singular, and itself leads to the determinantal equation

$$det\,(\mathbf{A} - \lambda\mathbf{I}_n) = 0, \tag{A.10}$$

which is a polynomial equation in λ of degree n. In this book, only eigenvalues and vectors of symmetric matrices will be considered any further, and this restriction leads to substantial simplifications. When the matrix $\mathbf{A}$ is symmetric, then there are n real roots to (A.10); given a root λ_j, the associated eigenvector $\mathbf{x}^{(j)}$ is in the nullspace of $(\mathbf{A} - \lambda_j\mathbf{I}_n)$. Again, when $\mathbf{A}$ is symmetric, if λ is a multiple root of (A.10), then the dimension of the nullspace (geometric multiplicity) matches the (algebraic) multiplicity of the polynomial equation. Eigenvectors associated with different eigenvalues are orthogonal; in the case of multiple eigenvalues, linearly independent eigenvectors spanning the nullspace of $(\mathbf{A} - \lambda_j\mathbf{I}_n)$ can be chosen to be orthogonal. Since eigenvectors have an undetermined scale, that is, if $\mathbf{x}$ is an eigenvector, $c\mathbf{x}$ is also an eigenvector for any nonzero constant c; usually eigenvectors are normalized to have unit Euclidean norm. In summary, when the matrix $\mathbf{A}$ of order n is symmetric, there are n real eigenvalues and their associated eigenvectors are orthogonal, or can be chosen to be so.

If the eigenvectors of a symmetric matrix $\mathbf{A}$ are stacked as columns of a matrix $\mathbf{Q}$, then the defining Equation (A.8) can be extended to form

$$\mathbf{A}\mathbf{Q} = \mathbf{Q}\Lambda \tag{A.11}$$

where Λ is a diagonal matrix of the eigenvalues, ordered in the same way the eigenvectors are stacked in $\mathbf{Q}$. Since the eigenvectors are linearly independent (see Exercise A.41), the matrix $\mathbf{Q}$ is nonsingular; moreover, since they are mutually orthogonal with unit norm, the matrix $\mathbf{Q}$ is an orthogonal matrix, whose inverse is its transpose:

$$\mathbf{Q}\mathbf{Q}^T = \mathbf{Q}^T\mathbf{Q} = \mathbf{I}_n.$$

Then (A.11) leads to the *spectral decomposition* of a symmetric matrix:

$$\mathbf{A} = \mathbf{Q}\Lambda\mathbf{Q}^T = \sum_j \lambda_j \mathbf{q}^{(j)}\mathbf{q}^{(j)T}.$$

Two useful consequences (see Exercises A.42 and A.43) of this decomposition are

$$|\mathbf{A}| = \lambda_1 \times \cdots \times \lambda_n$$

and

$$trace(\mathbf{A}) = \lambda_1 + \cdots + \lambda_n$$

where λ_j are the eigenvalues of $\mathbf{A}$.

Result A.19 *Let* $\mathbf{A}$ *be a symmetric* $n \times n$ *matrix. Then the rank of* $\mathbf{A}$ *is equal to the number of nonzero eigenvalues.*

Proof: See Exercise A.74. ∎

Example A.8: Eigenvalues and Eigenvectors
Let's find the eigenvalues and vectors for a simple 2×2 matrix; let

$$\mathbf{A} = \begin{bmatrix} 6 & -2 \\ -2 & 9 \end{bmatrix}$$

so that $\|\mathbf{A} - \lambda\mathbf{I}_n\| = (6-\lambda)(9-\lambda) - 4 = \lambda^2 - 15\lambda + 50$; setting this equal to zero yields the roots $\lambda_1 = 10$ and $\lambda_2 = 5$. For λ_1, $(\mathbf{A} - \lambda\mathbf{I}_n)\mathbf{q} = \mathbf{0}$ becomes

$$-4q_1 - 2q_2 = 0 \text{ and } -2q_1 - q_2 = 0$$

where the first equation is just twice the second. Solving for q_1, we have $q_1 = -(1/2)q_2$; solving for q_2 gives $q_2 = -2q_1$. These two routes yield the eigenvector

$$\begin{bmatrix} -(1/2) \\ 1 \end{bmatrix} q_2 \quad \text{or} \quad \begin{bmatrix} 1 \\ -2 \end{bmatrix} q_1$$

or, when normalized,

$$\frac{1}{\sqrt{5}} \begin{bmatrix} 1 \\ -2 \end{bmatrix}.$$

For λ_2, the equations are

$$q_1 - 2q_2 = 0 \quad \text{and} \quad -2q_1 + 4q_2 = 0$$

leading to the eigenvector

$$\frac{1}{\sqrt{5}}\begin{bmatrix}2\\1\end{bmatrix}.$$

Notice that the matrix of eigenvectors $\mathbf{Q}$ below is an orthogonal matrix:

$$\mathbf{Q} = \frac{1}{\sqrt{5}}\begin{bmatrix}1 & 2\\-2 & 1\end{bmatrix}.$$

See also Exercise A.47.

A.7 Definiteness and Factorizations

The multivariate generalization of a square is the quadratic form $\mathbf{x}^T\mathbf{A}\mathbf{x}$, which is composed of squares $A_{ii}x_i^2$ and cross-products $A_{ij}x_ix_j$. We are interested in characterizing matrices that have properties similar to a square.

Definition A.9 *A matrix* $\mathbf{A}$ *is* nonnegative definite *if and only if* $\mathbf{x}^T\mathbf{A}\mathbf{x} \geq 0$ *for all* $\mathbf{x}$.

Definition A.10 *A matrix* $\mathbf{A}$ *is* positive definite *if and only if* $\mathbf{x}^T\mathbf{A}\mathbf{x} > 0$ *for all* $\mathbf{x} \neq \mathbf{0}$.

A nonnegative definite matrix that is nonsingular will be positive definite (Exercise A.61). From the spectral decomposition we can see that a symmetric matrix is positive definite iff all of its eigenvalues are positive and nonnegative definite iff the eigenvalues are nonnegative.

Note that in this book we will only speak of quadratic forms in symmetric matrices. One motivation is that a quadratic form in a matrix that is not symmetric can be written in terms of one that is symmetric: $\mathbf{x}^T\mathbf{A}\mathbf{x} = \mathbf{x}^T\mathbf{A}^T\mathbf{x}$, so $\mathbf{x}^T\mathbf{A}\mathbf{x} = \mathbf{x}^T(\frac{1}{2}\mathbf{A} + \frac{1}{2}\mathbf{A}^T)\mathbf{x}$, since $(\frac{1}{2}\mathbf{A} + \frac{1}{2}\mathbf{A}^T)$ is symmetric. The other motivation is that all of the applications of quadratic forms that appear in this book involve symmetric matrices.

Nonnegative definite matrices also permit certain square root factorizations of a matrix. Usually the circumstance requires finding a matrix whose square or outer product produces a given matrix. In these situations the given matrix is positive definite, and so the problem is analogous to trying to find a number whose square is positive. There are two constructive solutions to this problem: *Cholesky factorization* and *spectral decomposition*.

Result A.20 *(Cholesky factorization) A square matrix* $\mathbf{A}$ *is positive definite if and only if there exists a nonsingular lower triangular matrix* $\mathbf{L}$ *such that* $\mathbf{A} = \mathbf{L}\mathbf{L}^T$.

Proof: One direction is easy: if $\mathbf{A} = \mathbf{L}\mathbf{L}^T$, then $\mathbf{x}^T\mathbf{A}\mathbf{x} = \|\mathbf{L}^T\mathbf{x}\|^2$, which is clearly nonnegative and only zero if $\mathbf{x}$ is zero since $\mathbf{L}$ is nonsingular. For the other direction, we will construct the matrix $\mathbf{L}$. Begin the induction with $L_{11} = \sqrt{A_{11}}$. Now let $L_{[k-1]}$ be the first $(k-1)$ rows and columns of $\mathbf{L}$, and form $\mathbf{A}_{[k-1]}$ similarly from $\mathbf{A}$. Then the kth step is

$$\mathbf{L}_{[k]}\mathbf{L}_{[k]}^T = \begin{bmatrix} \mathbf{L}_{[k-1]} & \mathbf{0} \\ \ell_k^T & L_{kk} \end{bmatrix} \begin{bmatrix} \mathbf{L}_{[k-1]}^T & \ell_k \\ \mathbf{0} & L_{kk} \end{bmatrix} = \begin{bmatrix} \mathbf{A}_{[k-1]} & \mathbf{a}_k \\ \mathbf{a}_k^T & A_{kk} \end{bmatrix} \begin{matrix} k-1 \\ 1 \end{matrix}$$

Matching up the upper-left blocks repeats $\mathbf{A}_{[k-1]} = \mathbf{L}_{[k-1]}\mathbf{L}_{[k-1]}^T$. The off-diagonal block yields the system of equations

$$\mathbf{L}_{[k-1]}\ell_k = \mathbf{a}_k$$

and the lower-right elements give $A_{kk} = L_{kk}^2 + \ell_k^T\ell_k$. Notice that in the system of equations above, both $\mathbf{L}_{[k-1]}$ and $\mathbf{a}_k$ are known, and the system can be solved for the vector ℓ_k. Once ℓ_k is found, we can find $L_{kk} = \sqrt{\mathbf{A}_{kk} - \ell_k^T\ell_k}$. This algorithm will work as long as we can compute this square root (see Exercise A.62). If the matrix $\mathbf{A}$ is positive definite, then $A_{kk} - \ell_k^T\ell_k$ will be positive and the algorithm will run to completion. □

For the alternative method, if $\mathbf{A}$ is symmetric, then the spectral decomposition of $\mathbf{A}$ is $\mathbf{A} = \mathbf{Q}\boldsymbol{\Lambda}\mathbf{Q}^T$ with the eigenvalues of $\mathbf{A}$ forming the diagonal matrix $\boldsymbol{\Lambda}$ and the eigenvectors stacked as columns in the orthogonal matrix $\mathbf{Q}$. If $\mathbf{A}$ is nonnegative definite, then the eigenvalues are all nonnegative, and forming the diagonal matrix of their square roots yields $\boldsymbol{\Lambda}^{\frac{1}{2}}$. The matrix $\mathbf{Q}\boldsymbol{\Lambda}^{\frac{1}{2}}\mathbf{Q}^T$ then is symmetric, and its square (or outer product) produces the nonnegative definite $\mathbf{A}$. The matrix $\mathbf{Q}\boldsymbol{\Lambda}^{\frac{1}{2}}\mathbf{Q}^T$ is sometimes called the *symmetric square root* of $\mathbf{A}$, sometimes denoted as $\mathbf{A}^{\frac{1}{2}}$. (See also Exercises A.64 and A.65.) Note also the following: if $\mathbf{A}$ is positive definite, then

$$\left(\mathbf{A}^{\frac{1}{2}}\right)^{-1} = \mathbf{A}^{-\frac{1}{2}} = \mathbf{Q}\boldsymbol{\Lambda}^{-\frac{1}{2}}\mathbf{Q}^T$$

Example A.9: Cholesky Factorization

Let us compute the Cholesky factorization for the following matrix:

$$\mathbf{A} = \begin{bmatrix} 4 & 2 & 2 & 4 \\ 2 & 5 & 7 & 0 \\ 2 & 7 & 19 & 11 \\ 4 & 0 & 11 & 25 \end{bmatrix}$$

Step $k = 1$: $L_{11} = \sqrt{A_{11}} = \sqrt{4} = 2$

Step $k = 2$: Solve $L_{11}L_{21} = A_{21} = 2L_{21} = 2$ so that $L_{21} = 1$, and

$$L_{22} = \sqrt{A_{22} - L_{21}^2} = \sqrt{5-1} = 2$$

Step $k = 3$: Solve $\mathbf{L}_{[2]}\ell_3 = \mathbf{a}_3$ or

$$\begin{bmatrix} 2 & 0 \\ 1 & 2 \end{bmatrix} \begin{bmatrix} L_{31} \\ L_{32} \end{bmatrix} = \begin{bmatrix} 2 \\ 7 \end{bmatrix}$$

to get $L_{31} = 1, L_{32} = 3$, and then

$$L_{33} = \sqrt{A_{33} - L_{31}^2 - L_{32}^2} = \sqrt{19 - 9 - 1} = 3$$

Step $k = 4$: Solve $\mathbf{L}_{[3]}\ell_4 = \mathbf{a}_4$ or

$$\begin{bmatrix} 2 & 0 & 0 \\ 1 & 2 & 0 \\ 1 & 3 & 3 \end{bmatrix} \begin{bmatrix} L_{41} \\ L_{42} \\ L_{43} \end{bmatrix} = \begin{bmatrix} 4 \\ 0 \\ 11 \end{bmatrix}$$

to get $L_{41} = 2, L_{42} = -1, L_{43} = 4$, and

$$L_{44} = \sqrt{A_{44} - L_{41}^2 - L_{42}^2 - L_{43}^2} = \sqrt{25 - 4 - 1 - 16} = 2$$

Example A.10: Symmetric Square Root
Let us compute the symmetric square root for the following matrix:

$$\mathbf{A} = \begin{bmatrix} 10 & 3 & 2 \\ 3 & 9 & 3 \\ 2 & 3 & 10 \end{bmatrix}$$

$$\begin{aligned} det(\mathbf{A} - \lambda\mathbf{I}) &= (10 - \lambda)(9 - \lambda)(10 - \lambda) + 18 + 18 - 9(10 - \lambda) \\ &\quad - 9(10 - \lambda) - 4(9 - \lambda) \\ &= -\lambda^3 + 29\lambda^2 - 258\lambda + 720 = -(\lambda - 15)(\lambda - 8)(\lambda - 6) \end{aligned}$$

So we have three eigenvalues, $\lambda_1 = 15, \lambda_2 = 8, \lambda_3 = 6$, and to get their eigenvectors, compute $\mathbf{A} - \lambda\mathbf{I}$ and find $\mathbf{q}$ so that $(\mathbf{A} - \lambda\mathbf{I})\mathbf{q} = \mathbf{0}$. In turn, we find the associated eigenvectors to be proportional to the following:

$$\mathbf{q}^{(1)} \propto \begin{bmatrix} 1 \\ 1 \\ 1 \end{bmatrix}, \quad \mathbf{q}^{(2)} \propto \begin{bmatrix} 1 \\ 0 \\ -1 \end{bmatrix}, \quad \mathbf{q}^{(3)} \propto \begin{bmatrix} 1 \\ -2 \\ 1 \end{bmatrix}$$

And so $\mathbf{A}$ can be factored as $\mathbf{A} = \mathbf{Q}\boldsymbol{\Lambda}\mathbf{Q}^T$ or

$$\mathbf{A} = \begin{bmatrix} 1/\sqrt{3} & 1/\sqrt{2} & 1/\sqrt{6} \\ 1/\sqrt{3} & 0 & -2/\sqrt{6} \\ 1/\sqrt{3} & -1/\sqrt{2} & 1/\sqrt{6} \end{bmatrix} \begin{bmatrix} 15 & 0 & 0 \\ 0 & 8 & 0 \\ 0 & 0 & 6 \end{bmatrix} \begin{bmatrix} 1/\sqrt{3} & 1/\sqrt{3} & 1/\sqrt{3} \\ 1/\sqrt{2} & 0 & -1/\sqrt{2} \\ 1/\sqrt{6} & -2/\sqrt{6} & 1/\sqrt{6} \end{bmatrix}$$

$$\mathbf{A}^{\frac{1}{2}} = \begin{bmatrix} 1/\sqrt{3} & 1/\sqrt{2} & 1/\sqrt{6} \\ 1/\sqrt{3} & 0 & -2/\sqrt{6} \\ 1/\sqrt{3} & -1/\sqrt{2} & 1/\sqrt{6} \end{bmatrix} \begin{bmatrix} \sqrt{15} & 0 & 0 \\ 0 & \sqrt{8} & 0 \\ 0 & 0 & \sqrt{6} \end{bmatrix} \begin{bmatrix} 1/\sqrt{3} & 1/\sqrt{3} & 1/\sqrt{3} \\ 1/\sqrt{2} & 0 & -1/\sqrt{2} \\ 1/\sqrt{6} & -2/\sqrt{6} & 1/\sqrt{6} \end{bmatrix}$$

which does not simplify any further (but see Exercise A.66).

A.8 Notes

• A linear space over the field of real numbers is any set that satisfies the following list of properties with $\mathbf{x}$, $\mathbf{y}$, and $\mathbf{z}$ elements of the set, and α, β real numbers:

1. $\mathbf{x} + \mathbf{y}$ is a unique element of the set.
2. $\mathbf{x} + \mathbf{y} = \mathbf{y} + \mathbf{x}$.
3. $(\mathbf{x} + \mathbf{y}) + \mathbf{z} = \mathbf{x} + (\mathbf{y} + \mathbf{z})$.
4. There exists a zero element $\mathbf{0}$ such that $\mathbf{x} + \mathbf{0} = \mathbf{x}$.
5. For every $\mathbf{x}$ there exists $-\mathbf{x}$ such that $\mathbf{x} + (-\mathbf{x}) = \mathbf{0}$.
6. $\alpha\mathbf{x}$ is an element of the set.
7. $\alpha(\beta\mathbf{x}) = (\alpha\beta)\mathbf{x}$.
8. $\alpha(\mathbf{x} + \mathbf{y}) = \alpha\mathbf{x} + \alpha\mathbf{y}$.
9. $1 \times \mathbf{x} = \mathbf{x}$.

In this text, we are dealing with elements that are the usual vectors or n-tuples, but the reader should be aware that many of the same concepts can be generalized, as in Exercises A.7 and A.8.

• The literature on generalized inverses (or pseudoinverses) is extensive, and there are different classes of generalized inverses, according to whether they satisfy certain requirements:

Definition A.11 *A generalized inverse* $\mathbf{G}$ *belongs to a class of generalized inverse according to whether it satisfies the sequential levels of conditions:*

1. $\mathbf{AGA} = \mathbf{A}$.

2. $\mathbf{GAG} = \mathbf{G}$.

3. $\mathbf{AG}$ *is symmetric.*

4. $\mathbf{GA}$ *is symmetric.*

We have been working with Definition (1), sometimes called an *inner* or *1-inverse* or *one-condition* generalized inverse. Note that some authors, for example, Rao [37] or Graybill [15, 16], define a generalized inverse as one satisfying both (1) and (2); see Exercise A.27. The generalized inverse found in Result 2.5, $\mathbf{G} = \left(\mathbf{X}^T\mathbf{X}\right)^g \mathbf{X}^T$, satisfies (1), (2), and (3), and leads in Theorem 2.1 to the resulting projection matrix $\mathbf{XG}$ being symmetric. A generalized inverse that satisfies (1),(2), (3), and (4) is called *the Moore-Penrose generalized inverse*, usually denoted as $\mathbf{A}^+$, and has some interesting properties. The use of the definite article "the" emphasizes the uniqueness of this generalized inverse. Pringle and Rayner [35] give an interesting early discussion of the field, as well as Boullion and Odell [4] and Rao and Mitra [38].

A.9 Exercises

* Exercises denoted by an asterisk should not be bypassed; some contain results that will be used in subsequent chapters.

A.1. * Show that a set of mutually orthogonal nonzero vectors are also linearly independent. (Hint: Suppose the contrary, then use (A.2) and form $\mathbf{x}^{(j)T}\mathbf{x}^{(j)}$.)

A.2. Find a vector orthogonal to $\begin{bmatrix} 1 \\ 1 \\ 1 \end{bmatrix}$ and $\begin{bmatrix} 1 \\ 0 \\ -1 \end{bmatrix}$.

A.3. Is the vector $\begin{bmatrix} 1 \\ 3 \\ -1 \\ -3 \end{bmatrix}$ linearly dependent on the vectors $\begin{bmatrix} 1 \\ 1 \\ 1 \\ 1 \end{bmatrix}$ and $\begin{bmatrix} 1 \\ 2 \\ 0 \\ -1 \end{bmatrix}$? If so, find the coefficients.

A.4. Show that $\{\, \mathbf{x} \in R^d : \sum_{i=1}^d x_i = 0\}$ is a vector space.

A.5. Is $\{\, \mathbf{x} \in R^d : \sum_{i=1}^d x_i = 1\}$ a vector space?

A.6. Is $\{\, \mathbf{x} \in R^d : \sum_{i=1}^d x_i^2 = 1\}$ a vector space?

See Section A.8 for the following two exercises:

A.7. Show that the set of polynomials of degree $\leq k$ forms a linear space over the field of real numbers.

A.8. Show that the set of real-valued functions on [0, 1] having a continuous first derivative forms a linear space over the field of real numbers.

A.9. For $\mathbf{a} \neq \mathbf{0}$ and $\mathbf{b} \neq \mathbf{0}$, what is the rank of the matrix $\mathbf{A} = \mathbf{ab}^T$?

A.10. For the matrix $\mathbf{A}$ in Example A.2, find bases for the following:

a. $\mathcal{C}(\mathbf{A})$ and $\mathcal{N}(\mathbf{A}^T)$
b. $\mathcal{C}(\mathbf{A}^T)$ and $\mathcal{N}(\mathbf{A})$

A.11. Repeat Exercise A.10 with a matrix whose columns are the three vectors from Exercise A.3.

A.12. * Show that the nullspace of a matrix is closed under addition and scalar multiplication.

A.13. Let $\mathbf{A} = \mathbf{11}^T$; describe the nullspace of $\mathbf{A}$ in words.

A.14. Find the rank and basis vectors for the nullspace of the following matrix:

$$\begin{bmatrix} 1 & 1 & 1 & 1 \\ 0 & 1 & 0 & -1 \\ 1 & 0 & 1 & 2 \end{bmatrix}.$$

A.15. Show that if $rank\,(\mathbf{BC}) = rank\,(\mathbf{B})$, then $\mathcal{C}(\mathbf{BC}) = \mathcal{C}(\mathbf{B})$.

A.16. * Let $\mathbf{E} = \mathbf{FA}$ with $\mathbf{F}$ a nonsingular matrix. Show that $\mathcal{N}(\mathbf{E}) = \mathcal{N}(\mathbf{A})$.

A.17. For $d = 4$, find a basis for the orthogonal complement of the vector space in Exercise A.4.

A.18. Find a set of basis vectors for the column space of A, $\mathcal{C}(\mathbf{A})$, where

$$\mathbf{A} = \begin{bmatrix} 1 & 2 & 1 \\ 1 & 1 & -2 \\ 1 & 1 & -2 \\ 1 & 2 & 1 \\ 1 & 1 & -2 \end{bmatrix}.$$

A.19. For the matrix $\mathbf{A}$ above, find a set of basis vectors for the nullspace of A^T, $\mathcal{N}(\mathbf{A}^T)$.

A.20. Show that your basis vectors in Exercises A.18 and A.19 are orthogonal.

A.21. Repeat Exercises A.18 to A.20 for the transpose; that is, find bases for $\mathcal{C}(\mathbf{A}^T)$ and $\mathcal{N}(\mathbf{A})$.

A.22. * Prove that if $\mathbf{G}$ is a generalized inverse of $\mathbf{A}$, then $\mathbf{G}^T$ is a generalized inverse of $\mathbf{A}^T$.

A.23. Prove that if $\mathbf{G}$ is a generalized inverse of a symmetric matrix $\mathbf{A}$, then $\frac{1}{2}\left(\mathbf{G}+\mathbf{G}^T\right)$ is a symmetric generalized inverse of $\mathbf{A}$.

A.24. Suppose the matrix $\mathbf{A}$ has the identity matrix as a generalized inverse. Find another generalized inverse of $\mathbf{A}$.

A.25. Find all solutions to the system of equations

$$\begin{bmatrix} 1 & 1 & 1 & 1 \\ 1 & 2 & 3 & 4 \end{bmatrix} \begin{bmatrix} x_1 \\ x_2 \\ x_3 \\ x_4 \end{bmatrix} = \begin{bmatrix} 0 \\ 1 \end{bmatrix}.$$

A.26. After the proof of Result A.13, the comment is made "by varying $\mathbf{z}$ over R^n, we can sweep out all possible solutions to a system of equations," but $\mathbf{z}$ has n dimensions and taking $\mathbf{c} = \mathbf{0}$, $\mathcal{N}(\mathbf{A})$ has dimension $n - r$. Is this comment correct? If so, what happened to the remaining r dimensions?

The following few exercises discuss further approaches for constructing a generalized inverse.

A.27. Consider the factorization of the $(m \times n)$ matrix $\mathbf{A}$ where $rank\,(\mathbf{A}) = r$ resulting from elementary row and column operations

$$\mathbf{A} = \mathbf{B} \begin{bmatrix} \mathbf{I}_r & \mathbf{0} \\ \mathbf{0} & \mathbf{0} \end{bmatrix} \mathbf{C}$$

where $\mathbf{B}$ is $m \times m$, $\mathbf{C}$ is $n \times n$, both nonsingular.

a. Show that the matrix $\mathbf{G}_1$ below is a (class 1) generalized inverse, satisfying $\mathbf{A}\mathbf{G}_1\mathbf{A} = \mathbf{A}$, for any matrices $\mathbf{E}_1, \mathbf{E}_2, \mathbf{E}_3$ with the proper shape.

$$\mathbf{G}_1 = \mathbf{C}^{-1} \begin{bmatrix} \mathbf{I}_r & \mathbf{E}_1 \\ \mathbf{E}_2 & \mathbf{E}_3 \end{bmatrix} \mathbf{B}^{-1}$$

b. Show that the matrix $\mathbf{G}_2$ below is a class 2 generalized inverse, satisfying $\mathbf{A}\mathbf{G}_2\mathbf{A} = \mathbf{A}$ and $\mathbf{G}_2\mathbf{A}\mathbf{G}_2 = \mathbf{G}_2$, for any matrices $\mathbf{E}_1$ and $\mathbf{E}_2$ with the proper shape.

$$\mathbf{G} = \mathbf{C}^{-1} \begin{bmatrix} \mathbf{I}_r & \mathbf{E}_1 \\ \mathbf{E}_2 & \mathbf{E}_2\mathbf{E}_1 \end{bmatrix} \mathbf{B}^{-1}$$

A.28. Apply the method in Exercise A.27 with $\mathbf{E}_j = \mathbf{0}$ to the matrix in Exercise A.14.

A.29. Apply the method in Exercise A.27 to the matrix in Exercise A.18.

For constructing a generalized inverse of a symmetric $n \times n$ matrix $\mathbf{A}$ with rank r, a case that arises often in this book, the method suggested by Result A.11 can be shortened to the following algorithm:

- Select a nonsingular symmetric $r \times r$ submatrix by deleting $n - r$ rows and columns of $\mathbf{A}$ (if row j is deleted, also delete column j).
- Invert the remaining submatrix.
- Replace the entries in the selected rows and columns of $\mathbf{A}$ with the corresponding entries from the inverted submatrix.
- Zero out the remaining entries of $\mathbf{A}$ (those previously deleted).

For example, consider the matrix $\mathbf{A}$ below. Deleting row/column 4 leads to the solution following Result A.10. But deleting row/column 2 leads to the submatrix below, whose inverse is given.

$$\mathbf{A} = \begin{bmatrix} 5 & 1 & 3 & 1 \\ 1 & 3 & 1 & 3 \\ 3 & 1 & 3 & 1 \\ 1 & 3 & 1 & 3 \end{bmatrix}, \quad \begin{bmatrix} 5 & 3 & 1 \\ 3 & 3 & 1 \\ 1 & 1 & 3 \end{bmatrix}^{-1} = \begin{bmatrix} 1/2 & -1/2 & 0 \\ -1/2 & 7/8 & -1/8 \\ 0 & -1/8 & 3/8 \end{bmatrix},$$

$$\text{so } \mathbf{A}^g = \begin{bmatrix} 1/2 & 0 & -1/2 & 0 \\ 0 & 0 & 0 & 0 \\ -1/2 & 0 & 7/8 & -1/8 \\ 0 & 0 & -1/8 & 3/8 \end{bmatrix}$$

A.30. Show that the matrix $\mathbf{A}^g$ above is a generalized inverse of $\mathbf{A}$.

A.31. Apply the method in Exercise A.30 to the matrix below.

$$\begin{bmatrix} 6 & 1 & 3 & -2 \\ 1 & 5 & -2 & 2 \\ 3 & -2 & 3 & -2 \\ -2 & 2 & -2 & 2 \end{bmatrix}$$

The Moore-Penrose generalized inverse, defined in Section A.8 as satisfying the four properties listed in (A.11), has some interesting sidelights:

A.32. If $\mathbf{A}$ has full-column rank, then $(\mathbf{A}^T\mathbf{A})^{-1}\mathbf{A}^T = \mathbf{A}^+$, the Moore-Penrose generalized inverse of $\mathbf{A}$.

A.33. If $\mathbf{A}$ has full-row rank, then $\mathbf{A}\left(\mathbf{A}\mathbf{A}^T\right)^{-1} = \mathbf{A}^+$.

A.34. Let $\mathbf{A} = \begin{bmatrix} 8 & 4 & 2 & 2 \\ 4 & 4 & 0 & 0 \\ 2 & 0 & 2 & 0 \\ 2 & 0 & 0 & 2 \end{bmatrix}$ and $c = \begin{bmatrix} 1 \\ -2 \\ -1 \\ 4 \end{bmatrix}$.

a. Show that $\mathbf{c}$ is in $\mathcal{C}(\mathbf{A})$. (Find a vector $\mathbf{x}$ such that $\mathbf{Ax} = \mathbf{c}$.)
b. Find two different generalized inverses for $\mathbf{A}$.
c. For one of your generalized inverses in (b), compute $\mathbf{AA}^g$.

d. Is $\mathbf{AA}^g$ from (c) idempotent? Symmetric?
e. Show that $\mathbf{AA}^g\mathbf{c} = \mathbf{c}$.
f. For your two generalized inverses in (b), compute $\mathbf{A}^g\mathbf{c}$ and show that each vector solves $\mathbf{Ax} = \mathbf{c}$.

A.35. For matrix $\mathbf{A}$ in Exercise A.34, find a generalized inverse

a. that is symmetric.
b. that is not symmetric.
c. that has rank 4 (and hence nonsingular).
d. so that $\mathbf{A}$ is a generalized inverse of it.
e. so that $\mathbf{A}$ is not a generalized inverse of it.

A.36. * If $\mathbf{P}$ is idempotent, show that $(\mathbf{I} - \mathbf{P})$ is also idempotent.

A.37. If $\mathbf{P}$ is symmetric and idempotent, show that the Pythagorean relationship holds:

$$\|\mathbf{y}\|^2 = \|\mathbf{Py}\|^2 + \|(\mathbf{I} - \mathbf{P})\mathbf{y}\|^2$$

A.38. Prove the results about the trace operator in Result A.17.

A.39. Show that the trace of the matrix $\frac{1}{n}\mathbf{1}_n\mathbf{1}_n^T$ is equal to its rank.

A.40. Find the trace and determinant for each of the following four matrices:

$$\begin{bmatrix} 13 & 4 \\ 4 & 7 \end{bmatrix} \quad \begin{bmatrix} 2 & 0 & 0 \\ 0 & 2 & 0 \\ 0 & 0 & -1 \end{bmatrix} \quad \begin{bmatrix} 12 & 8 \\ 8 & 12 \end{bmatrix} \quad \begin{bmatrix} 9 & 10 \\ 10 & 30 \end{bmatrix}$$

A.41. Show that the eigenvectors of a symmetric matrix with distinct eigenvalues are linearly independent.

A.42. Prove that for a symmetric matrix $|\mathbf{A}| = \lambda_1 \times \cdots \times \lambda_n$.

A.43. Prove that for a symmetric matrix, $trace(\mathbf{A}) = \lambda_1 + \cdots + \lambda_n$, where λ_j are the eigenvalues of $\mathbf{A}$.

A.44. Use Result A.18(e) to prove: if $\mathbf{A}, \mathbf{B}$ are $m \times n$, then $|\mathbf{I}_m + \mathbf{AB}^T| = |\mathbf{I}_n + \mathbf{B}^T\mathbf{A}|$.

A.45. Find the eigenvalues and eigenvectors of the following four matrices:

$$\begin{bmatrix} 13 & 4 \\ 4 & 7 \end{bmatrix} \quad \begin{bmatrix} 2 & 0 & 0 \\ 0 & 2 & 0 \\ 0 & 0 & -1 \end{bmatrix} \quad \begin{bmatrix} 12 & 8 \\ 8 & 12 \end{bmatrix} \quad \begin{bmatrix} 9 & 10 \\ 10 & 30 \end{bmatrix}$$

A.46. Find the eigenvalues and eigenvectors for a matrix of the following form:

$$\begin{bmatrix} a & b \\ b & a \end{bmatrix}$$

A.47. Compute the spectral decomposition of the matrix in Example A.8.

A.48. Find the eigenvalues of the (asymmetric) matrix $\mathbf{A} = \mathbf{a}\mathbf{b}^T$. (Hint: Use Exercise A.44.)

A.49. * Show that $\mathbf{A}\mathbf{A}^T$ and $\mathbf{A}^T\mathbf{A}$ have the same nonzero eigenvalues.

A.50. * The *singular value decomposition* of a matrix $\mathbf{A}$ arises from the relationship of the eigenproblems of $\mathbf{A}^T\mathbf{A}$ and $\mathbf{A}\mathbf{A}^T$. The spectral decompositions of $\mathbf{A}^T\mathbf{A}$ and $\mathbf{A}\mathbf{A}^T$, both nonnegative definite with nonnegative eigenvalues, lead to the expressions

$$\mathbf{A}^T\mathbf{A} = \mathbf{V}\begin{bmatrix} \mathbf{\Lambda}^2 & \mathbf{0} \\ \mathbf{0} & \mathbf{0} \end{bmatrix}\mathbf{V}^T \quad \text{and} \quad \mathbf{A}\mathbf{A}^T = \mathbf{U}\begin{bmatrix} \mathbf{\Lambda}^2 & \mathbf{0} \\ \mathbf{0} & \mathbf{0} \end{bmatrix}\mathbf{U}^T,$$

where $\mathbf{U}$ and $\mathbf{V}$ are orthogonal matrices with eigenvectors as columns, and $\mathbf{\Lambda}^2$ is the $(r \times r)$ diagonal matrix of nonzero eigenvalues with $rank\,(\mathbf{\Lambda}) = rank\,(\mathbf{A}) = r$. The diagonal elements of $\mathbf{\Lambda}$, square roots of eigenvalues of the inner and outer product matrices, are known as *singular values* of the matrix $\mathbf{A}$. Note that the blocks $\mathbf{0}$ above are sized to fit the appropriate dimensions.

a. Show that if $\mathbf{v}^{(j)}$ is a column of $\mathbf{V}$ and an eigenvector of $\mathbf{A}^T\mathbf{A}$, then $\mathbf{A}\mathbf{v}^{(j)}$ is an (unnormalized) eigenvector of $\mathbf{A}\mathbf{A}^T$.
b. Show that $\mathbf{AV} = \mathbf{U\Sigma}$ where $\mathbf{\Sigma}$ is some diagonal matrix.
c. Show that $\mathbf{\Sigma} = \begin{bmatrix} \mathbf{\Lambda} & \mathbf{0} \\ \mathbf{0} & \mathbf{0} \end{bmatrix}$.
d. Show that we can write the singular value decompositon as

$$\mathbf{U}^T\mathbf{AV} = \begin{bmatrix} \mathbf{\Lambda} & \mathbf{0} \\ \mathbf{0} & \mathbf{0} \end{bmatrix}$$

e. Show that the following is the Moore-Penrose generalized inverse for $\mathbf{A}$:

$$\mathbf{A}^+ = \mathbf{V}\begin{bmatrix} \mathbf{\Lambda}^{-1} & \mathbf{0} \\ \mathbf{0} & \mathbf{0} \end{bmatrix}\mathbf{U}^T$$

A.51. Find the eigenvalues and eigenvectors for the inverses of the matrices in Exercise A.45.

A.52. If the spectral decomposition of a nonsingular matrix $\mathbf{A}$ is $\mathbf{Q\Lambda Q}^T$, show that the decomposition of its inverse follows $\mathbf{A}^{-1} = \mathbf{Q\Lambda}^{-1}\mathbf{Q}^T$.

A.53. * Prove that the eigenvalues of a symmetric, idempotent matrix are zero or one.

A.54. * Prove the following result: if $\mathbf{P}$ is symmetric and idempotent, then $rank\,(\mathbf{P}) = trace(\mathbf{P})$.

Exercises A.53 and A.54 include the assumption of symmetry to make them easier to prove, since the eigenvalue decomposition is so precarious without symmetry. The following Exercises (A.55 to A.59) show that symmetry is not needed, but require a more solid understanding of eigenvalues:

A.55. Using the basic definition for eigenvalues (A.8), prove that the eigenvalues of an idempotent matrix are either zero or one.

A.56. Let $\mathbf{P}$ be idempotent, and let $\mathbf{p}^{(i)}$ be the ith column of $\mathbf{P}$. Show that $\mathbf{p}^{(i)}$ is an eigenvector of $\mathbf{P}$ with associated eigenvalue 1.

A.57. Let $\mathbf{P}$ be idempotent, and show that $dim\,(\mathcal{N}(\mathbf{P}-\mathbf{I})) = rank\,(\mathbf{P}) = r$.

A.58. Let $\mathbf{u}^{(i)}$, $i = 1, \ldots, r$ be a set of linearly independent columns of $\mathbf{P}$, let $\mathbf{u}^{(i)}$, $i = r+1, \ldots, n$ be a basis for $\mathcal{N}(\mathbf{P})$, and let $\mathbf{D}$ be a diagonal matrix with $D_{ii} = 1$ for $i = 1, \ldots, r$, and $D_{ii} = 0$ for $i = r+1, \ldots, n$. Show that $\mathbf{PU} = \mathbf{UD}$ and $\mathbf{U}$ is nonsingular, where the columns of $\mathbf{U}$ are the vectors $\mathbf{u}^{(i)}$, $i = 1, \ldots, n$.

A.59. Show that $rank\,(\mathbf{P}) = trace(\mathbf{D}) = trace(\mathbf{P})$.

A.60. * Prove that if a matrix is positive definite, then its inverse is positive definite.

A.61. Prove that a matrix that is nonsingular and nonnegative definite is positive definite.

A.62. The critical step in the Cholesky factorization algorithm is the square root of the quantity $A_{kk} - \ell_k^T \ell_k$. Show that if $\mathbf{A}$ is positive definite, this quantity will always be positive by showing that $A_{kk} - \ell_k^T \ell_k$ can be written as a quadratic form $\mathbf{x}^T\mathbf{A}\mathbf{x}$ where $\mathbf{x}$ is given below,

$$\mathbf{x} = \begin{bmatrix} \mathbf{L}_{[k-1]}^{-T}\ell_k \\ -1 \\ \mathbf{0} \end{bmatrix} \begin{matrix} k-1 \\ 1 \\ n-k \end{matrix}$$

So if $\mathbf{A}$ is positive definite, then the quadratic forms in these vectors $\mathbf{x}$ will be positive and the algorithm will proceed until completion.

A.63. Compute the Cholesky factorization for the following four matrices.

$$\begin{bmatrix} 1 & -1 & 2 \\ -1 & 5 & -4 \\ 2 & -4 & 0 \end{bmatrix} \quad \begin{bmatrix} 9 & -3 & -3 \\ -3 & 5 & 1 \\ -3 & 1 & 5 \end{bmatrix} \quad \begin{bmatrix} 9 & 6 \\ 6 & 20 \end{bmatrix} \quad \begin{bmatrix} 20 & 6 \\ 6 & 9 \end{bmatrix}$$

A.64. Compute the symmetric square root for the four matrices in Exercise A.63.

A.65. Show that the symmetric square root is unique.

A.66. Finish Example A.10 by computing $\mathbf{A}^{1/2}$ numerically.

The Cholesky factorization can be extended to the case where $\mathbf{A}$ is positive semidefinite, that is, $\mathbf{x}^T\mathbf{A}\mathbf{x} \geq 0$ for all $\mathbf{x}$, but $\mathbf{A}$ is singular, so $\mathbf{x}^T\mathbf{A}\mathbf{x} = 0$ for some nonzero $\mathbf{x}$.

A.67. If $\mathbf{A}$ is $p \times p$ and positive semidefinite, with $rank\,(\mathbf{A}) = r$, extend Result A.20 to show that $\mathbf{A} = \mathbf{L}\mathbf{L}^T$ where $\mathbf{L}$ is $p \times r$.

A.68. Find the Cholesky factorization for the positive semidefinite matrix

$$\begin{bmatrix} 4 & 2 & 2 & 2 \\ 2 & 5 & -3 & 7 \\ 2 & -3 & 5 & -5 \\ 2 & 7 & -5 & 19 \end{bmatrix}$$

A.69. * Find the inverse of the matrix $\mathbf{I}+\mathbf{a}\mathbf{b}^T$. Hint: Try the form $c\mathbf{I}+d\mathbf{a}\mathbf{b}^T$ and find c and d. What happens if $\mathbf{a}^T\mathbf{b} = -1$?

A.70. Let $\mathbf{D}$ be a diagonal matrix; find the inverse of $\mathbf{D} + \mathbf{1}\mathbf{1}^T$.

A.71. * Let $\mathbf{D}$ be a diagonal matrix; find the inverse of $\mathbf{D} + \mathbf{a}\mathbf{b}^T$.

A.72. Verify these results for the inverse of a partitioned matrix:

$$\begin{bmatrix} \mathbf{A} & \mathbf{B} \\ \mathbf{C} & \mathbf{D} \end{bmatrix}^{-1} = \begin{bmatrix} \mathbf{A}^{-1} + \mathbf{A}^{-1}\mathbf{B}\mathbf{E}^{-1}\mathbf{C}\mathbf{A}^{-1} & -\mathbf{A}^{-1}\mathbf{B}\mathbf{E}^{-1} \\ -\mathbf{E}^{-1}\mathbf{C}\mathbf{A}^{-1} & \mathbf{E}^{-1} \end{bmatrix}$$

$$= \begin{bmatrix} \mathbf{F}^{-1} & -\mathbf{F}^{-1}\mathbf{B}\mathbf{D}^{-1} \\ -\mathbf{D}^{-1}\mathbf{C}\mathbf{F}^{-1} & \mathbf{D}^{-1} + \mathbf{D}^{-1}\mathbf{C}\mathbf{F}^{-1}\mathbf{B}\mathbf{D}^{-1} \end{bmatrix}$$

where $\mathbf{E} = \mathbf{D} - \mathbf{C}\mathbf{A}^{-1}\mathbf{B}$ and $\mathbf{F} = \mathbf{A} - \mathbf{B}\mathbf{D}^{-1}\mathbf{C}$.

A.73. Let $\mathbf{V}$ be a symmetric $p \times p$ matrix with eigenvalues $\lambda_1, \cdots, \lambda_p$; show the following:

a. $|\mathbf{I}_p + t\mathbf{V}| = \prod_{i=1}^{p}(1 + t\lambda_i)$.

b. The derivative of $|\mathbf{I}_p + t\mathbf{V}|$ with respect to t is equal to $trace(\mathbf{V})$.

A.74. Prove Result A.19.

A.75. (Binomial inverse theorem) Verify the following for $\mathbf{A}, \mathbf{B}$ nonsingular:

$$(\mathbf{A} + \mathbf{U}\mathbf{B}\mathbf{V})^{-1} = \mathbf{A}^{-1} - \mathbf{A}^{-1}\mathbf{U}\mathbf{B}^{-1}(\mathbf{B}^{-1} + \mathbf{V}\mathbf{A}^{-1}\mathbf{U})^{-1}\mathbf{B}\mathbf{V}\mathbf{A}^{-1}$$

Appendix B

Lagrange Multipliers

B.1 Main Results

Lagrange multipliers are employed to solve optimization problems with equality constraints. We will first present the specific case of linear constraints, with a proof, then the general case.

Definition B.1 *A function* $f(\mathbf{x}) : R^n \to R$ *is* continuously differentiable *iff the gradient* $\frac{\partial f(\mathbf{x})}{\partial \mathbf{x}}$ *exists and is continuous.*

Result B.1 *Let* $f(\mathbf{x}) : R^n \to R$ *be continuously differentiable and let* $\mathcal{T} = \{\mathbf{x} : \mathbf{A}\mathbf{x} = \mathbf{c}\}$. *Then if* $\mathbf{x}^*$ *is a local minimum point of* f *over* $\mathcal{T}$, *then* $\frac{\partial f(\mathbf{x})}{\partial \mathbf{x}} \in \mathcal{C}(\mathbf{A}^T)$.

Proof: Any other point in $\mathcal{T}$ can be written as $\mathbf{x}^* + t\mathbf{d}$, where $\mathbf{d} \in \mathcal{N}(\mathbf{A})$. If $\mathbf{x}^*$ is a local minimum, then the directional derivative at $\mathbf{x}^*$ in the direction $\mathbf{d}$, given by $\mathbf{d}^T \frac{\partial f(\mathbf{x})}{\partial \mathbf{x}}$, must be zero for all $\mathbf{d} \in \mathcal{N}(\mathbf{A})$. Since this means that $\frac{\partial f(\mathbf{x})}{\partial \mathbf{x}}$ is orthogonal to $\mathcal{N}(\mathbf{A})$, then $\frac{\partial f(\mathbf{x})}{\partial \mathbf{x}}$ must be in its orthogonal complement. □

Since we have $\frac{\partial f(\mathbf{x})}{\partial \mathbf{x}} \in \mathcal{C}(\mathbf{A}^T)$, we can write $\frac{\partial f(\mathbf{x})}{\partial \mathbf{x}} = \mathbf{A}^T \lambda$, where λ is known as the Lagrange multiplier. The name arises from the effect of marginal changes in the constraint vector $\mathbf{c}$. If we write the points in $\mathcal{T}$ as

$$\mathbf{x}(\mathbf{c}, \mathbf{z}) = \mathbf{A}^g \mathbf{c} + (\mathbf{I} - \mathbf{A}^g \mathbf{A})\mathbf{z}$$

and now look at the gradient of $f(\mathbf{x}(\mathbf{c}, \mathbf{z}))$ as a function of the constraint vector $\mathbf{c}$, we find

$$\lambda = \frac{\partial f(\mathbf{x}(\mathbf{c}, \mathbf{z}))}{\partial \mathbf{c}}.$$

We can now state the general case with nonlinear constraints; we will dispense with the proof as it is beyond the scope of this course.

Definition B.2 *Let* $\mathbf{H} : R^n \to R^m$; *then* $\mathbf{x}^*$ *is a* regular point *of* H *if the Jacobian matrix of the mapping,* $\mathbf{J}_H(\mathbf{x}^*)$, *has rank* m. *(If* $\mathbf{H}$ *is affine, that is,* $\mathbf{H}(\mathbf{x}) = \mathbf{A}\mathbf{x} - \mathbf{c}$, *then* $\mathbf{J}_H(\mathbf{x}) = \mathbf{A}$ *and* $\mathbf{A}$ *has full-row rank.)*

Result B.2 *(Lagrange multiplier theorem) If* $\mathbf{H}$ *is regular at* $\mathbf{x}^*$, *the functions* f *and* H *continuously differentiable, and if* $f(\mathbf{x})$ *achieves a relative extremum at* $\mathbf{x}^*$ *subject to* $\mathbf{H}(\mathbf{x}) = \mathbf{0}$, *then*

$$\frac{\partial f(\mathbf{x}^*)}{\partial \mathbf{x}} \in \mathcal{C}(\mathbf{J_H}(\mathbf{x}^*)^T).$$

Corollary B.1 *If* $\mathbf{H}$ *is regular at* $\mathbf{x}^*$, *the functions* f *and* H *continuously differentiable, and if* $f(\mathbf{x})$ *achieves a relative extremum at* $\mathbf{x}^*$ *subject to* $\mathbf{H}(\mathbf{x}) = \mathbf{0}$, *then there exists* λ *such that the function* L *below ("the Lagrangian"),*

$$L(\mathbf{x}, \lambda) = f(\mathbf{x}) - \lambda^T \mathbf{H}(\mathbf{x}),$$

is stationary at $\mathbf{x}^*$.

Notice that taking the derivatives of the Lagrangian produces some familiar results:

$$\frac{\partial L}{\partial \mathbf{x}}|_{\mathbf{x}=\mathbf{x}^*} = \frac{\partial f(\mathbf{x}^*)}{\partial \mathbf{x}} - \mathbf{J_H}(\mathbf{x}^*)^T \lambda = \mathbf{0}$$

and

$$\frac{\partial L}{\partial \lambda}|_{\mathbf{x}=\mathbf{x}^*} = \mathbf{H}(\mathbf{x}^*) = \mathbf{0}.$$

Here the second part, the partial derivatives with respect to the Lagrange multipliers, just reproduces the constraint. The first part produces a similar result to the linear (affine) case, that the gradient is in the column space of a matrix arising from the constraints, where the vector λ, the Lagrange multipliers, solves the system of equations

$$\mathbf{J_H}(\mathbf{x}^*)^T \lambda = \frac{\partial f(\mathbf{x}^*)}{\partial \mathbf{x}}.$$

Notice that the statement of the constraint function allows some flexibility. That is, the constraint function $\mathbf{H}(\mathbf{x}) = \mathbf{0}$ is logically equivalent to $-2\mathbf{H}(\mathbf{x}) = \mathbf{0}$, so that sign changes and factors of two may be tossed in arbitrarily to simplify the algebra to the surprise of the unsuspecting reader.

Example B.1: Best Shape for a Can

Let's find the smallest area of a cylinder, in terms of height h and radius r, that has a specified volume. The application is minimizing the metal to form a can that holds a certain volume. Here the area is $f(r, h) = 2\pi rh + 2\pi r^2$, and the constraint is $H(r, h) = \pi r^2 h - 1$, taking care to set this constraint function to zero. The Lagrangian function then takes the form

$$L(r, h, \lambda) = f(r, h) - \lambda H(r, h) = 2\pi rh + 2\pi r^2 - \lambda(\pi r^2 h - 1),$$

and the derivatives are

$$\begin{aligned}\partial L/\partial r &= 2\pi h + 4\pi r - \lambda 2\pi r h\\ \partial L/\partial h &= 2\pi r - \lambda \pi r^2\\ \partial L/\partial \lambda &= \pi r^2 h - 1.\end{aligned}$$

Setting the last one to zero gives $h = 1/(\pi r^2)$; working on the middle one gives $\lambda = 2/r$. With these substitutions, the first becomes $2/r^2 + 4\pi r - 4/r^2$. Setting this to zero gives $r = 1/\sqrt[3]{2\pi}$. At the margin, increasing the volume by δ means an increase in metal of $\lambda\delta = \delta 2\sqrt[3]{2\pi}$.

Example B.2: Constrained Least Squares
Consider the least squares problem subject to the linear constraint $\mathbf{Ax} = \mathbf{c}$. Let $f(\mathbf{x}) = \mathbf{x}^T\mathbf{Bx} + \mathbf{b}^T\mathbf{x}$; then the Lagrangian becomes

$$L(\mathbf{x}, \lambda) = \mathbf{x}^T\mathbf{Bx} + \mathbf{b}^T\mathbf{x} - \lambda^T(\mathbf{Ax} - \mathbf{c})$$

and taking derivatives leads to the system of equations

$$\begin{bmatrix} 2\mathbf{B} & -\mathbf{A}^T \\ \mathbf{A} & \mathbf{0} \end{bmatrix} \begin{bmatrix} \mathbf{x} \\ \lambda \end{bmatrix} = \begin{bmatrix} -\mathbf{b} \\ \mathbf{c} \end{bmatrix}. \tag{B.1}$$

Example B.3: Eigenvalues
Let $f(\mathbf{x}) = \mathbf{x}^T\mathbf{Ax}$ and consider the scalar constraint $\mathbf{x}^T\mathbf{x} = 1$. The Lagrangian takes the form

$$L(\mathbf{x}, \lambda) = \mathbf{x}^T\mathbf{Ax} - \lambda(\mathbf{x}^T\mathbf{x} - 1)$$

and the partial derivative with respect to $\mathbf{x}$ gives

$$2\mathbf{Ax} - 2\lambda\mathbf{x}.$$

Setting this partial derivative to zero yields the familiar definition of eigenvectors and eigenvalues,

$$\mathbf{Ax} = \lambda\mathbf{x},$$

where the stationary points $f(\mathbf{x})$ occur at the eigenvectors where f takes the value of the eigenvalues. Setting the partial derivative of L with respect to λ reproduces the familar normalization constraint. See also Exercise 2.2.

B.2 Notes

The theorems are given in Section 9.2 of Luenberger [27] in terms of Frechet derivatives and Banach spaces; here we have just regular partial derivatives and Euclidean spaces.

B.3 Exercises

B.1. Maximize $f(\mathbf{x}) = \sum_i \log x_i$ subject to the constraint $\sum_i x_i = 1$. (Of course, we must have $x_i \geq 0$.)

B.2. Maximize $f(\mathbf{x}) = \sum_i n_i \log x_i$ again subject to the same constraint $\sum_i x_i = 1$. (And again, we must have $x_i \geq 0$.)

B.3. Minimize $f(\mathbf{x}) = 5x_1^2 + 2x_1x_2 + x_2^2$ subject to the constraint $x_1 + 2x_2 = 1$.

B.4. For the previous problem, plot the contours of the objective function $f(\mathbf{x})$ and the line that describes the constraint to verify your solution.

Bibliography

1. T. W. Anderson. The integral of a symmetric unimodal function over a symmetric convex set and some probability inequalities. *Proceedings of the American Mathematical Society* 6:170–76, 1955.

2. T. W. Anderson. *Introduction to Multivariate Statistical Analysis*. Wiley, Hoboken, NJ, 2003.

3. Patrick Billingsley. *Probability and Measure*. Wiley, New York, 1995.

4. Thomas L. Boullion and Patrick L. Odell. *Generalized Inverse Matrices*. Wiley, New York, 1971.

5. Peter J. Brockwell and Richard A. Davis. *Introduction to Time Series and Forecasting*, 2nd ed. Springer, New York, 2002.

6. George Casella and Roger Berger. *Statistical Inference*. Duxbury, Pacific Grove, CA, 2002.

7. Kai Lai Chung. *A Course in Probability Theory*, 3rd ed. Academic Press, San Diego, 2001.

8. D. Cochrane and G. H. Orcutt. Applications of least squares regression to relationships containing autocorrelated error terms. *Journal of the American Statistical Association* 44:32–61, 1949.

9. R. C. Elston and J. E. Grizzle. Estimation of time-response curves and their confidence bands. *Biometrics* 18:148–59, 1962.

10. W. A. Fuller. *Measurement Error Models*. Wiley, New York, 1987.

11. W. A. Fuller. *Introduction to Statistical Time Series*. Wiley, New York, 1995.

12. A. S. Goldberger. Best linear unbiased prediction in the generalized linear regression model. *Journal of the American Statistical Association* 57:369–75, 1962.

13. Gene H. Golub and Charles van Loan. *Matrix Computations*, 3rd ed. Johns Hopkins University Press, Baltimore, 1996.

14. J. H. Goodnight. A tutorial on the sweep operator. *The American Statistician* 33:149–58, 1979.

15. Franklin A. Graybill. *Theory and Application of the Linear Model*. Duxbury, North Scituate, MA, 1976.

16. Franklin A. Graybill. *Matrices with Applications in Statistics*. Wadsworth, Belmont, CA, 1983.

17. D. A. Harville. Maximum likelihood approaches to variance component estimation and to related problems. *Journal of the American Statistical Association* 72:320–38, 1977.

18. David A. Harville. Bayesian inference for variance components using only error contrasts. *Biometrika* 61:383–85, 1974.

19. David A. Harville and Alicia L. Carriquiry. Classical and Bayesian predication as applied to an unbalanced mixed linear model. *Biometrics* 48:987–1003, 1992.

20. A. J. Hayter. A proof of the conjecture that the Tukey-Kramer multiple comparisons procedure is conservative. *Annals of Statistics* 12:61–75, 1984.

21. James P. Hobert and George Casella. The effect of improper priors on Gibbs sampling in hierarchical linear models. *Journal of the American Statistical Association* 91:1461–73, 1996.

22. R. R. Hocking. *The Analysis of Linear Models*. Brooks/Cole, Monterey, CA, 1985.

23. H. Hotelling and H. Working. Application of the theory of error to the interpretation of trends. *Journal of the American Statistical Association* (*Supplement*) 24:73–85, 1929.

24. Jason C. Hsu. *Multiple Comparisons: Theory and Methods*. Chapman & Hall, New York, 1996.

25. Jan Kmenta. *Elements of Econometrics*. MacMillan, New York, 1971.

26. Charles L. Lawson and Richard J. Hanson. *Solving Least Squares Problems*. Prentice-Hall, Englewood Cliffs, NJ, 1974.

27. David G. Luenberger. *Optimization by Vector Space Methods*. Wiley, New York, 1969.

28. Eugene Lukacs. *Characteristic Functions*, 2nd ed. Hafner, New York, 1970.

29. K. V. Mardia, J. T. Kent, and J. M. Bibby. *Multivariate Analysis*. Academic Press, New York, 1979.

30. Alan J. Miller. *Subset Selection in Regression*, 2nd ed. Chapman & Hall, Boca Raton, FL, 2002.

31. George A. Milliken and Mohammed Albohali. On necessary and sufficient conditions for ordinary least squares estimators to be best linear unbiased estimators. *The American Statistician* 18:298–99, 1984.

32. John F. Monahan. *Numerical Methods of Statistics*. Cambridge University Press, New York, 2001.

33. J. Neyman and Elizabeth L. Scott. Consistent estimates based on partially consistent observations. *Econometrica* 16:1–32, 1948.

34. H. D. Patterson and R. Thompson. Recovery of inter-block information when block sizes are unequal. *Biometrika* 58:545–54, 1971.

35. R. M. Pringle and A. A. Rayner. *Generalized Inverse Matrices with Applications to Statistics*, Number 28. Hafner (Griffin's Monographs), New York, 1971.

36. C. R. Rao. Markoff's theorem with linear restrictions on parameters. *Sankhya* 7:16–19, 1945.

37. C. R. Rao. *Linear Statistical Inference and Its Applications*. Wiley, New York, 1973.

38. C. R. Rao and S. K. Mitra. *Generalized Inverse of Matrices and Its Applications*. Wiley, New York, 1971.

39. John O. Rawlings, Sastry G. Pantula, and David A. Dickey. *Applied Regression Analysis: A Research Tool*. Springer-Verlag, New York, 1998.

40. Alvin C. Rencher. *Methods of Multivariate Analysis*. Wiley, New York, 2002.

41. G. K. Robinson. That blup is a good thing: The estimation of random effects. *Statistical Science* 6:15–51, 1991.

42. Shayle R. Searle. *Linear Models*. Wiley, New York, 1971.

43. Robert H. Shumway and David S. Stoffer. *Time Series Analysis and Its Applications*. Springer, New York, 2006.

44. G. W. Stewart. *Introduction to Matrix Computations*. Academic Press, New York, 1973.

45. Stephen M. Stigler. *Statistics on the Table: The History of Statistical Concepts and Methods*. Harvard University Press, Cambridge, MA, 1999.

46. R. Thompson. The estimation of variance and covariance components with an application when records are subject to culling. *Biometrics* 22:527–50, 1973.

47. John W. Tukey. The problem of multiple comparisons. 1953.

48. Daniel T. Voss. Resolving the mixed models controversy. *The American Statistician* 53:352–56, 1999.

49. R. D. Wolfinger, R. D. Tobias, and J. Sall. Computing Gaussian likelihoods and their derivatives for general linear mixed models. *SIAM Journal on Scientific Computing* 15:1294–310, 1994.

50. Arnold Zellner. An efficient method of estimating seemingly unrelated regressions and tests for aggregation bias. *Journal of the American Statistical Association* 57:348–68, 1962.

Index

A

B

C

D

F

G

H

M

N

S

T

U

V

W

Z

Printed in the United States
by Baker & Taylor Publisher Services